Ground Characterization and Structural Analyses for Tunnel Design

Ground Characterization and Structural Analyses for Tunnel Design

Edited by
Benjamín Celada Tamames and
Z.T. Bieniawski von Preinl

With Contributions by
Mario Fernández Pérez
Juan Manuel Hurtado Sola
Isidoro Tardáguila Vicente
Pedro Varona Eraso
Eduardo Ramón Velasco Triviño

CRC Press
Taylor & Francis Group
Boca Raton London New York

CRC Press is an imprint of the
Taylor & Francis Group, an **informa** business

CRC Press
Taylor & Francis Group
6000 Broken Sound Parkway NW, Suite 300
Boca Raton, FL 33487-2742

First issued in paperback 2021

© 2020 by Taylor & Francis Group, LLC
CRC Press is an imprint of Taylor & Francis Group, an Informa business

No claim to original U.S. Government works

ISBN 13: 978-1-03-209042-9 (pbk)
ISBN 13: 978-0-8153-8662-9 (hbk)

Library of Congress Cataloging-in-Publication Data

Names: Celada, Benjamin, author. | Bieniawski, Z. T., author.
Title: Ground characterization and structural analyses for tunnel design /
Benjamin Celada and Z.T. Bieniawski.
Description: Boca Raton : Taylor & Francis, a CRC title, part of the Taylor &
Francis imprint, a member of the Taylor & Francis Group, the academic
division of T&F Informa, plc, [2019] | Includes bibliographical references
and index.
Identifiers: LCCN 2019011099| ISBN 9780815386629 (hardback) | ISBN
9781351168489 (ebook)
Subjects: LCSH: Tunnels--Design and construction. | Soil mechanics. |
Engineering geology.
Classification: LCC TA815 .C45 2019 | DDC 624.1/93--dc23
LC record available at https://lccn.loc.gov/2019011099

Visit the Taylor & Francis website at
http://www.taylorandfrancis.com

and the CRC Press website at
http://www.crcpress.com

Contents

In Memoriam

This is the English version of the Spanish book *Fundamentos del Diseño de Túneles*, which was edited in October 2016 by AGA Ediciones, as the result of a close and unique collaboration between Professor Z. T. Bieniawski von Preinl and Benjamín Celada Tamames, which started in 1997 during the Vienna Congress of the International Tunneling Association and ended in December 2017 when Professor Z. T. Bieniawski von Preinl passed away.

The approaches to tunnel design explained in this book are the result of wide experience gained during the design and site supervision of numerous underground works built in Spain between 1995 and 2012, and more recently in Santiago de Chile, Sao Paulo and Lima.

The editors' involvement with the ever-developing field of tunnel design goes back to the pioneering book *Rock Mechanics Design in Mining and Tunnelling* (1984), subsequently crystallizing ideas on the importance of a systematic tunnel design methodology, published as *Design Methodology for Rock Engineering* (1992).

This book is the culmination of all the mentioned topics and will benefit tunnel design professionals in civil, geological and mining engineering, challenging both practical engineers and current researchers on four topics studied in this book, originally introduced in the journal *Tunnels & Tunnelling International*. They are:

- **Development of the interactive structural design methodology** (*Diseño Estructural Activo*, DEA, in its original Spanish name)
- **Creation of the rock mass excavability (RME) index**
- **Creation of the index of elastic behavior** (*Índice del Comportamiento Elástico*, ICE, in its original Spanish name)
- **Improving the rock mass rating (RMR)** to update it to the modern practices, after 45 years of use

As this book deals with site characterization and structural analyses of tunnel design, which are two of the three pillars of the DEA, the aim of the living editor is to follow up with a new volume dedicated to engineering during construction, the third main pillar of the DEA.

It is also necessary to express our thanks to the people who participated in the drafting of this book, whose splendid collaboration has been essential to enable its realization.

Last but not least, this book shall be considered a due tribute to honor the memory of Professor Z. T. Bieniawski von Preinl for a whole life that was full of professional success and that was devoted to his wife Elizabeth. May he rest in the peace of God.

Benjamín Celada Tamames,
Madrid, Spain

Editors

With their technical and cultural collaboration of over 20 years going back to their first meeting at the World Tunnel Congress in Vienna in 1997, Professors Celada and Bieniawski combined American and European tunnel design and construction experience in their respective countries, sharing their achievements in many novel, practical and cost-effective tunnel design projects and publications. Both are unique in that having started as university professors and researchers, they have accumulated significant practical experience in both tunneling and mining. Both are also dedicated educators and excellent teachers, able to prepare their students to meet the practical challenges of tunnel engineering. They complement each other because of their special practical backgrounds such as the deep-level mining and tunneling expertise of Professor Bieniawski, and the extensive large-size urban highway tunneling expertise of Professor Celada.

Benjamín Celada Tamames, President and CEO Geocontrol, Spain Benjamín Celada Tamames was born in Zaragoza, Spain, studied at the Technical School of Mining Engineering in Oviedo, Asturias, Spain, and in 1979 he achieved a Ph. D. at the same university. During the 1970s he worked in Hunosa and Potasas de Navarra S.A. developing rock mechanics activities in Gresa as chief tunnel engineer.

In 1982 he founded the engineering company Geocontrol S.A. of which he is currently President and Chief Executive Officer. Between 1992 and 2011 he was Professor of Tunneling and Underground Excavations at the Technical School of Mining Engineering in Madrid, Spain, a position which he left in order to manage the international expansion of Geocontrol, S.A. in Latin America, establishing Geocontrol offices in Sao Paolo, Brazil, Santiago, Chile and Lima, Peru.

In between, he developed a unique tunnel design and construction methodology, called DEA, which has been used for many years on tunneling projects in Spanish speaking countries.

He is the author of more than 80 scientific publications and during his professional career in Geocontrol he has participated in the design of more than 342 tunnels totaling more than 1000 km.

Z. T. Bieniawski von Preinl, President, Bieniawski Design Enterprises, Prescott, AZ Professor Bieniawski obtained his first doctorate, D.Sc (Eng) at the University of Pretoria, South Africa, in 1967, when he was the Head of the Rock Mechanics Division of the Council for Scientific and Industrial Research (CSIR) until 1978. He and his family then emigrated to the United States by invitation.

Retiring after 20 years in mining and tunnel engineering education at the Pennsylvania State University, at the time of his death he was professor emeritus of mineral engineering at Penn State and adjunct professor in geoengineering at the AGH University of Science and Technology in Kraków, Poland. In 2001, he received a *Doctor Honoris Causa* degree from the Technical University of Madrid, and in 2010, he was awarded a third doctorate: a *Doctor Honoris Causa* degree from the AGH University of Science and Technology in Kraków, Poland, the city where he was born.

The author of 12 books, over 200 research papers (some translated into Spanish, German, Russian, Polish, Chinese and Korean), he guest-lectured in many countries and universities and held visiting professorships at Stanford University, Harvard University, the University of Cambridge, England, and consulted on over 50 tunneling projects in 66 countries, including the Channel Tunnel. A special honor was serving as Chairman of the U.S. National Committee on Tunneling Technology during 1984–86, and representing the United States on the ITA.

An American professor teaching Spanish graduate students – in their own language – was a whole new experience for all arising from the collaboration between these two professors, whose unusual teaching method was through interactive discussion and practical case studies.

Chapter 1

Typology of underground excavations and design issues

Benjamín Celada Tamames

> Everything that lives changes. We shouldn't be satisfied with some unaltered traditions.
>
> C. G. Jung

1.1 INTRODUCTION

In 1961 Karoly Szechy published the book titled *Alagutepitestan* which, in 1967, was translated into English as *The Art of Tunnelling*. This book was a memorable milestone in the synthesis of the knowledge that should be applied for tunnel design, but it had to live together with the obstacle derived from the limited knowledge about rock mechanics of those years.

In the year 1964 Leopold Rabcewicz published the first part of his work about the new Austrian tunneling method (NATM) which revolutionized the concepts of tunnel design and construction which had existed up to that time.

The key to the huge and positive influence that the NATM has had in tunnel design lies in recognizing the relevant role that ground behavior has in tunnel stabilization, which revolutionized the previous concepts based on the prevalence of the lining with respect to the excavated media.

Another prominent milestone in tunnel design was the publication, in 1980, of Evert Hoek and Edward Brown's book, *Underground Excavations in Rock*, in which they put the rock mass in the spotlight when constructing tunnels and achieve an excellent assemblage of the principal concepts that are present in tunnel design. Since the publication of this book, which still continues to be a reference point in tunnel design, 35 years have passed and during this tim huge development in tunnel design has occurred, as a result of the large tunneling construction activity in recent decades.

In Spain, during the period between 1995 and 2012, extensive activity in tunnel construction has been witnessed, including some particularly challenging constructions, such as never-seen-before performance in the use of Tunnel Boring Machines, both in soil and in rock.

This book updates the criteria and methodology for tunnel design based on important experiences occurring in the last decades, with the aim of being of help to the improved design of safe and economic tunnels.

1.2 TYPOLOGY OF UNDERGROUND EXCAVATIONS

The concept of underground excavation implies the construction under the ground surface but, also, this concept assumes that the construction work is done without removing the existing ground between the excavation and the surface.

In the case that the ground removal above the tunnel was necessary for its construction, it has to be kept in mind that it fully loses its resistant capacity, meaning the ground cannot play an active role in tunnel stabilization.

In fact, tunnels constructed by removing the ground up to the surface are referred to as cut-and- cover tunnels, to be distinguished from tunnels constructed with conventional procedures, whose typology is presented in the following sections.

1.2.1 Microtunnels

Microtunnels are those tunnels whose width is less than 3 m. This width substantially determines the construction methods which, in practice, are reduced to mini Tunnel Boring Machines and jacking pipes. Microtunnels are beyond the scope of this book.

1.2.2 Tunnels

Tunnels are underground excavations with an excavation width smaller than 17 m and which, in addition, have a minimum length of several hundred meters.

Tunnels are the subject of this book, and can be classified depending on their intended use, as shown in the following sections.

1.2.2.1 Hydraulic tunnels

Conceptually, most hydraulic tunnels are underground channels and, as such, they are probably the oldest tunnels constructed in the world. In Photograph 1.1, a view of the diversion tunnel, constructed by the Nabateans, to control the entry of water into Petra (Jordan) is shown.

Photograph 1.1 Water diversion tunnel constructed by the Nabateans in Petra (Jordan).

The length of hydraulic tunnels normally exceeds 1000 m and their cross section is often between 7 and 40 m², although the most common values range between 15 to 25 m².

Photograph 1.2 shows a view of the Cerro Azul Tunnel, constructed in the vicinity of Guayaquil (Ecuador), for irrigation water transportation; the tunnel cross section is 30 m² and its length is 6.45 km.

1.2.2.2 Mining tunnels

Mining tunnels compete in seniority with hydraulic ones, but have two distinct differences as these tunnels are very constrained by the mining activities and often have a service life shorter than 30 years.

In this book, only mining tunnels not affected by stress changes associated with mining activities will be considered. Mining tunnels often have cross sections between 15 and 50 m² with lengths of several kilometers.

Photograph 1.3 shows a view of an exploration tunnel, excavated by means of a Tunnel Boring Machine, in Los Bronces Mine (Chile), that has a circular cross section with diameter of 4.5 m and a length of 8.12 km.

1.2.2.3 Railway tunnels

Railway tunnels began being constructed in the mid-19th century and have had two distinct phases of growth; the first one until the beginning of the Second World War, while the

Photograph 1.2 Cerro Azul Tunnel, Guayaquil (Ecuador).

Photograph 1.3 Exploration tunnel in the South Seam from Los Bronces Mine (Chile).

Photograph 1.4 Huancayo–Huancavelica railway tunnel constructed with the rock exposed (Peru).

second one was associated with subway and high-speed railway lines, whose construction was intensified at the end of the 20th century.

The first railway tunnels had moderate sections, around 20 m² if one track was accommodated, and between 35 and 60 m² if it had two tracks.

Conventional railway tunnels, mainly those of only one track, frequently had stretches in which rock was left exposed when the construction was finished. Photograph 1.4 shows one of the tunnels of the Huancayo–Huancavelica railway (Peru), put into service in 1933.

The low technological level of the machinery used for the first railway tunnels construction did not restrict the construction of long length tunnels, mainly in the Swiss Alps.

The Simplon Tunnel, constructed between Switzerland and Italy, has two tubes with a length of 19.8 km and the first tube was put into service in the year 1906. The Simplon Tunnel was the world's longest tunnel until 1982, when the Daishimizu Tunnel (Japan) was inaugurated.

Tunnels for high-speed railways are larger than conventional ones, because speeds up to 300 km/h are reached during the train operation, which demands an enlargement of the cross section in order to mitigate the dynamic effects on passengers.

Table 1.1 Longest railway tunnels

No	Name	Country	Length (km)	Service start-up
1	Gotthard Base	Switzerland	57.1	2016
2	Seikan	Japan	53.8	1988
3	Eurotunnel	France–England	50.5	1994
4	Lötschberg	Switzerland	34.6	2007
5	Guadarrama	Spain	28.4	2007
6	Taihang	China	27.8	2007
7	Hakkoda	Japan	26.4	2005
8	Iwate Ichinohe	Japan	25.8	2002
9	Dai-Shimizu	Japan	22.2	1982
10	Wushaoling	China	21.1	2007

Currently, high-speed railway tunnels which accommodate two tracks have an excavated cross section of between 125 and 170 m², while the cross section of those with one track is around 80 m².

The strong inclination limitation that high-speed trains have to overcome makes such tunnels be constructed the longest in the world, exceeding 20 km easily, as shown in Table 1.1.

Leading the world's longest railway tunnels is the Gotthard Base Tunnel, 57.1 km long, which was put into service in June 2016. In Photograph 1.5 a view of the Gotthard Base Tunnel during construction is shown, taken from the tail of one of the boring machines used for its construction.

Subway tunnels have very limited dimensions as, if they accommodate only one track, the excavated area often ranges between 35 and 40 m², while the tunnel section with two tracks ranges between 60 and 70 m².

Photograph 1.6 shows a view of the construction of the Line 6 Santiago de Chile Subway tunnel designed for two tracks, whose straight stretch has an excavated cross section of 62 m².

The Line 9 Barcelona Subway is an exception to the above mentioned, as in this line the tunnel has an excavated area of 113 m², to allow the location of two platforms with two tracks each.

Thus, in the stations, the platforms are incorporated inside the tunnel, as shown in Figure 1.1 and in Photograph 1.7.

Photograph 1.5 Gotthard Base Tunnel construction (Switzerland).

Photograph 1.6 Line 6 Santiago Subway tunnel (Chile).

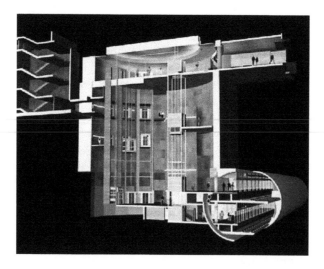

Figure 1.1 Line 9 tunnel of the Barcelona Subway (Spain).

1.2.2.4 Highway tunnels

Road tunnel construction has increased in parallel with motorized vehicle production growth, since the beginning of the 20th century.

Road tunnels often have shorter lengths than high-speed railway tunnels, as roads allow larger inclinations and higher radius of alignment than railway lines.

A standard road tunnel with two lanes often has an excavated area between 60 and 80 m²; depending on the lanes, shoulders and sidewalk width to be accommodated, varying substantially depending on the existing legislation in each country.

Figure 1.2 shows the cross section of one of the tubes from the San Cristóbal Tunnels, Santiago de Chile, while Photograph 1.8 shows one of the portals of these tunnels.

The San Cristóbal tunnels are 1,588 m long and each tube has an excavated area of 67 m².

In order to accommodate three traffic lanes, the excavated area often ranges between 100 and 130 m², as shown in Figure 1.3, which corresponds to the Bracons Tunnel cross section

Photograph 1.7 Platform in the Line 9 station in Barcelona Subway (Spain).

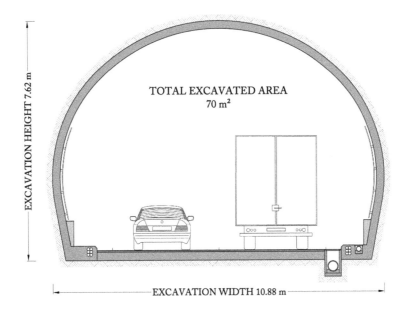

Figure 1.2 Cross section of one of the tubes from the San Cristóbal tunnels, Santiago de Chile.

(Barcelona, Spain), which has a length of 4.5 km and three lanes, two of them inclined upwards.

In Table 1.2 data from the world's largest road tunnels is shown.

Since its inauguration in 1980, the Gotthard Road Tunnel, constructed between Switzerland and Italy, had been the world's longest road tunnel until the Laerdal Tunnel in Norway was put into service in 2000.

The Laerdal Tunnel is an exceptional tunnel not only due to its length of 24.5 km, but also for having a longitudinal ventilation system, supported with electrostatic filters, and roundabouts so that vehicles can change running direction inside the tunnel, as illustrated in Photograph 1.9. Obviously, this design has been possible due to the low traffic flows inside the tunnel.

Photograph 1.8 Portal of the San Cristóbal tunnels, Santiago de Chile.

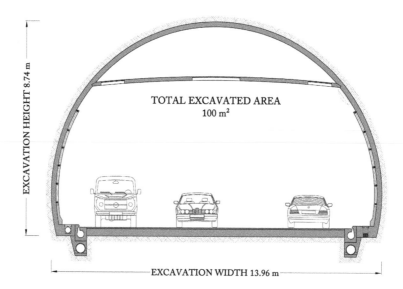

Figure 1.3 Bracons Tunnel cross section (Barcelona, Spain).

Table 1.2 World's longest road tunnels in 2016

No	Name	Country	Length (km)	Service start-up
1	Laerdal	Norway	24.5	2000
2	Zhongnanshan	China	18.0	2007
3	Gotthard	Switzerland–Italy	17.0	1980
4	Arlberg	Austria	14.0	1978
5	Hsuehshan	Taiwan	12.9	2006
6	Fréjus	France–Italy	12.9	1980
7	Maijishan	China	12.2	2009
8	Mont Blanc	France–Italy	11.7	1965
9	M-30 South	Spain	11.5	2007
10	Gudvanga	Norway	11.4	1991

Photograph 1.9 Roundabout inside Laerdal Tunnel (Norway).

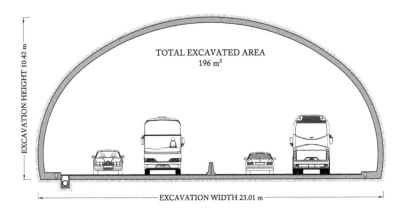

Figure 1.4 Manquehue I Tunnel cross section (Chile).

1.2.3 Large underground excavations

Large underground excavations are those whose sections exceed 200 m², as happens in road tunnels with four lanes, subway and railway underground stations, and hydroelectric power plant caverns.

In the following sections, the most relevant characteristics of these excavations are summarized.

1.2.3.1 Four-lane tunnels

Four-lane tunnels, arranged in the same level plan, have excavation widths that exceed 20 m, as is the case of the Manquehue I Tunnel, part of the Northeastern Access to Santiago de Chile, with a section of 196 m² and an excavated width of 23 m.

Figure 1.4 and Photograph 1.10 show, respectively, the cross section of Manquehue I Tunnel and a side view of its state after its service start-up.

The excavated portal of this tunnel is relatively small for a four-lane tunnel and this design was possible due to the good rock mass quality, which allowed the projection of a very lowered arch.

Photograph 1.10 Side view of Manquehue I Tunnel (Chile).

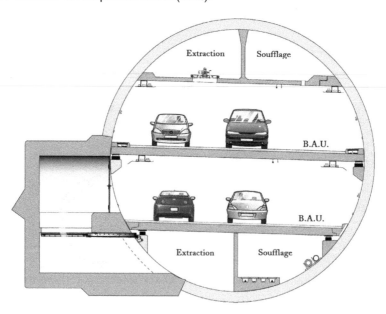

Figure 1.5 A-86 West Tunnel cross section in Paris (France).

At the beginning of the 20th century a new four-lane tunnel typology began to be proposed, characterized by the accommodation of a double-deck road and the use of Tunnel Boring Machines for its construction.

The first tunnel constructed with this typology was the highway A-86 West, in the vicinity of Versailles (Paris). As shown in Figure 1.5, this tunnel has an excavated diameter of almost 12 m, which allows the accommodation of two lanes in each deck, with a limiting clearance of 2.5 m in height. The tunnel is around 7.5 km long and does not allow truck traffic. Photograph 1.11 shows an illustration of the A-86 Tunnel in service.

Photograph 1.11 Illustration of the A-86 West Tunnel in Paris.

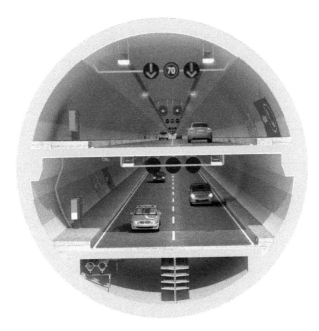

Figure 1.6 Eurasia Tunnel cross section (Turkey).

Probably the most emblematic mega tunnel being constructed at present, is the one which will link Asia to Europe under the Bosphorus strait (Turkey), called the Eurasia Tunnel.

This tunnel will have a length of 5.4 km and two overlapped lanes, so it is being excavated with a 13.7 m diameter Tunnel Boring Machine.

In Figure 1.6 the cross section of this tunnel is depicted and in Photograph 1.12 an already constructed stretch of this tunnel is shown.

1.2.3.2 Subway stations

Underground subway stations have excavated areas ranging from 140 to 220 m², depending on the platform width, which often varies between 3.5 and 5 m.

Photograph 1.12 A stretch of the Eurasia Tunnel, during its construction at the end of 2015.

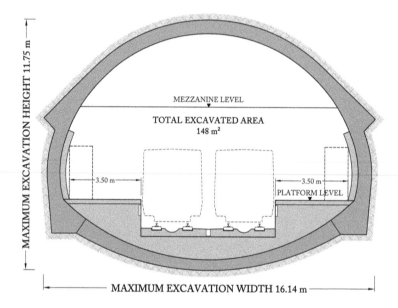

Figure 1.7 Subway station with 3.5 m wide platforms.

Figure 1.7 shows the cross section of a subway station which belongs to Line 6 of the Santiago de Chile Subway with platforms of 3.5 m wide, which has an excavated area of 148 m².

In Los Leones Station, also of Line 6, the excavated area reaches 184 m², as the platforms are 5 m wide. Photograph 1.13 shows a view of the already constructed subway station.

1.2.3.3 Caverns

Caverns are underground excavations whose excavated cross section exceeds 300 m². Traditionally, standard caverns were represented by the machine halls in the hydroelectric plants, of around 25 m wide, 60 m high and had a length that could be up to 200 m depending on the installed number of turbines.

Photograph 1.14 shows the construction of the cavern which will accommodate the turbines of Ituango Hydroelectric Plant (Colombia), which will have an installed power capacity of 2,400 MW.

Photograph 1.13 Subway station with 5 m wide platforms.

Photograph 1.14 Main cavern of Ituango Hydroelectric Plant (Colombia).

Photograph 1.15 Winter Olympic Games inauguration in 1994, inside Gjøvik Cavern (Norway).

Photograph 1.16 CERN ATLAS cavern construction (France, Switzerland).

This cavern has a length of 240 m, a width of 23 m and a height of 49 m, so its excavated area is 1,127 m².

Also remarkable are some caverns constructed for public use, such as the Gjøvik underground stadium (Norway), which has a capacity of 5,500 people and was inaugurated for the Winter Olympic Games in 1994.

Photograph 1.15 shows a view of the inauguration of the Gjøvik Olympic stadium (Norway), which is 61 m wide, 25 m high and 91 m long. Its excavated area is 1,525 m², larger than Ituango cavern and, even more important, with a much larger width.

The CERN nuclear physics laboratory, in the vicinity of Geneva (Switzerland) has underground excavations with similar sections as to those of the Gjøvik cavern, as four caverns have been excavated, at around 160 m below ground surface. The largest of these excavations is called ATLAS; its dimensions are 35 m wide, 56 m long and 42 m high, which leads to an excavated area of 1,380 m², which is quite similar to the Ituango cavern section. Photograph 1.16 shows the ATLAS cavern during its construction.

1.3 BASIC ASPECTS OF TUNNEL DESIGN

Having introduced the scope of underground excavations, the following sections present the basic aspects related to tunnel design.

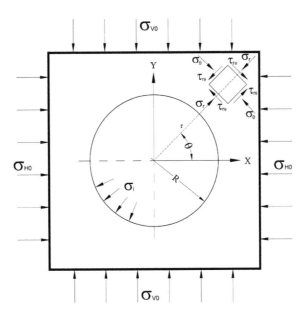

Figure 1.8 Kirsch problem approach.

1.3.1 Tunnel stabilization

In a simplified and direct way, tunnel construction can be synthesized as the substitution of a ground volume by air and by the support elements.

As air does not have significant strength, the immediate issue is to know the tunnel stabilization: the supporting elements or the excavated ground itself.

To understand this issue, the equations established by Kirsch (1898) are helpful, as they determine the stresses around a circular hole made in an elastic media.

In Figure 1.8 the Kirsch approach is exposed, in polar coordinates.

The parameters that define the Kirsch problem are the following:

σ_r = Stress component acting in the direction toward the excavation center (**radial stress**)
σ_θ = Stress component acting perpendicular to that of the excavation center (**tangential or circumferential stress**)
$\tau_{r\theta}$ = Shear stress acting on the surface
σ_{H_0} = Horizontal stress acting on the ground (**field horizontal stress**)
σ_{V_0} = Vertical stress acting on the ground (**field vertical stress**)

If $k = \sigma_{H_0}/\sigma_{V_0}$ (stress redistribution coefficient) and $\sigma_{H_0} = \sigma_{V_0}$, the solution given by Kirsch is defined by the equations

$$\sigma_r = \sigma_0 \frac{1+k}{2}\left(1 - \frac{R^2}{r^2}\right) - \sigma_0 \frac{1-k}{2}\left(1 - 4 - \frac{R^2}{r^2} + 3\frac{R^2}{r^2}\right)\cos 2\theta$$

$$\sigma_\theta = \sigma_0 \frac{1+k}{2}\left(1 + \frac{R^2}{r^2}\right) + \sigma_0 \frac{1-k}{2}\left(1 + 3\frac{R^2}{r^2}\right)\cos 2\theta \qquad (1.1)$$

$$\tau_{r\theta} = \sigma_0 \frac{1-k}{2}\left(1 + 2\frac{R^2}{r^2} - 3\frac{R^4}{r^4}\right)\text{sen}\,2\theta$$

In the perimeter of excavation, $r = R$, is obtained:

$$\sigma_r = 0$$

$$\sigma_\theta = \sigma_0 \left[(1 + k) + 2(1 - k) \cos 2\theta \right] \qquad (1.2)$$

$$\tau_{r\theta} = 0$$

The maximum tangential stress value in the excavation perimeter is obtained by choosing $\cos 2\theta = 1$ is:

$$\sigma_{\theta MAX} = (3 - K)\sigma_0 \qquad (1.3)$$

In order to maintain the ground in elastic regime, as is assumed in the Kirsch problem, it will be necessary that $\sigma_{CM} > \sigma_{\theta MAX}$, in other words:

$$\sigma_{CM} > (3 - K)\sigma_0 \qquad (1.4)$$

where σ_{CM} is the uniaxial compressive strength of the ground.

An interesting particular case, of the problem solved by Kirsch, corresponds to an unsupported tunnel, excavated in an elastic ground and in a hydrostatic stressed state.

By assigning $K = 1$ in the Expression 1.1 it is obtained that

$$\sigma_r = \sigma_0 \left(1 - \frac{R^2}{r_2} \right)$$

$$\sigma_\theta = \sigma_0 \left(1 - \frac{R^2}{r_2} \right) \qquad (1.5)$$

$$\tau_{r\theta} = 0$$

Figure 1.9 represents the values of σ_r and σ_θ as a function of r.

From Figure 1.9 some important conclusions can be deduced:

1. At a distance large enough from the excavation center, the values σ_θ and σ_r, match with the hydrostatic field stress value (σ_θ). In particular, for a distance to the tunnel center of six times its radius, both stresses only differ from the field stress by 2.8%.

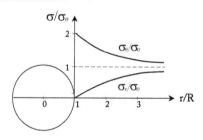

Figure 1.9 Stress distribution around a circular tunnel, excavated in an elastic media, without supporting elements and $k_0 = 1$.

2. In the tunnel perimeter the tangential stress value doubles the field stress value and the radial stress is zero; this makes sense as, by definition, no supporting elements that could counter the radial stress (σ_r) have been placed.
3. At any point on the tunnel perimeter the ground is subjected to uniaxial compressive stress, whose value is $2\sigma_0$, as illustrated in Figure 1.10.
4. In a circular tunnel excavated in ground with elastic behavior it is not necessary to place any kind of support elements in order to reach the stress–strain equilibrium.
5. The largest value of the tangential stress in an excavation with elastic behavior is at its perimeter.
6. The required condition so that the ground surrounding an excavated circular tunnel behaves in an elastic way, demands that its uniaxial compression strength be bigger than the double of the field stress.

From the above conclusions, it can be stated **that in a circular tunnel excavated in ground with elastic behavior it is not necessary to place any supporting elements for its stabilization; this means that in tunnels excavated in grounds with elastic behavior the ground itself ensures the stress–strain stability.**

In case that when excavating a tunnel the ground does not keep in the elastic domain and yields, Kirsch equations cannot further be used and have to be substituted by others which take into account ground yielding.

In Figure 1.11 the stress distribution in the ground when excavating a tunnel in yielding conditions and in a hydrostatic stress state is depicted, which was obtained by Kastner (1962).

Between the stress distributions from Figures 1.9 and 1.11 two very important differences exist:

1. **The largest value of the tangential stress is no longer in the perimeter of the excavation.** This is a direct consequence of the ground yielding at the excavation perimeter, this leads to a decrease in its strength and reaches the maximum tangential stress value inside the excavated ground, under triaxial conditions.

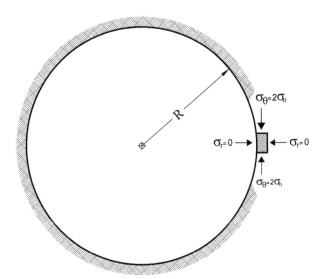

Figure 1.10 Uniaxial compressive state on a tunnel perimeter excavated in an elastic media.

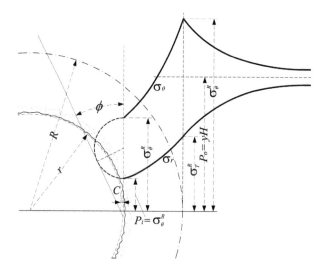

Figure 1.11 Stress distribution around a circular tunnel, excavated in a plasticized ground and subjected to hydrostatic stresses.

Source: Kastner, 1962.

According to this, an aureole of yielded ground is created around the tunnel, defined by the yielding radius (R), measured from the tunnel center.

Between 1960 and 1980, the concept of a pressure arch around the tunnel was used, which can be defined as the geometric location of the points, placed near the tunnel crown, in which the tangential pressure is maximum.

In a circular tunnel excavated in homogenous ground and with a hydrostatic stress field, the pressure arch is a circular arch, which, in the case that that ground remains in elastic regime after excavation, it matches exactly with the perimeter of the excavation as was said previously.

As ground plasticization grows, the position of the pressure arch moves away from the excavation perimeter and the pressure arch will always separate those grounds with an elastic behavior from those grounds that yield after the tunnel excavation.

2. **In order to stabilize the tunnel it is necessary that the support elements provide a radial stress (P_i)** to equilibrate the ground radial stress value on tunnel perimeter (σ_0^r).

The radial stress provided by the support elements of the tunnel is much lower than the tangential pressure value in the ground, which has to be supported by the ground.

In accordance with the above, as an answer to the issue raised at the beginning it turns out that **in tunnels excavated in grounds with elastic behavior, the ground is responsible for the stabilization process** and if the ground yields when the tunnel is excavated, it will continue bearing most of the stress increment that is produced, so the support elements must apply a much lower stress than the one produced by the ground in the stabilization process.

1.3.2 Rock mass and intact rock

In the previous section, the term ground was used to designate the media in which the tunnel is excavated and whose uniaxial compressive strength has been referred to as the key parameter to know if the tunnel is excavated in a ground with elastic behavior.

This creates an important problem, as the uniaxial compressive strength of the natural ground proves to be difficult to measure in an accurate way, due to the large dimensions that samples to be tested should be so that, during the test, the effect of the natural discontinuities present in the rock mass could be taken into account.

This problem has been solved by making laboratory tests with ground samples without discontinuities, which corresponds to what is known as "intact rock". In Figure 1.12, taken from Hoek and Brown (1980), the concept of intact rock and rock mass is clearly illustrated.

In this figure, it can be seen that the transition from the intact rock, which has no discontinuities, to the rock mass, can be obtained by enlarging the volume of the sample to be tested and the consequence is that as the sample volume is enlarged, the strength value decreases, a phenomenon known as the scale effect.

Rarely, *in situ* tests are carried out with samples having a volume larger than several cubic meters, as such tests are extremely expensive. It is usual to test samples of intact rock in the laboratory which normally have a sample volume of around one liter or less.

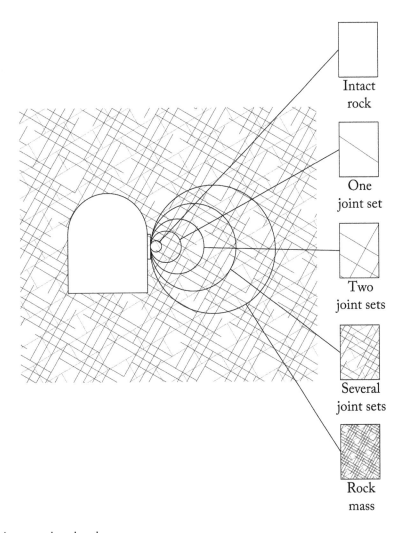

Figure 1.12 Intact rock and rock mass.

Source: Hoek and Brown, 1980.

To estimate the rock mass strength, which is always much lower than that of the intact rock, many years ago the intact rock uniaxial compressive strength used to be divided by a parameter varying from zero to ten depending on the ground degree of fracturation.

Subsequently, the strength reduction for taking into account the scale effect has been taken into account using the Rock Mass Rating (RMR), established by Bieniawski in 1973, as described in Section 8.6.1.3.

1.3.3 Evaluation of the ground elastic behavior

In Section 1.3.1 the influence of the tunnel support was discussed as depending on whether the ground excavated continues in the elastic domain or yields.

To evaluate if the ground behaves elastically when excavating an underground structure, the Index of Elastic Behavior (ICE, in Spanish acronym) was proposed by Celada (Bieniawski et al., 2011) which was found to be of great utility.

The ICE is defined by the following equations:

$$\text{For } K_0 < 1 \quad ICE = \frac{3704 \cdot \sigma_{ci} \cdot e^{\frac{RMR_C - 100}{24}}}{(3 - K_0) \cdot H} \cdot F \tag{1.6}$$

$$\text{For } K_0 \geq 1 \quad ICE = \frac{3704 \cdot \sigma_{ci} \cdot e^{\frac{RMR_C - 100}{24}}}{(3K_0 - 1) \cdot H} \cdot F \tag{1.7}$$

where:

K_0 = Natural stress distribution coefficient
σ_{ci} = Intact rock uniaxial compressive strength (MPa)
RMR_c = Rock Mass Rating corrected depending on the discontinuities orientation
H   = Overburden from the tunnel crown (m)
F   = Shape factor, which takes the following values:

- Circular tunnels with diameter 6 m: F = 1.3
- Circular tunnels with diameter 10 m: F = 1.0
- Conventional tunnels with diameter 14 m: F = 0.75
- Caverns 25 m wide and 60 m high: F = 0.55

The ICE has been defined so that a value of 100 corresponds to the excavation elastoplastic limit. However, as this index encompasses parameters known with some uncertainty, it is considered that an excavation will have a behavior between the elastic and plastic limit if its ICE is in the range from 70–130.

The criteria used to classify the stress–strain response of a tunnel as a function of its ICE are shown in Table 1.3.

The ICE can be calculated in a quick way, using parameters which are basic to define the stress–strain behavior of the ground, that provides an efficient estimation of the difficulties in the tunnel construction.

In Figure 1.13 the distribution of yielded elements is shown, obtained by solving a finite element geomechanical model, for two tunnels of conventional section with ICE values of 79 and 26.

Table 1.3 Ground stress behavior after underground excavations

ICE	Stress–strain behavior
> 130	Fully elastic
70–130	Near elastic
40–69	Moderated yielding
15–39	Intense yielding
< 15	Very intense yielding

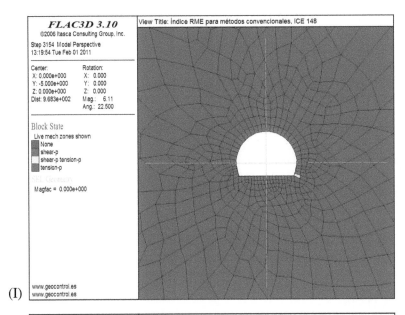

(I)

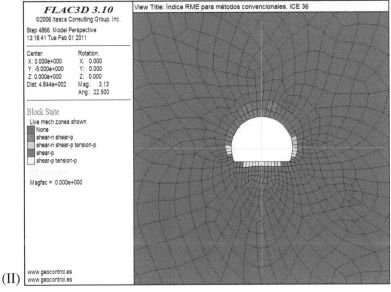

(II)

Figure 1.13 Distribution of yielded elements around two tunnels excavated in grounds with ICE values of 79 and 26. I. ICE = 79. II. ICE = 26.

1.3.4 Effect of water during tunnel construction

The presence of water in the ground always leads to an added difficulty during tunnel construction; but the precise estimation, during the detailed design, of the effect that water presence carries during tunnel construction still remains an unsolved issue.

In the following sections, two basic aspects in the design of tunnels excavated below the water table are presented.

1.3.4.1 Water infiltration into the tunnel

Infiltrated water flow inside the tunnel mainly depends on the permeability of the ground surrounding the tunnel. Grounds with permeability lower than 10^{-6} m/s, which is often greater than the permeability shown by clayey ground, can be considered as impermeable and tunnels excavated in this ground will not have water infiltrations.

An illustrative example is the Eurotunnel, whose geological profile is shown in Figure 1.14. Out of the 50.5 km of the Eurotunnel, 23.5 km were constructed below the English Channel with a maximum cover above the tunnel crown of 125 m, of which 50 m correspond to the water surface.

The tunnel was excavated in a marly chalk layer called the Chalk Marl, which is practically impermeable, and on which a chalk layer called Grey Chalk rested, which allowed the construction of the tunnel without water infiltration.

However, the most common situation when constructing a tunnel under the water table is that the excavation has a draining effect which could capture very important amounts of water and create construction problems.

The worst situation is the one produced by the water entry inside the tunnel when it is excavated downwards, as a pumping system is needed to keep the excavation face in good working condition. When the pumping system is insufficient or fails, there is no other remedy than to stop the construction until the excavation face is free of water. In Photograph 1.17 the excavation face from La Breña Tunnel in La Palma (Spain) is shown, with water infiltration of around 115 l/s.

When construction is done upwards, water entry inside the tunnel creates fewer problems, as gravity helps water to flow out of the excavation face.

When the tunnel is placed under the water table there are two situations that have to be foreseen during the construction phase, due to the great consequences that can be incurred: the flow of fine particles and infiltrations with large flows.

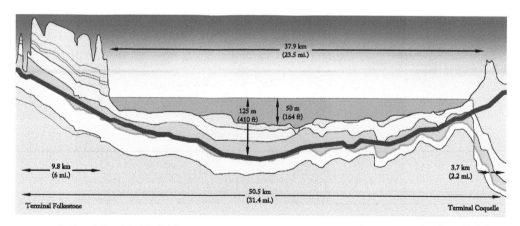

Figure 1.14 Eurotunnel geological profile (France, England).

Photograph 1.17 Water infiltration inside La Breña Tunnel, La Palma (Spain).

Water entry through the tunnel excavation face with fine particles is a highly dangerous situation, as, if it is not stopped, it can lead to excavation collapse due to the voids created next to it.

To prevent this risk, the probability of this happening has to be evaluated during the detailed design and, in the case of it being significant, a methodology has to be planned in order to control water entry and improve the ground before resuming the excavation.

In tunnels excavated upward, it is certain that water creates great problems when its flow exceeds 100 m³/s. In Photograph 1.18 water entry inside a Tunnel Boring Machine with a calculated flow of around 250 l/s is shown.

These situations can be foreseen in the detailed design, but it is very difficult to establish specific measures to solve problems caused by large infiltrated water flows, as the efficiency of these measures depends a lot on the way in which water is infiltrated at the excavation face and this can only be done at the working site.

The situation illustrated in Photograph 1.17 was solved by drilling holes, of 100 mm diameter and around 20 m long, behind the Tunnel Boring Machine, with the aim of collecting the water before its infiltration in the excavation face.

Photograph 1.18 Tunnel Boring Machine constructing a tunnel with a water entry calculated of around 250 l/s.

1.3.4.2 Degradation of rock due to water

Rocks with high strength are not affected by the water but in rocks with medium to low strength, water can produce important undesirable effects, as in saline rocks, anhydrites and, in general, soft rocks.

Saline rocks, of which the most common is the salt rock, are fully soluble in water, which simply stops the construction of tunnels in these grounds if during construction infiltrations with continuous flows are produced.

Tunnel excavation through anhydrite layers is very dangerous due to the elevated pressures generated during its hydration. In the second half of the last century, many tunnels were destroyed after construction as they were excavated in anhydrite, without taking adequate measures into account.

Anhydrite is usually a layered rock with a uniaxial compressive strength around 60 MPa, which consists of anhydrous calcium sulfate, very common in Keuper grounds at depths greater than 60 m. If anhydrite comes in contact with water it is hydrated and becomes gypsum, through a very slow chemical reaction that generates a significant increase in volume.

When anhydrite is confined, as happens when excavating a tunnel, the increase in volume produced generates swelling pressures reaching 7 MPa, which results in a long-term overload on the support elements and the lining which, if not taken into account when dimensioning them, can destroy them completely.

Much more common than the two previous events is the loss of strength caused by water infiltration into a tunnel when soft rock moisture content is present. In Figure 1.15 the uniaxial compressive strength variation as a function of the loamy chalk moisture content in which the Eurotunnel was excavated is shown.

In Figure 1.15 it can be seen that the uniaxial compressive strength of the loamy chalk from the Eurotunnel exceeds 25 MPa for a moisture content of 2%, but decreases to 8 MPa if the moisture increases to 8%.

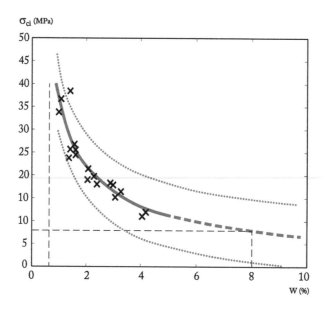

Figure 1.15 Uniaxial compressive strength of the marly chalk from the Eurotunnel as a function of its moisture content.

A similar situation was found during the San Formerio Tunnel design, located in Burgos (Spain), as illustrated in Figure 1.16.

The uniaxial compressive strength of these loams was also evaluated at around 25 MPa with a moisture content of 2%, but it decreased to 5 MPa if moisture rose to 8%.

Figure 1.17 jointly represents the strength data of the loamy chalk from the Eurotunnel, the loams from the San Formerio Tunnel and of other similar rocks.

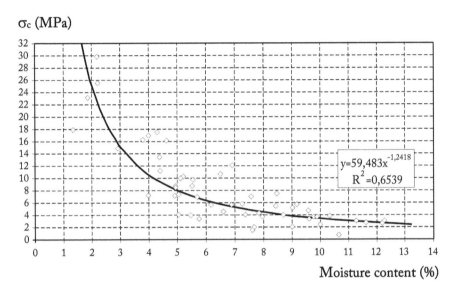

Figure 1.16 Uniaxial compressive strength of San Formerio Tunnel loams, as a function of the moisture content.

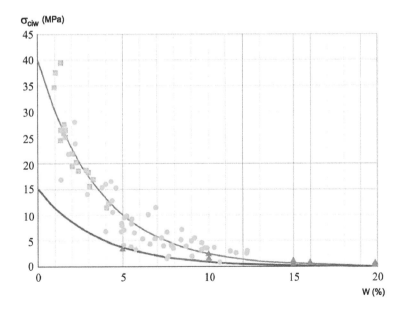

Figure 1.17 Uniaxial compressive strength from some soft rocks as a function of its moisture content.

From the data contained in Figure 1.17 (Celada, 2011) the following correlation was established:

$$\sigma_{ci\omega} = \sigma_{ciseca} \cdot 10^{\frac{-\omega}{8}} \tag{1.8}$$

where:

$\sigma_{ci\omega}$ = uniaxial compressive strength with moisture content ω
σ_{ciseca} = uniaxial compressive strength from a dry sample

With this correlation, the uniaxial compressive strength of soft rocks for a particular moisture content can quite accurately be calculated, which is very useful in order to know the uniaxial compressive strength of rocks sensible to moisture variations, in real moisture conditions.

Water is always used when exploration boreholes are drilled unless triple tube samplers are used, therefore the obtained samples will have a higher moisture content than in the ground. Under these conditions, if dry strength with the natural moisture content is not corrected, most likely much lower strength than the real one will be assigned to the ground and the obtained designs will be unrealistic.

1.3.5 Tunnels and structures

Tunnels and structures are engineering works in which ground plays a very different role, as for tunnels the ground is the strength medium itself, but in structures it represents just a dead load, as illustrated in Figure 1.18.

From this, it is deduced that the concepts and methodologies applied in structures calculation are not valid in tunnel design, except for two cases in which there can be some similarity between tunnels and structures: shallow tunnels and the dimensioning of the lining.

1.3.5.1 Shallow tunnels

In tunnel portals, in order to restore the original topography, it is very common to construct some structures and, then, cover them with soils. They are known as cut-and-cover tunnels. In Photograph 1.19 one such tunnel construction is shown.

Cut-and-cover tunnels are actually underground structures in which ground acts just like a dead load, so they must be dimensioned by applying the criteria used for structures calculation.

Differential feature	Structure	Tunnel
Construction material	Artificial and well defined	Natural soil
Loads definition	A priori well defined	Have to be calculated

Figure 1.18 Different features of an open air constructed structure and a tunnel.

Photograph 1.19 Cut-and-cover tunnel construction.

Photograph 1.20 Gallery access to Ñuñoa Station, Line 6 in the Santiago de Chile Subway.

Shallow tunnels, defined as those in which the cover above the tunnel crown is less than the tunnel width, constitute a limit case between tunnels and structures, which have to be analyzed in detail to apply the correct design criteria to enable taking advantage of the ground strength.

In Photograph 1.20 a view of the gallery access to Ñuñoa Station, of Line 6 in the Santiago de Chile Subway is shown, which has been constructed in gravel with an overburden above the tunnel crown of 67.2% the gallery width.

To make the decision about the design method to be applied for a shallow depth tunnel, it is necessary to know if the creation of a pressure arch around the tunnel is possible.

If a pressure arch is possible, the ground will play the role of being the strength element and it would not be necessary to consider the overburden as a dead load.

Underground works in Ñuñoa Station, shown in Photograph 1.20 have been designed with tunnel engineering criteria, considering the existence of a pressure arch, and have been constructed without any problem.

1.3.5.2 Dimensioning of the lining

Normally in underground excavations, there is a distinction between the support elements and the lining, also referred to as primary lining and secondary lining in some countries.

The support elements of an underground excavation are the set of strength measures which are placed in order to reach excavation stabilization and have to be assessed with the tunnel engineering criteria that are presented in Chapters 9 and 10.

The lining of an underground excavation is normally composed of a cast in place concrete or shotcrete layer, which is placed when the excavation has been stabilized, as a complementary long-term protection element.

A tunnel can be subjected to time-dependent disturbance, such as earthquakes, water table variations, fires or soil degradation by the water effect.

Photograph 1.21 Cast in place lined tunnel.

Photograph 1.22 Tunnel lined with shotcrete and reflective plates.

If these overloads can be quantified, the tunnel lining must be designed taking them into account and applying the criteria usually applied for concrete structures.

When there are no time-dependent effects there is no reasonable method for dimensioning the tunnel lining and, in these cases, it is usual to adopt a thickness of 30 cm if cast in place concrete is used or of 15 cm for shotcrete. In the Photograph 1.21 a cast in place lined tunnel is shown and Photograph 1.22 shows a tunnel lined with shotcrete and reflective plates.

BIBLIOGRAPHY

Bieniawski, Z.T., Aguado, D., Celada, B., Rodríguez, A., "Forecasting tunnelling behaviour", *Tunnels and Tunnelling International.*August. 2011, pp. 39–42.

Celada, B., "Caracterización de rocas sensibles al agua", *Ingeopres.* No. 209. 2011, pp. 14–24.

Hoek, E., Brown, E.T., *Underground Excavations in Rocks.* Institution of Mining and Metallurgy, London. 1980.

Kastner, H., *Statik des Tunnel-und Stollenbaues auf der Grundlagen geomechanischer Erkenntnisse.* Springer, Berlin/Göttingen. 1962.

Kirsch, E.G., "Die Theorie der Elastizität und die Bedürfnisse der Festigkeitslehre", *Zeitschrift des Vereines deutsher Ingeniere.* No. 42. 1898, pp. 797–907.

Kovari, K., Fechtig, R., *Percements historiques de tunnels alpins en Suisse.* Société pour l'art de l'Ingénieur Civil, Zurich, Switzerland. 2000.

Széchy, K., *The Art of Tunnelling.* Akademiai Kiadó, Budapest, Hungary. 1970.

Chapter 2

Tunnel design methodologies

Z.T. Bieniawski von Preinl and Benjamín Celada Tamames

Scientists discover what is, engineers create what has never been.

Theodore von Kármán, 1911

2.1 INTRODUCTION

The engineer's professional activity culminates in *design*, the word that comes from the Latin *designare*. It can be defined as the socioeconomic activity in which scientific and performance criteria are applied with ability, imagination and common sense to create processes, systems or things with an economic and aesthetic benefit and that are environmentally acceptable.

This definition contains an important message that the designer needs not only the knowledge to design but also the knowledge about the process of design itself.

As indicated by the *Austrian Society for Geomechanics* (2010), a good design product is the result of an intelligent combination of technical expertise, objectives, tools, analysis and management.

Design is one of the oldest human endeavors going back to prehistoric times when mankind conceived hunting implements, shelters and clothing. These and later inventions preceded the development of the sciences by many centuries. However, until recently, design was more an art rather than a science: experience was more important than formal education.

Tunnel design and its construction methods have had an impressive evolution over the last two centuries, especially since the second half of the 20th century, made possible by the contributions of a wide variety of disciplines, such as geology, soil and rock mechanics, civil engineering, geological engineering and mining engineering.

The aim of this chapter is to introduce the methodology of the design process, its guiding principles and the design methods currently in use in tunnel engineering.

2.2 HISTORICAL PERSPECTIVE

In the early millennia of human civilization explanations for many ordinary events were rooted in superstitions or unusual beliefs, which instead of providing reasonable explanations, limited the development of human knowledge.

Greek civilization brought about a major step forward which was the observation of everyday life situations, leading to natural explanations of many formerly inexplicable events.

Probably one of the best examples of progress through observation was by Hippocrates of Kos who managed to implement scientific medicine, separating it from religious mysticism, through the use of intelligence and the senses to diagnose diseases.

Also of note at that time was Aristotle, who is considered to be the father of biology, zoology, botany and anatomy, among other sciences, which use the observation and subsequent classification of data as a method.

In the 16th century Galileo Galilei astounded the world with numerous discoveries achieved when applying a **hypothetic-deductive methodology** composed of four stages:

- Problem observation
- Elaboration of an explanatory hypothesis
- Mathematical deduction of the solution
- Experimental verification of the solution

For the selection of a design method it is also useful to mention **empiricism**, a philosophical concept that reached its peak during the 17th and 18th centuries, and advocated that **experience was the basis for the origin of knowledge**, although the merit of this approach was gradually diluted in the 19th century with the emergence of the scientific method.

2.3 DESIGN METHODOLOGIES IN TUNNEL ENGINEERING

During the 20th century, tunnel design has progressed much slower than other branches of the engineering discipline, probably due to two specific difficulties:

- Experimental verification of the solutions to be applied is only possible during the tunnel construction process.
- The mathematical basis of the solutions to be applied is extremely difficult due to the lack of representative data about the ground behavior and of proper calculation tools.

In the evolution of the tunnel design process, a very significant event was the participation of Karl von Terzaghi in the Chicago Subway construction in the United States in the late 1930s, which, despite its importance, is not well known.

Terzaghi took part as a consultant for the Chicago Subway construction from late 1938 to mid-1941 and had the help of an able collaborator, Ralph B. Peck.

Chicago Subway had to be constructed in soft clayey soils, under the water table, with anticipated major construction problems.

To characterize the ground along the project layout, boreholes spaced 100 m apart were executed, whose length reached 3 m below the tunnel invert, which enabled the extraction of 2-inch-diameter drill cores. With these cores more than 10,000 uniaxial compressive tests were carried out, whose results ranged between 50 and 20 KPa (0.5 to 0.2 Kp/cm^2).

Once the works began, subsidence on the surface reached 300 mm, resulting in significant damage to some buildings and it was necessary to rethink the project in order to ensure subsidence of less than 100 mm.

To address the problem, Terzaghi proposed to measure the ground displacement at full scale by making topographic measurements, both on the surface and under the ground, measuring the displacements of the forepoling made by rammed rebar spiles, as well as the displacements of the tunnel perimeter.

The result of these tests enabled definition of the advance length during the tunnel construction, which was associated with the surface displacements of the ground. Surface surveying results were transmitted in real time, by telephone, to the excavation equipment so that the operators could modify, if necessary, the length of advance. Furthermore, the construction process and the elements of support used were varied to better control the earth pressures.

With the application of this methodology, subsidence was substantially reduced to below 100 mm and the Chicago Subway construction was completed without major problems.

A few years later, in 1948, Terzaghi and Peck published their well-known book *Soil Mechanics in Engineering Practice*, in which they introduced the observational method, based on their experiences in the Chicago Subway, whose aim was to save cost in construction by avoiding ultra-conservative designs.

In 1969 Peck published an article about the "Advantages and Limitations of the Observational Method in Applied Soil Mechanics" defining the following stages:

1. Exploration sufficient to define ground behavior
2. Estimation of the most likely ground behavior and the evaluation of the most unfavorable conditions that might be encountered in the works
3. Create the design considering the most likely ground behavior
4. Selection of the variables that have to be controlled during construction and forecasting their range
5. Recalculations of the design performed with the most likely behavior and also considering the most unfavorable behavior of the ground, to establish the limits of the control to be performed during construction
6. Early selection of the measurements to be taken during construction if the ground behavior changes from the most likely to the most unfavorable case
7. Monitor variable measurements during construction
8. If necessary, modify the design in order to adapt it to the real behavior of the ground

The observational method suggested by Peck has had many detractors in the nearly five decades since its formulation, due to two very different reasons:

1. The need for the zero-risk support, that is impossible to achieve, to justify highly conservative ground characteristics.
2. The great difficulties existing in the 1960s and 1970s to make tunnel designs with adequate scientific support, due to the lack of appropriate calculation tools. These tools, only fully operational in the tunneling field with the advent of computer modeling since the beginning of the 21st century, are essential to narrowing the variables and to performing the back-analyses necessary to modify the initial designs.

Indeed, the progress in tunnel design methodology has had to wait until 1980, the year in which Hoek and Brown published their book *Underground Excavations in Rock*, which consolidated most of the knowledge about tunnel designs available at that time.

Shortly afterward, Bieniawski (1984) in his book *Rock Mechanics Design in Mining and Tunnelling*, proposed the following stages for design:

1. Identification of the need to solve a problem
2. Problem statement, including identification of the objectives to be achieved in terms of stability, security and economy
3. Collection of the most relevant problem information
4. Conceptual formulation, research about the method to be applied, models and hypotheses
5. Analysis of the problem components, including the use of heuristics.

6. Summary of alternative solutions
7. Ideas evaluation and test of the solutions
8. Optimization
9. Solution specification
10. Implementation

Brown (1985), following the steps from Peck's observational method (1969), established the following stages of design:

1. Ground characterization
2. Geotechnical model formulation
3. Design of the solution
4. Ground behavior monitoring
5. Retrospective analysis, based on monitoring data

The International Tunnelling Association (ITA) established the activity diagram for tunnel design shown in Figure 2.1 (Duddek 1988) which emphasizes the importance of ground characterization, but does not include the concept of feedback in the design.

More recently, Bieniawski (1992) in his book *Design Methodology in Rock Engineering* proposed six principles for tunnel design:

1. **Clarity of the objectives:** the design objectives have to be clearly established, in terms of security, stability and economy.
2. **Decrease of the geological uncertainty:** the best design is the one based on the smallest geological uncertainties; there must be enough geotechnical data and it must be representative to achieve the design objectives.
3. **Simplicity in the design components:** the complexity of any design solution can be minimized by subdividing the design system into the lowest possible number of components.
4. **Utilization of the state of the art:** the best design will be that maximizing the transference of technology between the state of the art and the industrial practice.
5. **Optimization:** a good design should include an evaluation of the different alternatives, including cost analyses.
6. **Constructibility:** the best design enables efficient construction, defining the appropriate construction method, which also facilitates an efficient construction contract.

Special recognition should be given to the principle of constructibility as this principle is a unique innovation in rock engineering. In mechanical engineering, the concept of "design for manufacture" has led to highly streamlined production processes. The "design for construction" concept advocated here encompasses innovation as the essence of engineering practice. Figure 2.2 groups the principles and design stages as suggested by Bieniawski (1992), grouped into the three standard activities in tunnel engineering:

1. Prefeasibility studies (conceptual studies or profile studies, among others) in which several solutions are proposed and analyzed for one to be selected.
2. Feasibility studies, in which the one alternative previously selected in the prefeasibility study is developed.
3. Detailed studies or construction engineering, in which the final design for the construction is developed.

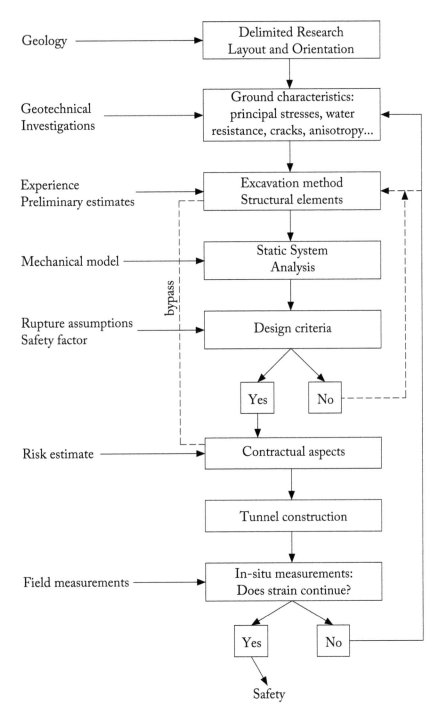

Figure 2.1 Activity diagram for tunnel design.

Source: Duddek, 1988.

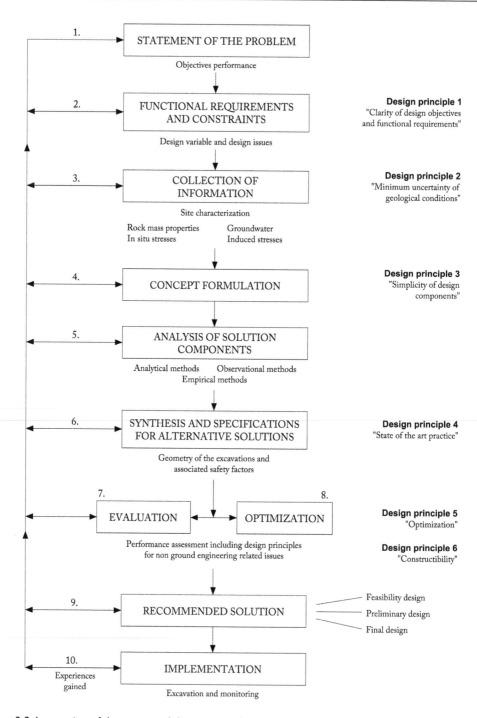

Figure 2.2 Integration of the stages and design principles into an overall methodology.

Source: Bieniawski, 1992 by permission of Taylor & Francis/Balkema.

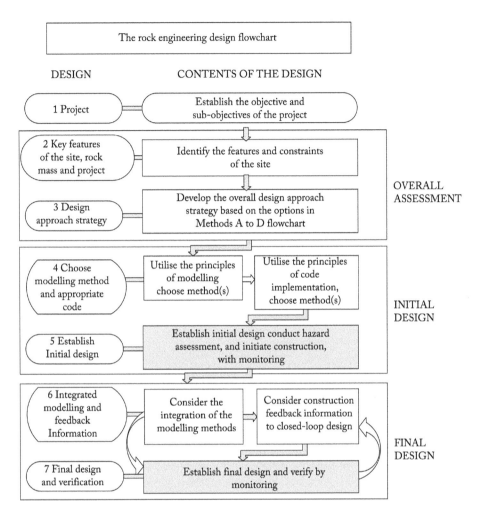

Figure 2.3 Flow chart for a rock engineering design process.

Source: Feng and Hudson 2011, by permission of Taylor & Francis.

Based on the above concept of Bieniawski, but nearly 20 years later, Feng and Hudson developed a similar chart, adding rock mechanics modeling as part of the process, as presented in Figure 2.3.

2.4 TUNNEL DESIGN PRACTICE

The following sections describe the methods and practice for tunnel design of relevance in the past and those in use today.

2.4.1 Early developments

During the first half of the 20th century there was a wide proliferation of methods to calculate the tunnel support elements, which were based on models to calculate the loads imposed on the tunnel support; among these those of Protodiakonov and Terzaghi stand out.

In the second half of the 20th century the so-called New Austrian Tunneling Method (NATM) emerged strongly, and some years later the so-called method of convergence–confinement characteristic curves emerged, which was considered the scientific support for the NATM.

2.4.1.1 Protodyakonov and Terzaghi's methods

Protodyakonov and Terzaghi were two practicing engineers in the first half of the 20th century; the former, much younger than Terzaghi, had a great influence in countries affiliated with the Soviet Union and became more specialized in rock mechanics, while Terzaghi has been a great figure on a global scale in the West, particularly in the United States and is considered the father of soil mechanics.

The calculation method of Protodyakonov assumes the presence, in the ground surrounding the tunnel, of an arch where the rock is subjected to compression, a concept similar to the relieving arch of the maximum values of σ_θ, presented in Section 1.3.1.

Protodyakonov considers that this arch of compressed rock can be approximated through a parabolic arc as illustrated in Figure 2.4.

The load to be borne by the tunnel support is given by the weight of the ground below the parabolic relieving arch, calculated as:

$$P = \frac{1}{3} \cdot \gamma \cdot \frac{b^2}{\tan\phi} \tag{2.1}$$

where:

    γ = ground specific weight
    b = tunnel width
    ϕ = ground friction angle

In the case of Terzaghi, he initially developed his method to calculate the loads on the support elements for granular soils; but, later, this method was extended to cohesive soils.

In Figure 2.5 the model assumed by Terzaghi is presented, which assumes the behavior of a collapsible ground around a cavity whose width is defined by the equation:

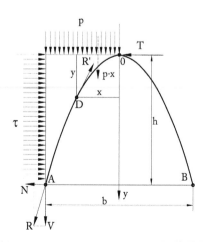

Figure 2.4 Definition of relieving arch by Protodyakonov.

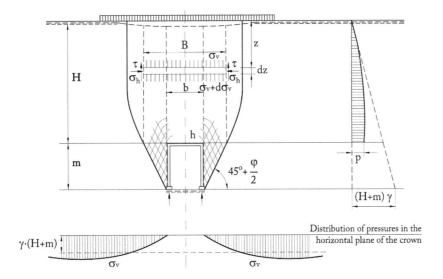

Figure 2.5 Terzaghi model for shallow depth tunnels.

$$B = 2\left[\frac{b}{2} + m \cdot \tan\left(45° - \phi/2\right)\right] \tag{2.2}$$

where:

   b = tunnel width
   m = tunnel height
   ϕ = ground friction angle

In this case, the pressure to be borne by the support elements is calculated by the following expression:

$$p_v = \frac{B\gamma}{2K\tan\phi}\left(1 - e^{-K\tan\phi\frac{2H}{B}}\right) \tag{2.3}$$

where:

   K = ratio between the horizontal and the vertical pressure, empirically set between 1.0 and 1.5
   H = ground cover above the tunnel crown

For deep tunnels, Terzaghi assumes that the relieving arch cannot reach the surface and the calculation model is the one indicated in Figure 2.6.

The pressure on the support elements is calculated through the expression:

$$p_v = \frac{B\gamma}{2K\tan\phi}\left(1 - e^{-K\tan\phi\frac{2H}{B}}\right) + \gamma H_1 e^{-K\tan\phi\frac{2H}{B}} \tag{2.4}$$

If the depth is quite substantial, the former expression can be simplified and it transforms into:

$$P_{max} = \frac{\gamma \cdot B}{2K\tan\phi} \tag{2.5}$$

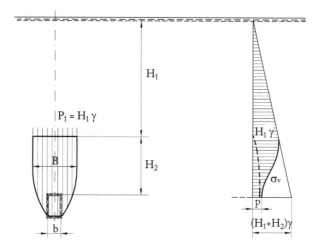

Figure 2.6 Terzaghi model for deep tunnels.

It is interesting to notice that both the expressions of Protodyakonov and Terzaghi show that the ground pressure on the support elements is proportional to the tunnel width.

This conclusion endorses the practical rule that, in poor quality grounds, the best way to construct a tunnel is by dividing the section to be excavated in several phases, with a width smaller than the final tunnel width. From this consideration could also be deduced that the wider the tunnel the greater the pressure to be carried by the support elements and, therefore, the tunnel construction will be more difficult and expensive.

However, this conclusion is no longer valid for tunnel segments with a clearly elastic behavior after excavation.

If a tunnel segment with a fully elastic behavior is considered which has, for example an ICE = 250, the ground surrounding the tunnel withstands a maximum shear stress well below its elastic limit. Under these conditions, if the tunnel width is doubled loads will also double; but as they are well below its elastic limit, it is highly likely that the new loads will not plasticize the ground, so the tunnel will remain self-supporting.

The methods of Protodyakonov and Terzaghi, like all the ones based on an assumed failure model lead to over-dimensioned designs, which only make sense when the tunnel strongly plasticizes and, in addition, the real ground plasticized extension matches the one imposed by the calculation model.

2.4.1.2 *New Austrian Tunneling Method*

At the 13th symposium of the International Society of Rocks Mechanics, hosted in Salzburg (Austria) in 1962, Ladislaus von Rabcewicz presented the basis of what he called the New Austrian Tunneling Method, known worldwide by its acronym NATM.

Two years later, Rabcewicz (1964) published three consecutive articles in the journal *Water Power* titled "The New Austrian Tunnelling Method" which was a remarkable milestone in tunnel design and construction.

Initially, Rabcewicz introduced NATM as a new method in tunnels construction versus those used until then that had one characteristic in common: the excavation and tunnel support were fully dissociated; as once excavated, the support elements were placed, made of brick or masonry which did not get loaded until the excavated perimeter moved away far enough.

This approach led to, in tunnels excavated in heterogeneous grounds with significant displacements, irregular load distribution on the support elements and its failure well after the tunnel excavation.

As a result of the support failures observed in several European tunnels, Rabcewicz proposed to modify the traditional tunneling method by adopting the following criteria:

1. Construct tunnels with almost circular sections, which are statically more favorable than the habitual ones with a horseshoe shape.
2. Place a provisional support as close as possible to the excavation face, composed of a thin shotcrete layer, with thicknesses of 10 to 15 cm, complemented by bolts.
3. Monitor the tunnel stabilization process and, if necessary, reinforce the provisional support.
4. Place the final tunnel lining, once stabilization is achieved.

The NATM generated controversy for a long time, but did gain considerable support as, although the Rabcewicz initial formulation was too descriptive and mixed theoretical concepts with practical experience, in a few years an enormous quantity of information was generated that enhanced the NATM extension.

Probably the most relevant input to the NATM were the recommendations for the definition of the tunnel support sections, shown in Figure 2.7, elaborated by Lauffer (1958).

Looking at Figure 2.7, a question arises as a shortcoming of the NATM, namely: just how does one arrive at a rock mass class? In fact, this is not being done rationally and – in Austria and Germany – the task is simply left to a decision by a "specialist consultant" on any given project. Moreover, in Europe, the NATM serves mainly as a basis for payment purposes, to facilitate understanding of contract documents and to define unit prices for tunneling activities; in America, it is considered a tunnel design method and no monetary amounts are allocated to rock mass classes.

Anyhow, 11 years after the first presentation of the NATM Rabcewicz himself, with the collaboration of Golser (1973), published the work *Principles of Dimensioning the Supporting System for the New Austrian Tunnelling Method*.

When making a detailed assessment about the impact that NATM had in tunnel design, the dominant role that the NATM gave to the rock mass in tunnel support, especially in the segments with the worst behavior, has to be highlighted as of unquestionable value.

In regard to the criticisms to the NATM, it should be noted that Rabcewicz himself in his third article about the NATM (1965) stated:

The new method is a little bit sensitive in its application, especially in soils below the water table [...] The design and dimensioning of the provisional and permanent support elements should be made exclusively by engineers with not only tunnel experience, but also with a wide knowledge about Rock Mechanics.

Unfortunately, the reality experienced in the last forty-five years is that a large proportion of engineers involved in the NATM application have had little knowledge about rock mechanics.

In short, the NATM, in its initial approach, has been mainly an observational and empirical tunnel design method which, due to the limited knowledge and calculation tools for tunnel design existing in the 1960s, did not have enough scientific support to achieve its full development.

ROCK CLASSES	I FROM STABLE TO SLIGHTLY BRITTLE	II VERY BRITTLE	III UNSTABLE TO VERY UNSTABLE	IV SQUEEZING	Va VERY SQUEEZING	Vb LOOSE MATERIAL
CHARACTERISTICS	COMPACT MATERIAL, SLIGHT TO MEDIUM FISSURING	HEAVY DIVISION INTO STRATA AND FRACTURING, SINGLE FISSURES ARE FULL OF CLAYEY MATERIAL; SCHISTOSE INTERCALATIONS	VERY HEAVY DIVISION INTO STRATA AND FRACTURING ON SEVERAL PLANES; FISSURES ARE FULL OF CLAYEY MATERIAL	VERY WEATHERED ROCK: FOLDED AND SCHISTOSE; BANDS OF FAULTS; WELL CONSOLIDATED, COHESIVE, LOOSE MATERIAL	COMPLETELY MYLONITIZED AND WEATHERED REDUCED TO SCREE, NOT CONSOLIDATED, SLIGHTLY COHESIVE	LOOSE MATERIAL, NON-COHESIVE
CHARACTERISTICS	UNI AXIAL COMPRESSIVE STRENGTH σ_c; TANGENTIAL STRESS σ_t; PERMANENT CONDITIONS OF EQUILIBRIUM OR GUARANTEED BY: MEASURES OF LOCAL PROTECTION	σ_{pd} IS GREATER THAN THE REINFORCEMENT OF THE RING OF LOAD BEARING ROCK IN THE CROWN	THE LIMIT STRENGTH OF THE ROCK IS REACHED AND EXCEEDED AROUND THE CROSS SECTION. SUPPORTS AND THE CREATION OF A RING OF LOAD BEARING ROCK ARE NECESSARY	THE TANGENTIAL STRESSES EXCEED THE STRENGTH OF THE ROCK. THE MATERIAL HAS PLASTIC BEHAVIOUR AND TENDS TO MOVE INTO THE CAVITY REDUCING THE CROSS SECTION; INTENSITY OF THE PHENOMENON: MEDIUM STRONG LATERAL THRUSTS AND RAISING OF THE FLOOR. THE MOVEMENTS ARE WITHSTOOD BY THE FULLY CLOSED LOAD BEARING RING		SEE CLASS Va
INFLUENCE OF WATER	NONE	UNIMPORTANT	MAINLY ON THE CAVITY, OF THE FISSURES	FAIR	EVEN STRONG (THE MATERIAL TENDS TO BECOME SOAKED)	
EXCAVATION	FULL FACE	FULL FACE	TOP HEADING AND BENCH	DIVISION OF FACE: I – IV	DIVISION OF FACE: I – VI	DIVISION OF FACE: I – VI

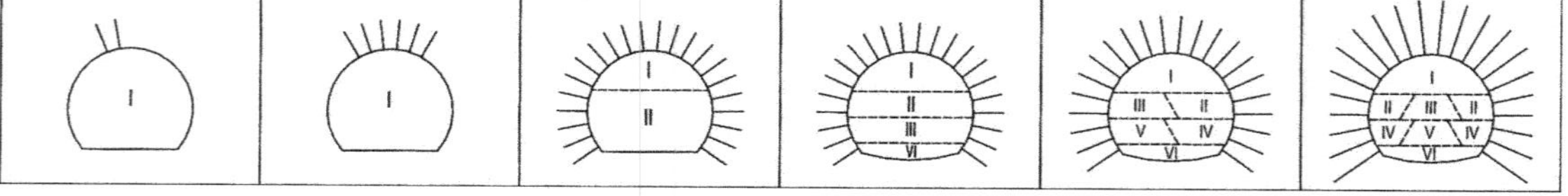

Figure 2.7 Rabcewicz–Pacher classification.

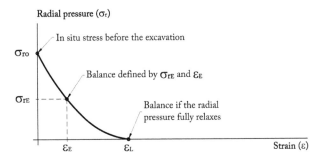

Figure 2.8 Generic characteristic curve of an excavation.

2.4.1.3 Convergence–confinement characteristic curves method

The tunnel characteristic curve is defined as the locus of infinite balance states that can be achieved by limiting the natural radial pressure relaxation when the support elements are placed after the excavation. The assumed characteristic curve of an excavation is shown as a function of the radial pressure on the excavation perimeter and the unit strain that occurs to reach the balance as illustrated in Figure 2.8.

The characteristic curve of an excavation is limited by two extreme points:

- **The natural stress state,** defined by $\sigma_r = \sigma_{ro}$ y $\varepsilon = 0$
- **The equilibrium state when fully relaxing the radial pressure,** where it is fulfilled that $\sigma_r = 0$, $\varepsilon = \varepsilon_L$

The characteristic curve of the tunnel support is normally defined as the linear part of its stress–strain curve, that corresponds to a straight line defined by the equation:

$$\sigma_{rs} = K_s \cdot \varepsilon \tag{2.6}$$

where k_{s0} is the tunnel support stiffness and ε is the tunnel support unit strain for each load step.

In Figure 2.9 the generic characteristic curve of a tunnel support is illustrated, which is limited by the maximum radial pressure value that can be withstood by the support elements ($\sigma_{rs\,max}$).

To define the state of equilibrium achieved in a tunnel by placing the support elements, the characteristic curves of both the excavation and the tunnel support have to be included, as illustrated in Figure 2.10.

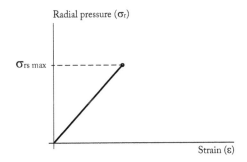

Figure 2.9 Generic characteristic curve of a tunnel support.

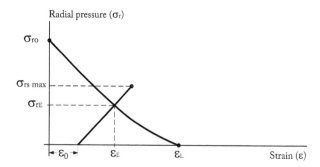

Figure 2.10 Equilibrium when placing the support elements.

At the balance point, the radial pressure on the excavation perimeter matches with the pressure that can be withstood by the support elements when a unit strain is produced on the perimeter ε_E.

The deployment of the characteristic curves enables the definition of the tunnel support safety factor though the following relationship:

$$SF = \frac{\sigma_{rs\,max}}{\sigma_{rE}} \qquad (2.7)$$

The characteristic curves were already used by Pacher (1964) as scientific support to NATM, as illustrated in Figure 2.11.

Since 1984 the Association Francaise de Travaux en Souterrain (AFTES) adopted the characteristic curves method calling it *convergence–confinement*, with the collaboration of Marc Panet (1974 and 1995) as the main scientific driver of this method.

Tunnel convergence is defined as the relative displacement between two reference points placed facing each other on the tunnel perimeter.

According to Sarukai (1997), the strain in the tunnel perimeter is related to the convergence through the expression:

$$\varepsilon = \frac{Convergence}{Tunnel\,width} \qquad (2.8)$$

The characteristic curves are grouped into the three types shown in Figure 2.12.

In a tunnel segment with elastic behavior, produced when ICE>130, the characteristic curve is a straight line and its maximum strain, caused if no lining is placed, is normally below 1%.

If the excavated segment has medium to intense plastification, which occurs if 15<ICE<69, equilibrium without support elements is produced with strains close to 4%.

When plastification is very intense, which occurs when ICE<15, equilibrium is not possible without support elements and the strains before the collapse reach 8%, according to Hoek and Marinos (2000).

The characteristic curves method has two very important limitations; one derives from the difficulty to accurately determine the strains in the tunnel after placing the support elements, ε_o, in Figure 2.10, and the other imposed by the fact that the characteristic curves can only be analytically calculated for tunnels with symmetry in respect to its axis, which implies that the tunnel has to be circular and cannot be shallow.

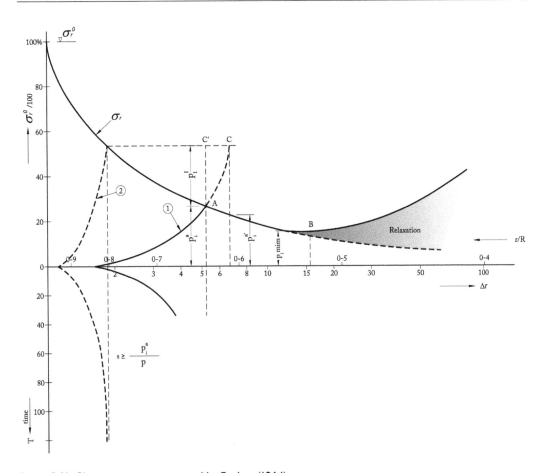

Figure 2.11 Characteristic curves used by Pacher (1964).

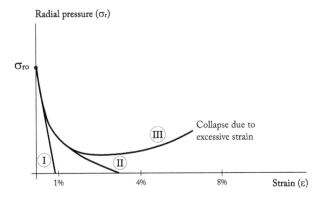

Figure 2.12 Typical characteristic curves.

The difficulty in determining the ground strain after placing the tunnel support, at least in the design phase, can be solved by applying Panet theory about the tunnel excavation face effect, Celada (2003); but the impossibility of calculating the characteristic curves for non-circular sections, can only be overcome by using numerical modeling methods. This has been the main reason for the obsolescence of the characteristic curves method, which

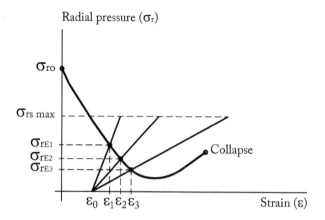

Figure 2.13 Tunnel support stiffness effect.

currently is only of educational interest, for explaining the interaction of ground-tunnel support, as illustrated in Figure 2.13.

In this figure the effect derived from placing tunnel supports with variable stiffness can qualitatively be appreciated, which means that the less stiff the tunnel support is, the lower the pressure needed to achieve the balance ($\sigma_{\varepsilon 1}$, $\sigma_{\varepsilon 2}$ and $\sigma_{\varepsilon 3}$) and the greater the allowed strain in the tunnel perimeter.

This supports Rabcewicz's idea to place yielding supports very close to the excavation face, as in them, the soils quickly confine and their resistant properties can be kept during the stabilization process. Figure 2.14 illustrates the effect of placing tunnel supports, with the same stiffness and strength, at increasing distances from the tunnel face for the initial strains ε_{01}, ε_{02} and ε_{03}.

Following on from this, in cases where the stabilization process is time dependent, the best solution is to reinforce the provisional tunnel support with a stiffer one, for example, with highly accelerated shotcrete.

The delay of the tunnel support installation with respect to the excavation face, has theoretically the benefit of reducing the radial pressure at which the balance is reached; but if this delay is excessive it will lead to the excavation collapse without reaching the limit load of the tunnel support.

Theoretically, the potential of the convergence–confinement characteristic curves method was very high but the limitation of having to accept symmetry with respect to the

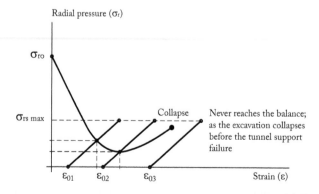

Figure 2.14 Effect of delaying the tunnel support with respect to the excavation face.

tunnel axis, prevents the analysis of the behavior of different points on the tunnel perimeter, which led in the late 20th century to its substitution by the stress–strain analysis methods.

2.4.2 Stress–strain analysis

The stress–strain analysis approach is based on the deployment of approximate numerical calculation methods, which allow symmetries of any kind and little geometrical constraints.

To apply the stress–strain analysis in tunnel design, a numerical model is made, including the ground surrounding the tunnel of the region to be analyzed and with dimensions of six times the tunnel width.

In such models, normally referred to as geomechanics solutions or colloquially 'numerical methods', all types of ground can be included, with or without water, with the desired excavation shapes and considering any kind of tunnel support. These models are divided into elements ensuring that their size is sufficiently small in the parts of the model where more abrupt stress changes are expected.

Figure 2.15 is a reproduction of an example of a tunnel geomechanical model.

Once the numerical model is produced, the initial boundary conditions are defined, which are then altered to become unbalanced, simulating the tunnel construction process in the model, and the displacements and pressures which lead to a new equilibrium state in the model are calculated.

As a result of the calculations, the stress and strain distributions inside the numerical model are obtained, which provides a superior analysis capacity.

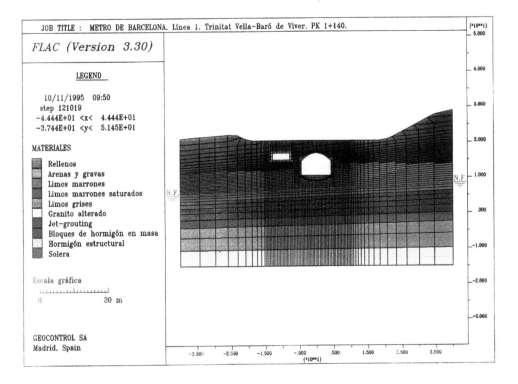

Figure 2.15 Tunnel geomechanical model.

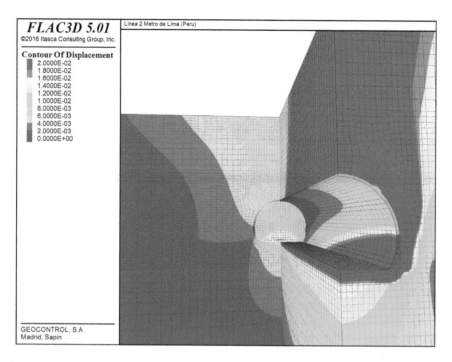

Figure 2.16 Strains around the tunnel.

Figure 2.16 shows ground displacement distribution around the tunnel in the construction final stage.

The stress–strain analysis techniques were already known in the 1960s (Zienkiewicz and Cheung, 1967), but their deployment was extraordinarily time consuming, both due to the high cost of the existing computers and their limited performance.

The emergence of the personal computer (PC) in the 1980s dramatically changed the situation, as the running costs became affordable but, until the end of the 20th century, most of the commercial calculation programs only worked in two dimensions.

With the turn of the century, the PC performances reached an excellent level of performance and three-dimensional calculation programs were developed, which enabled the solving of three-dimensional tunnel models in a few hours.

Currently, the stress–strain analysis applied to tunnel design is well advanced, but calculation limitations are imposed by the lack of reliable input data of the properties that characterize the ground and its behavior under load that, often, does not have the adequate determination.

2.4.3 Other design approaches

Other design approaches in tunnel design feature primarily empirical methods, which are solely based on previous experience. They provide recommendations which are particularly suitable in prefeasibility studies, but are insufficient in detail studies.

2.4.3.1 Empirical recommendations based on the Q index

The Q index was proposed by Barton et al. (1974) to classify the ground depending on its quality and is presented in detail in Chapter 6.

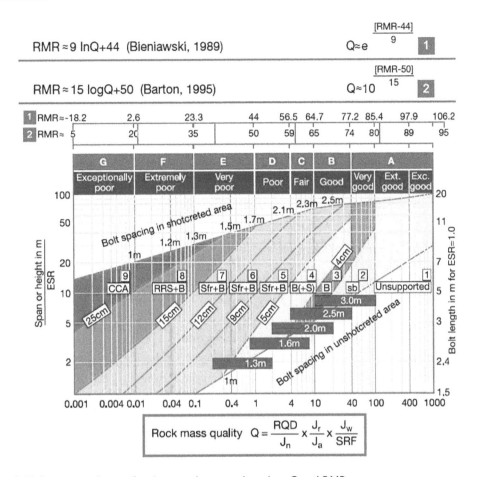

Figure 2.17 Recommendations for the tunnel support based on Q and RMR.

Source: Barton and Bieniawski 2008.

Since its presentation, the Q index was clearly focused on the definition of the tunnel supports to be used in underground excavations and 38 different types of support were defined. Later Grimstad and Barton (1993) summarized these recommendations on an abacus which, fifteen years later, was updated by Barton and Bieniawski (2008) and it is illustrated in Figure 2.17.

In order to use the abacus in Figure 2.17, besides knowing the value of the Q index of the rock mass where the tunnel is going to be designed, the excavation width or height have to be known, whichever is most critical, which is divided by a factor called the excavation support ratio (ESR), which depends on the intended excavation use, as shown in Table 2.1.

With this data it is possible to select the support elements to use among nine predefined types that combine shotcrete, bolts and steel arches.

2.4.3.2 Empirical recommendations based on the ICE

The Elastic Behavior Index (Índice de Comportamiento Elástico [ICE] to give it its Spanish acronym), which was introduced in Section 1.3.3, enables knowing the stress–strain behavior in a tunnel segment and can also be used to set recommendations about the construction process and the tunnel support to be used in tunnels 14 m wide, as it is presented in Table 2.2.

Table 2.1 Excavation support ratio (ESR) values

Excavation type	ESR
Temporary mining excavations.	3–5
Circular vertical shafts.	2.5
Permanent mining excavations, hydraulic tunnels, pilot tunnels, inclined surfaces, large section excavations.	1.6
Storage caverns, water treatment plants, road and railway tunnels with medium section.	1.3
Hydroelectric caverns, large tunnels, military excavations, tunnel portals.	1.0
Nuclear facilities, railway stations and industrial facilities.	0.8

From the data in Table 2.2 recommendations about the following aspects in tunnel design can be extracted:

1. Full section or sequential excavation
2. Type of support elements
3. Complementary support elements
4. Types of lining

2.4.3.3 Empirical recommendations by the Eurocode

The Eurocode was an attempt by the European Union to unify the design criteria for engineering structures. It is divided into 10 parts, and in the Eurocode 7 the criteria to design "geostructures" are defined.

Generally, it must be noted that the Eurocode specifies that the use of the limit state method can be suitable to designing shallow retaining structures, such as piles or slurry walls; but it is not suitable to design tunnels where there is an interaction between the ground and the tunnel support, as was described in Section 1.3.5.

2.4.3.4 Empirical recommendations of the Austrian Society for Geomechanics

The Austrian Society for Geomechanics established some geotechnical design criteria in 2001 for underground structures excavated with traditional methods, that were updated in 2008 and published in English in 2010. It must be highlighted that these recommendations did not mention anything about the NATM.

These recommendations are divided into two parts: one is focused on the design and the other refers to the verifications during construction.

The activities to be performed in the design phase are shown in Figure 2.18 and are grouped into the following seven steps:

1. Determination of the ground type
2. Determination of the ground behavior
3. Selection of the construction concept
4. Selection of the excavation system
5. Selection of the tunnel support
6. Selection of the excavation and tunnel support plan
7. Determination of the excavation types

Table 2.2 Constructive recommendations for tunnels 14 m wide, as a function of the ICE

Ice	Excavation behavior	Section excavation	Orientative support elements	Special support elements	Lining
> 130	Fully elastic	Full section	Bolts $\begin{cases} E_T = 2.5\,m \\ E_L = 2.5\,m \end{cases}$ $L = 4.5\,m$ Shotcrete: 5 cm	None	Crown and walls of hydraulic concrete or shotcrete. Bottom slab on natural ground.
70–130	In the elasto-plastic limit		Bolts $\begin{cases} E_T = 2.0\,m \\ E_L = 2.5\,m \end{cases}$ $L = 4.5\,m$ Shotcrete: 10 cm		Crown and walls of hydraulic concrete or shotcrete. Concrete invert with a vertical deflection of 0.1 × excavation width.
40–69	Moderated yielding		Bolts $\begin{cases} E_T = 2.0\,m \\ E_L = 2.0\,m \end{cases}$ $L = 4.5\,m$ Shotcrete: 15 cm		Vault and facings of hydraulic concrete or shotcrete. Concrete invert with a vertical deflection of 0.2 × excavation width.
15–39	Intense yielding	In phases	Steel arches TH-29, 1 m spacing + Shotcrete: 25 cm	Elefant foot to support the heading arches	
< 15	Very intense yielding		Steel arches HEB-180, 1 m spacing + Shotcrete ≥ 30 cm	Elefant foot. Heavy umbrellas. Bolts on the excavation face. Underpinning micropiles.	Almost circular lining of concrete, reinforced in the invert.

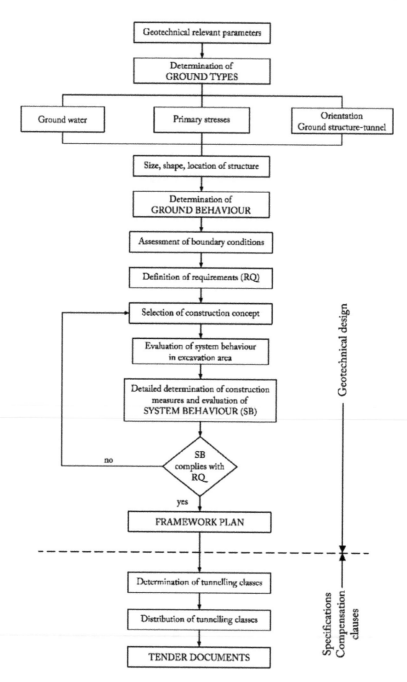

Figure 2.18 Recommendations of the Austrian Society for Geomechanics for the tunnel design.

Source: Austrian Society for Geomechanics 2010.

It must be emphasized that the design phase is aimed at establishing the contractual documents which anticipate the differences observed during construction with regard to the design.

These recommendations draw from the premise that, in many cases, the real ground behavior during the tunnel construction cannot be defined with the required accuracy in the design phase and, therefore, it is necessary to make adjustments in the initial design during the construction.

Figure 2.19 presents the activities to carry out during the tunnel construction, according to the Austrian Society for Geomechanics, which are grouped into the following four stages:

1. Identification of ground and prediction of its characteristics
2. Establishment of the ground behavior in the excavation zone
3. Determination of the excavation process and support elements and ground behavior prediction
4. Verification of the tunnel behavior as built

These recommendations are based on the idea that in order to achieve a safe and economical tunnel construction it is necessary, during construction, to upgrade the geomechanical design according to the new information obtained during the evolution of the works.

2.4.3.5 Empirical recommendations based on the Rock Mass Rating

The Rock Mass Rating (RMR) system was proposed by Bieniawski (1973) and is described in detail in Chapter 6. It is an engineering classification of rock masses, particularly suitable for empirical design of rock tunnels.

The tunnel support guidelines based on RMR were provided originally in the form of a table (Bieniawski 1989) that provides support recommendations for a tunnel span/diameter of 10 meters. In view of the improving technology for rock bolting, shotcrete and steel ribs, it was left to tunnel designers to modify these guidelines for other tunnel sizes, which served its purpose well.

Today, after 45 years of use, practical tunnel designers find it useful to have charts for the selection of rock support as a function of *both* tunnel size and rock mass quality. Accordingly, this was done and Figure 2.20 is a reproduction of a chart, developed by Lowson and Bieniawski (2013). This enables dimensioning of the shotcrete thickness and the bolts length as a function of tunnel width and the RMR values of the rock mass involved.

As shown in Table 2.3, the rock support measures for each rock mass quality include a combination of the various support types. Since, for example, two support methods are additive to some extent, determination of support requirements for individual types, such as rock bolts, shotcrete and steel ribs have been determined.

An important question is what are the current needs of practical tunnel designers. It should be noted, with respect to modeling versus empirical assessments based on accumulated experience, that using continuum models often gives unreliable results for support, particularly at shallow depths, although it is useful for cases where squeezing effects are present and rock mass plasticity is extensive. For most purposes, a practical tunnel engineer needs design charts and simple aids to pragmatic design. It is not just an issue of difficulty or complexity; numerical modeling does not currently have a good way of modeling the support effect from interlocking blocks, unless one uses discontinuum modeling software such as UDEC. However in this case, one faces the problem that the range of joint parameters that comes from laboratory tests is so wide that at one end any excavation is stable and at the other nothing is stable. Experienced modelers can produce convincing results but the modeling process is arguably rather subjective.

Accordingly, the practical tunnel engineer is interested in design rules for both shotcrete and rock bolts as a function of tunnel size as well as rock mass quality, for assessing support requirements.

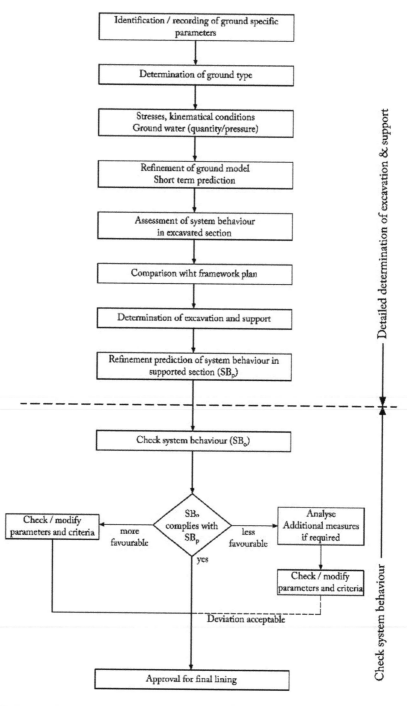

Figure 2.19 Recommendations of the Austrian Society for Geomechanics for the tunnel construction.

Source: Austrian Society for Geomechanics, 2010.

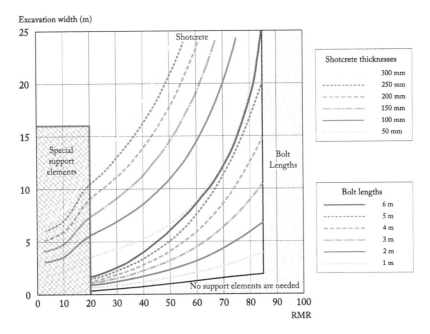

Figure 2.20 Rock bolts length and shotcrete thickness as a function of the excavation width and the RMR.

Source: Lawson and Bieniawski, 2013.

2.4.4 Interactive Structural Design

Interactive Structural Design (DEA in its Spanish acronym, *Diseño Estructural Activo*) was proposed in 1997 in the first edition of the *Tunnels and Underground Works Manual* and was updated in the second edition (Celada 2011).

DEA is a methodology that includes the **design** and **construction** phases, by combining the scientific method with the feedback observational method in three phases:

- **Phase I. Ground characterization:** aims to minimize the geological risks and to evaluate, as well as possible, the stress–strain behavior of the ground.
- **Phase II. Structural design:** aims to select the construction method, define the tunnel support sections, minimize the design risks and estimate the ground displacements during construction.
- **Phase III. Engineering during construction:** aims to find the solution for the construction problems which appear during the works, measure ground displacements to check the accuracy of the calculations and, if necessary, optimize the tunnel construction process taking advantage of the excavated ground behavior.

Concerning ground characterization, it must be noted that, currently, there are no technical problems to accurately characterize the stress–strain behavior of any ground, but in practice many times this target is not reached, due to either the lack of time or lack of economic means.

It is quite common to have uncertainties about the ground where the tunnel is going to be built until the construction is well advanced.

Table 2.3 Original guidelines for support of rock tunnels based on the RMR system (Bieniawski 1989)

Rock mass class	Excavation	Support		
		Rock bolts (20-mm Dia, fully grouted)	Shotcrete	Steel sets
Very good rock I RMR: 81–100	Full face 3-m advance	Generally, no support required except for occasional spot bolting		
Good rock II RMR: 61–80	Full face 1.0–1.5-m advance. Complete support 20 m from face	Locally, bolts in crown 3 m long, spaced 2.5 m, with occasional wire mesh	50 mm in crown where required	None
Fair rock III RMR: 41–60	Top heading and bench 1.5–3-m advance in top heading. Commence support after each blast Complete support 10 m from face	Systematic bolts 4 m long, spaced 1.5–2 m in crown and walls with wire mesh In crown	50–100 mm in crown and 30 mm in sides	None
Poor rock IV RMR: 21–40	Top heading and bench 1.0–1.5-m advance in top heading. Install support concurrently with excavation 10 m from face	Systematic bolts 4–5 m long, spaced 1–1.5 m in crown and wall with wire mesh	100–150 mm in crown and 100 mm in sides	Light to medium ribs spaced 1.5 m where required
Very poor rock V RMR: <20	Multiple drifts 0.5–1.5-m advance in top heading. Install support concurrently with excavation. Shotcrete as soon as possible after blasting	Systematic bolts 5–6 m long, spaced 1–1.5 m in crown and walls with wire mesh. Bolt invert	150–200 mm in crown, 150 mm in sides, and 50 mm on face	Medium to heavy ribs spaced 0.75 m with steel lagging and fore-poling if required. Close invert

Shape: horseshoe; width: 10 m: vertical stress: <25 MPa: construction: drilling and blasting.

Figure 2.21 illustrates the process about the soil behavior uncertainty as a function of time, during the 15 years that can elapse from when the decision of making a 10 km long tunnel is made until its service start-up.

As a starting point it must be highlighted that the ground behavior uncertainty rarely reaches 100% as, via the Internet, we can have access to substantial information about the site where the tunnel is going to be built.

Thus, it is considered that when a prefeasibility study is initiated, the ground behavior uncertainty about the area where the tunnel is going to be excavated can be up to 80%.

As the prefeasibility studies, feasibility studies and detailed engineering are carried out, the ground behavior uncertainty decreases and it is estimated that at the beginning of the tunnel construction the typical uncertainty will be 10%.

A relevant fact derived from Figure 2.21 is that the ground behavior around a long tunnel being excavated will not be fully known until several years after the start of its construction.

Under these conditions, despite using accurate calculation methods for the tunnel design, it is obvious that the designs will have a certain margin of error; which makes it absolutely necessary to observe the real ground behavior during construction and, if necessary, correct the initial design through back-calculations.

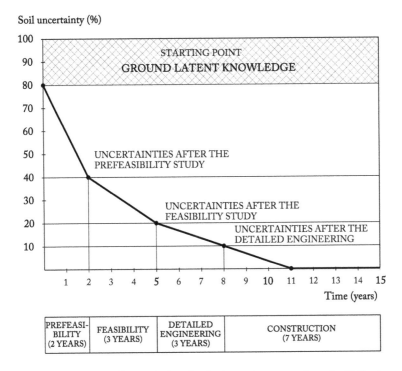

Figure 2.21 Ground uncertainty evolution during the design and construction of a 10 Km long tunnel.

DEA methodology follows the principle of ground behavior observation during the works, but, in case it is necessary to modify the design, it includes the feedback of the calculation models with the ground displacements measures observed in the works.

After several decades of experience in monitoring the tunnel stabilization process, it has been observed that the measurements which best allow controlling the stabilization process are the convergence and the movement of the tunnel crown, because the measurements of the stresses are very inaccurate and have very high costs.

Accordingly, Figure 2.22 shows an activity flow chart that can be used to implement DEA. The activities necessary to develop the ground characterization phase are covered in Chapter 3; although the ground stress–strain behavior must be found by calculating the ICE, which is described in Section 1.3.1.

Upon completion of the ground characterization phase, the tunnel geomechanical profile is obtained which will be divided into ground segments having similar stress–strain properties. This will indicate, for each segment, the stress–strain parameters that characterize the ground behavior.

Chapter 10 introduces the currently available techniques for tunnel structural designs based on numerical methods simulating the stress–strain behavior during the construction process.

As a result of the structural design phase of DEA, a tunnel construction profile is obtained whose length will be divided into segments with the same tunnel support sections, identifying the most relevant construction features.

The activities to carry out in the engineering phase during DEA construction are based on the ground displacement measurements and their comparison with the calculations performed in the design phase.

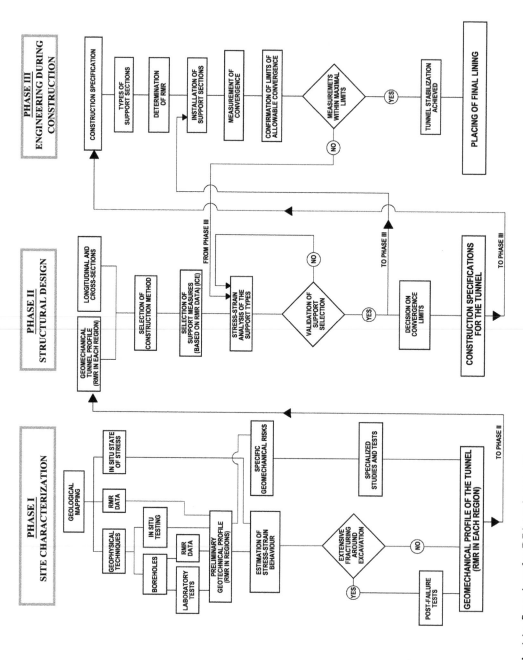

Figure 2.22 Activity flow chart for DEA implementation.

In essence, DEA methodology uses the scientific method for the ground characterization and structural design stages and uses the observational method as a feedback loop to the structural design during tunnel construction.

DEA methodology can be adapted to any design and construction method. When Tunnel Boring Machines are used it is necessary to make some changes because the convergence data cannot be used as the basic parameter to control the stabilization process.

The most recent and informative example of DEA application has been the design and construction of the 16.5 Km of Line 6 in the Santiago de Chile Subway, where DEA proved to be the decisive factor in increased efficiency achieved (Celada et al. 2015 and 2016).

BIBLIOGRAPHY

AFTES, *Recommendations on the Convergence-Confinement Method*. Versión 1. AFTES, Paris. 1984.

Austrian Society for Geomechanics, *Guideline for the Geotechnical Design of Underground Structures with Conventional Excavation*. Translation from Version 2.1. Salzburg, Austria. 2010.

Barton, N., "TBM performance estimation in rock using Q_{TBM}", *Tunnels and Tunnelling International*. September. 1999, pp. 30–34.

Barton, N., Bieniawski, Z.T., "RMR and Q: Setting the Record Straight", *Tunnels and Tunnelling International*. February. 2008, pp. 26–29.

Barton, N., Grimstad, E., "The Q-System following twenty years of application in NMT support selection", *Proceedings of the 43rd Geomechanic Colloquy*. Salzburg, Austria. 1994.

Barton, N., Lien, R., Lunde, J., "Engineering classification of rock masses for the Q design of tunnel support", *Rock Mechanics*. Springer Verlag. Vol. 6. 1974, pp. 189–236.

Bieniawski, Z.T., "Engineering classification of jointed rock masses", *The Civil Engineer in South Africa*. Vol. 15. 1973, pp. 335–343.

Bieniawski, Z.T., *Rock Mechanics Design in Mining and Tunnelling*. A. A. Balkema, Rotterdam, the Netherlands. 1984.

Bieniawski, Z.T., *Engineering Rock Mass Classifications: A Complete Manual*. John Wiley and Sons, New York. 1989.

Bieniawski, Z.T., *Design Methodology in Rock Engineering*. A. A. Balkema, Rotterdam, the Netherlands. 1992.

Bieniawski, Z.T., Celada, B., Galera, J.M., Álvarez, M., "Rock Mass Excavability (RME) index", *ITA World Tunnel Congress*. Seoul, Korea. 2006.

Brown, E.T., "From theory to practice in rock engineering", *Transactions of the Institution of Mining and Metallurgy*. Vol. 94. 1985, pp. A67–A83.

Bruland, A., "The NTNU prediction model for TBM performance", *Norwegian Tunnelling Society*. No. 23. Oslo. 2014.

Celada, B., "Concepto y Diseño del Sostenimiento de túneles", *Manual de Túneles y Obras Subterráneas*. U.D. Proyectos. E.T.S.I. Minas, Universidad Politécnica de Madrid. Capítulo 23. Madrid. 2011.

Celada, B., Fernández, M., "Consideraciones del efecto del frente de excavación de un túnel en los cálculos bidimensionales", *Ingeopres*. Madrid. May 2003.

Celada, B., Adasme, J., González, H., "Experiences in the construction of the interstation tunnels of L-6 Santiago Subway (Chile)", *Proceedings of the World Tunnel Congress*. Dubrovnik. 2015.

Celada, B., Adasme, J., González, H., "Design and engineering during the Construction of Los Leones Station. Line 6. Santiago Subway, Chile", *Proceedings of the World Tunnel Congress*. San Francisco. 2016.

Celada, B., Tardáguila, I., Varona, P., Rodríguez, A., Bieniawski, Z.T., "Innovating tunnel design by an improved experience-based RMR system", *Proceedings of the World Tunnel Congress*. Brazil. 2014.

Celada Tamames, B., "Consideraciones del efecto del frente de excavación de un túnel en los cálculos bidimensionales", *Ingeopres*. May 2003.

Deere, D.U., "Technical description of rock cores for engineering purposes", *Felsmechanic und Ingenieur Geologie*. Vol. 1. 1963, pp. 16–22.

Duddeck, H., "Guidelines for the Design of Tunnels", ITA Working Group on General Approaches to the Design of Tunnels, *Tunnelling and Underground Space Technology*. Vol. 3. No. 3. 1988.

Feng, X., Hudson, J. *Rock Engineering Design*. CRC Press, Boca Raton, FL. 2011.

Fenner, R., "Untersuchungen zur Erkenntnis des Gebirgsdruckes", *Glückauf*. Vol. 74. 1938, pp. 681–695.

Geocontrol, S.A., *Actualización del Índice Rock Mass Rating (RMR) para mejorar sus prestaciones en la caracterización del terreno*. Centro para el Desarrollo Técnico Industrial (CDTI). Proyecto: IDI-20120658. Madrid, Spain. 2012.

Grimstad, E., Barton, N., "Updating of the Q-System for NMT", *Proceedings of the International Symposium on Sprayed Concrete*. Norwegian Concrete Association, Oslo, Norway. 1993.

Hoek, E., Brown, E.T., "Practical estimates of rock mass strength", *International Journal of Rock Mechanics and Mining Sciences*. Vol. 34. No. 8. 1997, pp. 1165–1186.

Hoek, E., Carranza-Torres, C., Corkum, B., "Hoek-Brown failure criterion, 2002 edition", *Proceedings of the 5th North American Rock Mechanics Symposium*. Toronto, Canada. 2002.

Hoek, E., Marinos, P., "Predicting tunnel squeezing problems in weak heterogeneous rock masses", *Tunnelling and Underground Space Technology*. Vol. 12. No. 4. Elsevier, Oxford, UK. 2000.

Hoek, E., Kaiser, P.K., Bawden, W.F., *Support of Underground Excavations in Hard Rock*. Taylor & Francis, London/New York. 1995.

Lauffer, H., *Gebirgsklassifizierung in Sollenbace*. Geologie und Bauwesen. Jg. 24. 1958.

Lowson, A.R., Bieniawski, Z.T., "Validating the Yudhbir-Bieniawski rock mass strength criterion", *Proceedings of the World Tunnel Congress*. ITA, Bangkok, Thailand. 2012.

Lowson, A.R., Bieniawski, Z.T., "Critical assessment of RMR based tunnel design practices: A practical engineer's approach", *Rapid Excavation and Tunnelling Conference*. Washington, DC. 2013, pp. 180–198.

Lunardi, P., *Design and construction of Tunnels. Analysis of Controlled Deformation in Rock and Soils (ADECO-RS)*. Springer-Verlag, Berlin Heidelberg. 2008.

Pacher, F., "Deformations messungen in Versuchsstollen als Mittel Zur Erforschung des Gebirgs verhaltens und zur Bewessung des Ausbanes", *Felsmechanik und Ingenieurgeologie*. Suplementum IV. 1964.

Palmström, A., Singh, R., "The deformation modulus of rock masses. Comparisons between in situ tests and indirect estimates", *Tunnelling and Underground Space Technology*. Vol. 15. No. 3. 2001.

Panet, M., *Le calcul des tunnels par la méthode convergence-confinement*. Presses de l'Ecole National Ponts et Chaussèes, París. 1995.

Panet, M., Guellec, P., "Contribution á l'étude du soutenement derrière de front de taille", *Proceedings of the 3rd Congress International Society Rock Mechanics*. Vol. 12. Part B. Denver, CO. 1974.

Peck, R.B., "Advantages and limitations of the observational method in applied soil mechanics", *Geotechnique*. Vol. 19. 1969, pp. 171–187.

Priest, S.D., Brown, E.T., "Probabilistic stability analysis of variable rock slopes", *Transactions of the Institution of Mining and Metallurgy*. Vol. 92. 1983.

Rabcewicz, L., "The New Austrian tunnelling method", *Water Power*. November, December 1964 and January 1965.

Rabcewicz, L., Golser, J., "Principles of dimensioning the supporting system for the New Austrian Tunnelling Method", *Water Power*. March. 1973.

Sakurai, S., "Lessons learned from field measurements in tunnelling", *Tunnelling and Underground Space Technology*. No. 22. Oxford, UK. 1997.

Teale, R., "The concept of specific energy in rock drilling", *International Journal of Rock Mechanics and Mining Sciences*. Vol. 2. 1965, pp. 57–73.

Terzaghi, K., "Rock defects and loads on tunnel support", *Rock Tunnelling with Steel Supports*. Eds. R.V. Proctor and T. White. Commercial Shearing Co., Youngstown, OH. 1946.

Zienkiewicz, O.C., Cheung, P., *The Finite Element Method in Structural and Continuum Mechanics*. University of Wales, Swansea, UK. 1967.

Chapter 3

Site investigations

Eduardo Ramón Velasco Triviño and Juan Manuel Hurtado Sola

You always pay for a site investigation; whether you have one or not.

Littlejohn (1994)

3.1 INTRODUCTION

Ground characterization is the core of Phase 1 of the active structural design methodology, introduced in the previous chapter.

Ground characterization is so important that it is covered in four chapters: the present chapter deals with site investigation; Chapter 4 covers in situ stresses determination; Chapter 5 is dedicated to laboratory tests and Chapter 6 focuses on the geotechnical classifications, which combine the knowledge derived from chapters 3, 4 and 5.

3.2 PLANNING GROUND INVESTIGATIONS

The general aim of a geotechnical investigation campaign for tunnel design is to identify and quantify the characteristics of the ground which influence the cost and the tunnel construction time.

The scope of the investigations to be **executed** has to correspond to the project phase and to the quantity and quality of the information previously available.

The development of a tunnel project is structured, from its design until its operation, in the following stages:

- Prefeasibility studies
- Feasibility studies
- Finite designs
- Construction
- Operation

Figure 3.1 shows a graph which relates the ground investigation relative cost with the geological–geotechnical level of knowledge gained, for each tunnel design phase.

Important information can be extracted from Figure 3.1: at the end of the site investigation campaign, the level of knowledge about the ground will never be 100%. This implies that, during the tunnel construction, there are factors which could vary and negatively affect the cost and work deadlines.

The geological risk factors have to be identified and, if possible, quantified in the tunnel construction contract.

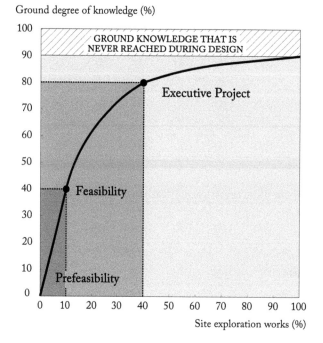

Figure 3.1 Evolution of the ground knowledge as a function of the site investigations performed.

Ground investigation campaigns are normally structured in three stages:

- Preliminary studies
- Geophysical surveys
- Borehole drilling

During the ground investigation campaign, especially when borehole drilling, in situ tests are carried out to quantify the ground properties, as described in this chapter. However, the actual procedures for laboratory tests are dealt with in Chapter 5.

3.3 PRELIMINARY STUDIES

The aim of the preliminary studies is to obtain as much information as possible about the ground where the tunnel is to be designed, using the elementary and low costs techniques described next.

3.3.1 Review of the available data

In many countries there are suitable maps about geological, geotechnical, hydrogeological and geomorphological aspects and mining activities, among others, that could provide very interesting preliminary information about the area where the tunnel is going to be designed. Nowadays, the majority of these publications are easily accessible through the Internet.

Figure 3.2 reproduces the sheet called "El Espinar" from the Spanish Geological Map, at scale 1:50.000, edited by the Geological and Mining Institute of Spain.

On the webpage One Geology (www.onegeology.org) there is access to the geological databases of many countries.

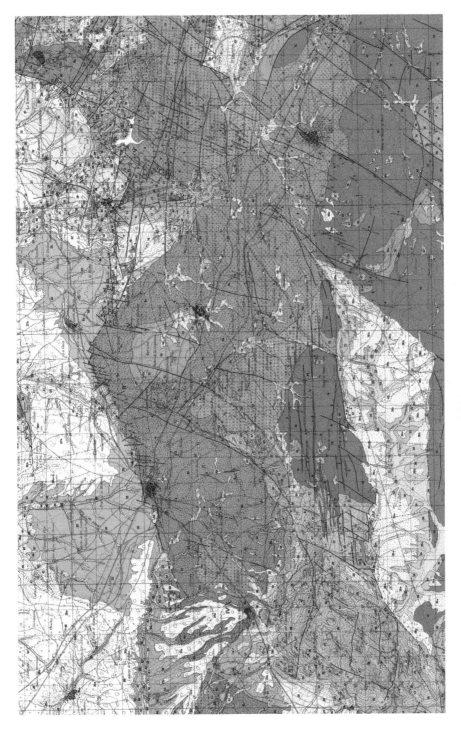

Figure 3.2 Sheet "El Espinar" from the Spanish Geological Map at 1:50.000 scale.

Source: Geological and Mining Institute of Spain.

3.3.2 Photo interpretation

Aerial photo interpretation allows observation of the ground features with regard to the morphology, geological structure, lithology and hydrology.

In the geological photo interpretation, photograph sets are taken orthogonally to the ground, which are overlapped with one another at around 60% in the longitudinal direction. Each picture in turn overlaps with the previous one at approximately 20% so that the whole area of interest is covered and, thus, two frames of the same area (stereoscopic pair) can be obtained which enable viewing the ground in three-dimensions by the use of stereoscopes.

Photograph 3.1 shows one of these devices and Figure 3.3 shows an example of photo interpretation.

The main advantages of photo interpretation are as follows:

- It is a quick and cheap technique, in comparison with other ground investigation methods.
- Allows the study of large surfaces, so it is an extremely helpful tool for the evaluation of different tunnel alignment alternatives.
- In some cases, aerial photographs disclose geological information which is hardly perceptible on the ground. For example, moderate vegetation does not mask the ground features, observable in an aerial photograph, in contrast to what happens in a tour across the ground.
- Aerial photographs are a useful tool to plan the ground investigation campaigns, as well as identifying potentially complex areas from the geotechnical point of view. This allows assessing the accessibility to the points of interest.

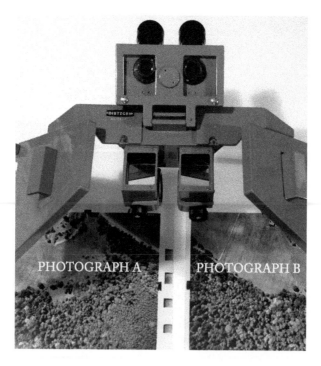

Photograph 3.1 Mirror stereoscope.

Figure 3.3 Example of geological photo interpretation.

3.3.3 Geotechnical cartography

Geotechnical cartography should be done after the analysis of the previous information about the area and making its geological photointerpretation. It is also essential to have a topographic base of the area, to translate the field observations. The scale of the topographic base should be appropriate with the study phase at which the project is and, therewith, with the required level of definition, as indicated in Table 3.1.

For each case, all the information considered of interest should be represented in the plans, where the following items should be highlighted:

- Topography
- Lithological distribution of the different existing units
- Presence of soils and/or altered rocks, with its thickness estimation
- Geological structure: layer strike and dip, presence of discontinuities and faults, alignments, folds, etc.
- Hydrogeological conditions: springs, floodable areas, rivers, ponds, etc.
- Geomorphological conditions
- Ground movements: landslides, creeping, etc.
- Other geological risks
- Previous investigations if existing

Table 3.1 Scales according to the design phases

Design phase	Work scales
Prefeasibility	1:25,000/1:10,000
Feasibility	1:5,000/1:2,000
Project execution	1:1,000/1:500

Figure 3.4 shows an example of the geotechnical cartography performed in the tunnel area.

3.3.4 Geomechanic stations

A geomechanic station is an aerial inspection of a rocky outcrop on the ground surface, which allows identification of the types of rocks constituting the rock mass, the discontinuities and estimates of some strength properties.

To get an accurate and effective description it is recommended to follow the methodological guide *Suggested Methods Quantitative Description of Discontinuities in Rock Masses*, Ulusay and Hudson (2007), published by the International Society for Rock Mechanics (ISRM).

For each of the identified joint sets, the following parameters are established:

- **Type of discontinuity:** stratification, joint, schistosity, fault, etc.
- **Discontinuity orientation:** obtained by measuring the discontinuity dip and dip direction. Figure 3.5 shows a sketch with the meaning of these terms, while Photograph 3.2 shows an example of field data collection.

  Each discontinuity is presented in a stereographic diagram and the result is statistically grouped with the aim of obtaining the average orientation of each set, in terms of the dip and dip direction. Figure 3.6 shows an example of this type of data processing through the software DIPS, of Rocscience (2015).
- **Spacing:** is the perpendicular distance between adjacent discontinuities. Normally it is expressed as the joint set average spacing.

  The studies carried out by Priest and Hudson (1976) indicate that the distribution function of the spacing follows an exponential distribution, like:

$$f(x) = \left(\frac{1}{\mu}\right) e^{-\frac{x}{\mu}}$$

  where μ is the average spacing of a number of measures large enough.

Figure 3.7 reproduces one of the examples presented by Priest and Hudson (1976) about 1,828 measures taken along a 60 m long tunnel and with an average spacing of 3.3 cm.

- **Continuity:** is the maximum joint length along its plane.

  Table 3.2 shows the classification of this parameter according to the ISRM (Ulusay and Hudson, 1977).
- **Aperture:** is the perpendicular distance between joint lips. Figure 3.8 illustrates the concept of joint aperture and filling thickness, if present. According to the ISRM, the classification of this parameter is shown in Table 3.3.
- **Filling:** The most common fillings or gauge consist of mineralizations (calcite, quartz, pyrite, etc.), clay, silts, sands or the decomposed rock itself.

  The strength of a filled joint, except for hard fillings (calcite, quartz, etc.), is less than that of a closed joint.
- **Roughness:** This term refers to the discontinuity waviness and the roughness of the joint surface. Joint roughness is not easy to quantify, as this parameter is very sensitive to the measuring scale, as illustrated in Figure 3.9. Barton and Choubey (1977) have classified the joint roughness in ten groups, by associating the Joint

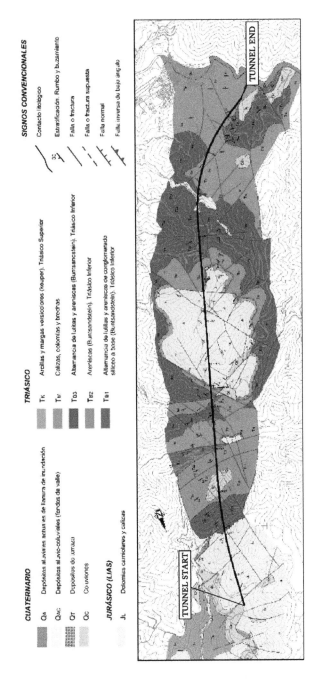

Figure 3.4 Geotechnical cartography of the tunnel alignment.

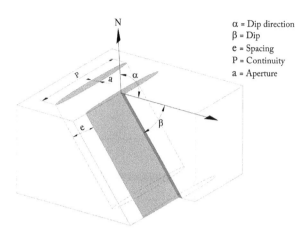

α = Dip direction
β = Dip
e = Spacing
P = Continuity
a = Aperture

Figure 3.5 Geometrical definition of a joint set.

Photograph 3.2 Joint data collection.

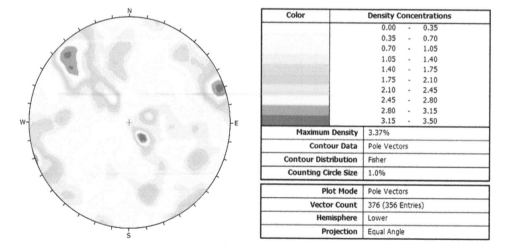

Color	Density Concentrations	
	0.00	-	0.35
	0.35	-	0.70
	0.70	-	1.05
	1.05	-	1.40
	1.40	-	1.75
	1.75	-	2.10
	2.10	-	2.45
	2.45	-	2.80
	2.80	-	3.15
	3.15	-	3.50

Maximum Density	3.37%
Contour Data	Pole Vectors
Contour Distribution	Fisher
Counting Circle Size	1.0%

Plot Mode	Pole Vectors
Vector Count	376 (356 Entries)
Hemisphere	Lower
Projection	Equal Angle

Figure 3.6 Statistical treatment of the discontinuities measured.

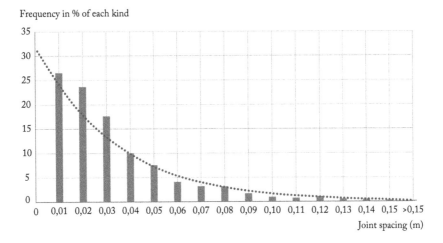

Figure 3.7 Histogram of the joint spacing in lutites.

Source: Priest and Hudson, 1976.

Table 3.2 Description of the joint continuity

Continuity	Trace length (m)
Very low	< 1
Low	1–3
Medium	3–10
High	10–20
Very high	>20

Source: Ulusay and Hudson, 1997.

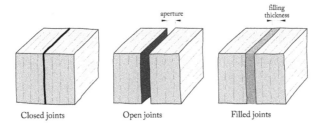

Figure 3.8 Definitions suggested by the ISRM (1977) for the joint aperture and filling thickness.

Roughness Coefficient parameter (JRC) to each of them, as illustrated in Figure 3.10. To enable the roughness profile evaluation, a Barton's profilometer is used as shown in Photograph 3.3.

- **Alteration:** is normally expressed according to the ISRM criteria, shown in Table 3.4.
- **Water presence:** dry joint, wet joint, dripping joint, joint with continuous flow, etc.; if the flow is high, its flow has to be estimated.

Additionally, the uniaxial compressive strength of the rock material is often evaluated using the ISRM criteria, as shown in Table 3.5.

Table 3.3 Joint aperture classification

Description	Aperture (mm)	Classification
Very closed	<0.1	Closed joints
Closed	0.1–025	
Partially opened	0.25–0.5	
Opened	0.5–2.5	Spaced joints
Moderately spaced	2.5–10	
Spaced	>10	
Very spaced	10–100	Open joints
Extremely spaced	100–1000	
Cavernous	>1000	

Source: Ulusay and Hudson, 1997.

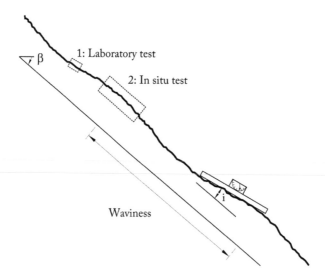

Figure 3.9 Different scales of the joint roughness.

Source: Ulusay and Hudson, 1997.

   The information obtained from a surface point, called geomechanics station, is presented in a synthesized manner, similarly to what is shown in Figure 3.11.

3.4 GEOPHYSICAL SURVEYS

Geophysical prospecting techniques are based on the variations of a physical parameter, correlating these changes with its geological features. They are non-destructive techniques and, generally, are very low cost when compared with the prospecting methods using boreholes. Therefore, this type of survey is very useful in the early stages of tunnel design.

   Geophysical prospecting methods are classified depending on the physical parameter measured, as indicated in Table 3.6. Although the theoretical bases of these methods are

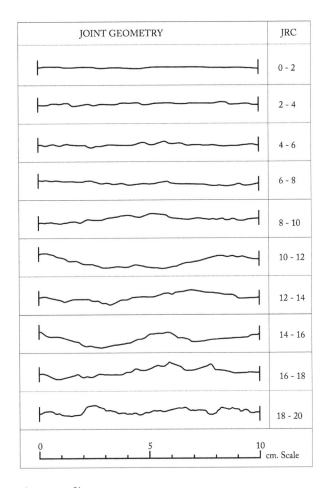

Figure 3.10 Typical roughness profiles.

Source: Barton and Choubey, 1977.

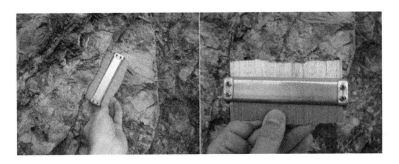

Photograph 3.3 Barton's profilometer.

Table 3.4 Rock weathering degree

Weathering degree	State	Description
1	Sound rock	No visible sign of rock material weathering.
2	Slightly weathered	Rock discoloration.
3	Moderately weathered	Less than half of the rock is decomposed or disintegrated to soil. The sound or discolored rock is present as a discontinuous framework.
4	Highly weathered	More than half of the rock material is decomposed or disintegrated to soil. The sound or discolored rock is present as a discontinuous framework.
5	Completely weathered	All rock material is decomposed or disintegrated to soil. The original rock mass structure is still largely intact.
6	Residual soil	All rock material is decomposed or disintegrated to soil. The original rock mass structure and material fabric are destroyed.

Source: Ulusay and Hudson, 1981.

Table 3.5 Rock uniaxial compressive strength estimation

Grade	Description	Field identification	Uniaxial compressive strength (MPa)
R6	Extremely strong rock	The hammer only splinters the rock.	> 250
R5	Very strong rock	Specimen requires many blows of a geological hammer to fracture it.	100–250
R4	Strong rock	Specimen requires more than one blow of a geological hammer to fracture it.	50–100
R3	Medium strong rock	Cannot be scraped or peeled with a pocket knife, specimen can be fractured with single firm blow of a geological hammer.	25–50
R2	Weak rock	Can be peeled by a pocket knife with difficulty, shallow indentations made by firm blow with the point of a geological hammer.	5–25
R1	Very weak rock	Crumbles under firm blows with the point of a geological hammer, can be peeled by a pocket knife.	1–5
R0	Extremely weak rock	Indented by thumbnail as soil-like.	0.25–1

Source: Ulusay and Hudson, 1981.

shared, a distinction is made between those carried out on the ground surface or inside the boreholes. With the latter approach, some of the techniques commonly used for tunnel design are described below.

3.4.1 Surface geophysics

Within the geophysical techniques carried out on the ground surface, the most used ones in tunnel projects are seismic refraction, seismic reflection, passive seismic and electrical tomography.

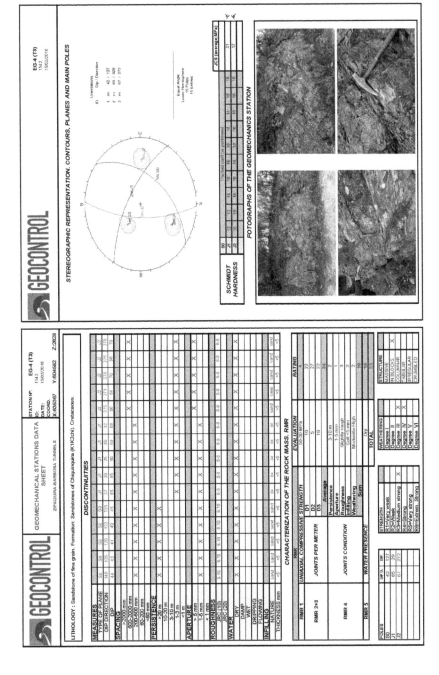

Figure 3.11 Information obtained from a geomechanics station.

Table 3.6 Geophysical methods

Methods	Specific technique	Measured physical parameter
Electrical	Electrical pits	Resistivity/Conductivity
	Vertical electrical boreholes	
	Electrical tomography	
	Spontaneous potential	
Electromagnetic	Electromagnetic boreholes	Electromagnetic field
	Georadar	
Seismic	Seismic refraction	Seismic wave velocity
	Seismic reflection	
	Passive seismic (REMi, MASW, MAM)	
Gravimetric	Gravimetry	Density
	Microgravimetry	

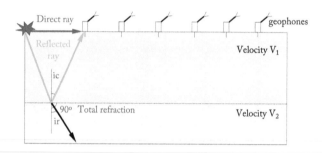

Figure 3.12 Seismic refraction performance.

3.4.1.1 Seismic refraction

Seismic refraction is based on the measure of the velocity of propagation of the compression waves through the ground, which are the fastest among those generated when applying a mechanical impulse on the ground.

When the compression waves propagate in stronger ground, their speed increases, causing refraction and reflection which allows measurement of the velocities of propagation; as is shown in Figure 3.12

The impulse generation is achieved by releasing energy to the ground, generally using a hammer, 8 kg in weight, that hits a metal plate placed on the ground surface. The mallet itself is instrumented with a sensor (trigger) that detects the moment when the blow takes place from which one can calculate the wave arrival time to each geophone.

The geophones are arranged according to longitudinal profiles, equally spaced and the profile length usually ranges between 60 m and 120 m.

For each profile, a series of impacts are carried out whose minimum number must be three, at the beginning, in the middle and at the end of the profile.

Figure 3.13. shows an example of a seismic record, obtained with this technique.

Once the distance-time curves are obtained, it is necessary to identify the points belonging to the same refractor, known as dromocrone. Figure 3.14 shows an example of dromocrones.

Seismic refraction is a technique that is easy to apply which requires also relatively simple equipment, making it extremely economical.

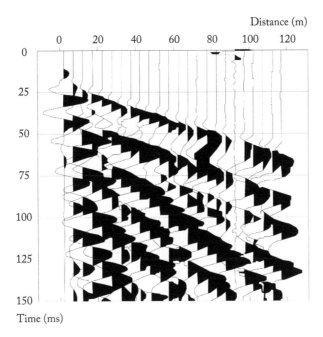

Figure 3.13 Seismic record.

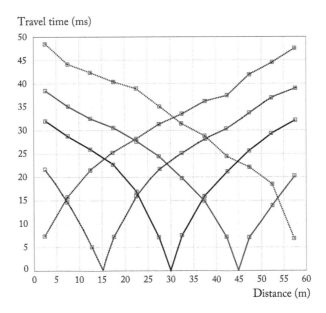

Figure 3.14 Example of dromocrones.

The disadvantages of this method are as follows:

- Its penetration is limited, depending mainly on the energy source used for the wave generation and the full profile extent. For a 100 m long profile, a penetration depth of about 30 m is often achieved.

- It is necessary to have an increase of the wave propagation velocity with depth, because if there is an intermediate layer with lower speed than the overlying layer, the method gives wrong values.
- When the ground has very inclined structures, the interpretation of the result is problematic.

Seismic refraction is often used to determine the thickness of soil or the ground altered in the tunnel portals.

3.4.1.2 Seismic reflection

In seismic reflection, the arrival times of the seismic waves to the geophones are measured, after being reflected in the interface between different lithological units, faults or other types of discontinuities. In general, the greater the relationship between the longitudinal wave propagation velocities of two adjacent materials, the better the reflector will be detected.

The devices used in seismic reflection are similar to those used in seismic refraction, but to increase the energy of the impulses and reach a greater research depth, the impact is replaced by a controlled explosion. Figure 3.15 presents an example of a seismic reflection profile.

3.4.1.3 Passive seismic

Passive seismic is a particular case of seismic refraction, which takes advantage of the environmental "noise" existing in the area of study as an energy source for wave generation. Currently, there are several variants of this technique, the most common ones being the multichannel analysis of surface waves (MASW) and the microtremor array measurements (MAM).

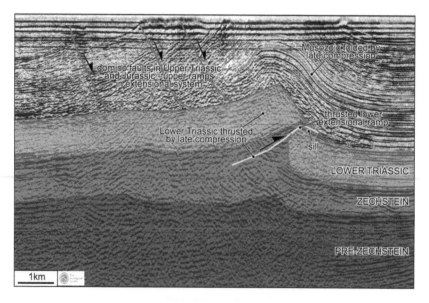

Figure 3.15 Example of interpretation of a seismic reflection profile.

Source: The Virtual Seismic Atlas (www.seismicatlas.org), ©The Geological Society London (https://www.geolsoc.org.uk/) reproduced with permission.

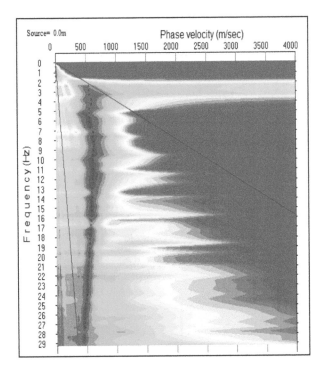

Figure 3.16 Example of the dispersion curve in a MAM test.

The MASW test consists in the interpretation of the surface waves (Rayleigh waves) in a multichannel record; obtaining the shear wave velocity profile (Vs) in the central point of that line.

The MAM method consists the of monitoring of environmental vibrations with devices similar to the previous ones and, through dispersion analysis, determining the shear wave velocity profile.

Both methods have the same theoretical bases, as, in both, the record interpretation allows obtaining a dispersion curve consisting of a graph that relates the phase velocity of the surface waves and the frequency, as shown in Figure 3.16.

Depending on the shear velocities, ground type can be classified in the groups indicated in Table 3.7.

Figure 3.17 shows a ground profile interpreted from a MAM test.

The main difference between both methods lies in the penetration achieved and the resolution obtained. Thus, through the MASW approach, investigation depths of around 20 or 30 m can be reached.

Table 3.7 Ground classification depending on the shear waves

Group	Shear wave velocity (m/s)	Ground description
I	< 180	Soft soils
II	180–380	Medium soils
III	380–750	Resistant soils
IV	750–1500	Strongly resistant soils or soft rock
V	> 1500	Rock

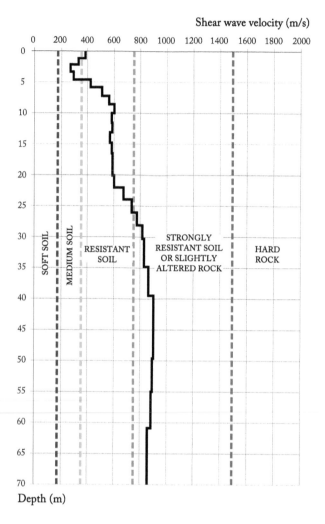

Figure 3.17 Profile of a MAM test.

The microtremors used in the MAM approach are typically of low frequency (1–30 Hz), with wavelengths ranging from a few kilometers in natural sources, to a few tens or hundreds of meters in artificial sources.

This allows obtaining the shear velocity profiles up to depths of 80 m.

These methods are widely used for ground identification in urban areas where, normally, tunnels are constructed at shallow depths.

3.4.1.4 Electrical tomography

The electrical tomography is based on the ground resistivity measurement, which depends on several factors such as the lithology, the internal structure and, fundamentally, on the water content.

To carry out the electrical tomography, potential measuring electrodes and others for intensity measurement are needed.

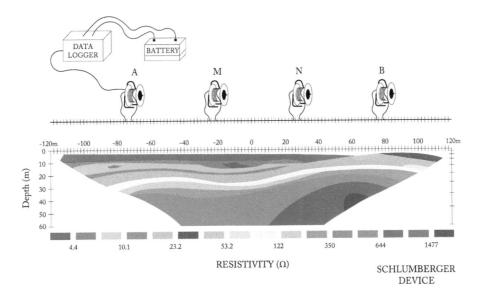

Figure 3.18 Scheme of an electrical tomography test.

Potential measuring electrodes (MN) are located laterally to the intensity ones (AB) and aligned with them. Keeping the dipole AB fixed, the MN is successively moved. Afterward, the process is repeated, moving a step toward the dipole AB.

The existing devices perform this process automatically, so a physical exchange of the electrodes is not necessary. Figure 3.18 illustrates a scheme of one of these tests.

From the resistivity values obtained it is possible to make a geological characterization of the ground estimating the fault planes, fracture zones, dikes and alteration state or weathering of the rock mass. Figure 3.19 shows an example of a resistivity profile.

The depth which can be reached by this technique depends on the profile total length; however, it must be kept in mind that, the larger the electrode spacing, the lower the accuracy obtained would be. Table 3.8 shows several penetration examples, as a function of the selected layout.

During recent years, the continuous improvement in the computer systems and the processing software has allowed a substantial increase in the use of the tridimensional electrical tomography.

This method starts from the same principles as the two-dimensional tomography and for its three-dimensional extension currently two trends exist: electrode arrangement in square or rectangular meshes and computer processing of profiles taken according to parallel profiles. Figure 3.20 shows an example of a three-dimensional electrical tomography.

The main disadvantage of the electrical tomography is the need of excessive subjective interpretation as a fault zone can provoke a resistivity decrease in saturated ground. Thus, this method has to be supplemented with others when used to enable meaningful comparison.

3.4.2 Cross-hole geophysics

The application of geophysical techniques between boreholes enabled overcoming the limitation of these methods in terms of its penetration depth; but, to obtain efficient results, the

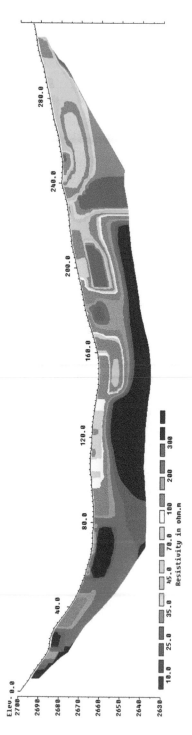

Figure 3.19 Example of an electrical topography profile.

Table 3.8 Example of different layouts and the depth reached

No. of electrodes	Spacing between electrodes	Total length	Max. approximated depth
48	5 m	240 m	48 m
	10 m	480 m	96 m
72	5 m	360 m	72 m
	10 m	720 m	144 m

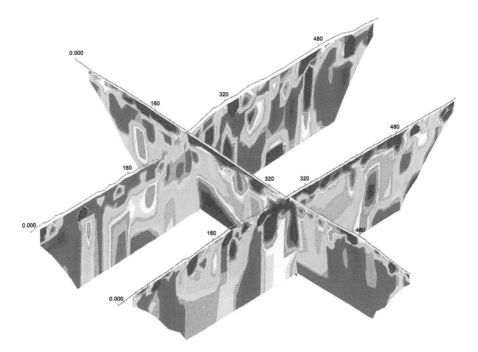

Figure 3.20 Example of a 3D electrical tomography.

boreholes have to be strictly parallel and separated from each other by only a few tens of meters.

In tunnel projects, the most common techniques are seismic between boreholes (cross-hole seismic tomography) and electrical tomography between boreholes (cross-hole electrical tomography). The theoretical bases of both techniques are the same as those of their counterparts on the ground surface so these aspects will not be emphasized here. Both techniques are briefly described below.

3.4.2.1 Cross-hole seismic tomography

In cross-hole seismic tomography, a triaxial probe is placed inside one of the boreholes, previously lined, to record the arrival times of the longitudinal (P) and shear waves (S) from which the propagation velocities can be calculated. In another borehole, the instrument that produces the impact and generates the waves is placed.

The elements which define the resolution of this geophysical technique are:

- The distance between the emission and reception boreholes. The greater the distance between them, the smaller the resolution obtained will be.

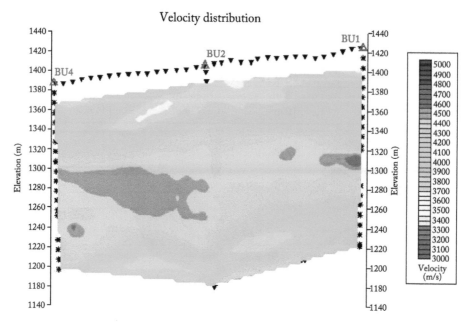

Figure 3.21 Cross-hole seismic tomography.

- The distance among geophones or hydrophones and the distance among the impact points. The smaller the distance between the receivers the better the tomogram resolution. On the other hand, the greater the impact point number, the more available combinations the source-receiver will obtain and therefore, more data to define a particular structure.

Figure 3.21 shows the interpretation of a cross-hole seismic tomography.

3.4.2.2 Cross-hole electrical tomography

Through cross-hole electrical tomography, data about the existing resistivity distribution in the rock volume bounded by the position of two adjacent and parallel boreholes is obtained.

To make the system operational the conditions indicated in the previous section must be met in regard to the required boreholes characteristics.

Figure 3.22 shows an example of two electrical tomograms obtained among three boreholes.

3.5 PROBE BOREHOLES

Probe boreholes with continuous sampling extraction are the most efficient means of ground identification but also the most expensive one, because the drilling campaign has to be carried out when important geological information is available and has to be planned in detail.

3.5.1 Sampling

In the case of boreholes involving core recovery, the ground is drilled with a diamond or tungsten carbide bit, which allows the extraction of a rock core, as the drill proceeds deeper.

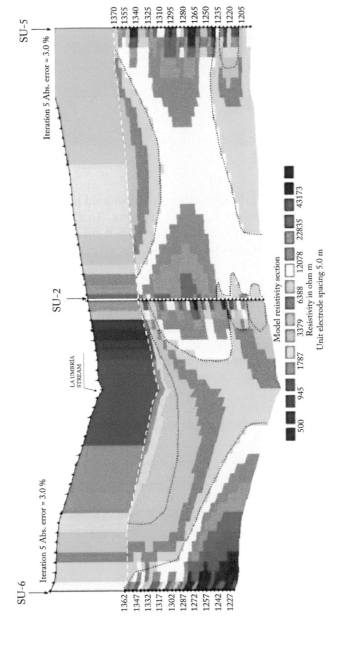

Figure 3.22 Profile with two electrical tomographies developed among three consecutive boreholes.

Photograph 3.4 Double tube sampler.

The drilling progress is achieved through the push transmitted by the machine to the drill string which, at the same time, has a rotational movement. To ease the drilling, the bit is lubricated with a drilling fluid, normally water. The bit, as it goes deeper, recovers the core that penetrates into a tube threaded to the bit, called the core barrel. The set of bit, couplings, extractor and core barrel is called a sampler (soil or rock coring).

Sampling can be by a single, double or triple tube. In single tube samplers, the drilling fluid washes the full core surface, so in soils and soft rocks there is a clear disturbance of the recovered sample.

To partly mitigate this effect, double tube sampling is used, in which water flows down through the contact between the two tubes and the water contact with the core is only produced at its bottom. Both tubes are separated by bearings, which avoids the rotation transfer of the inner tube from the drill string. Photograph 3.4 shows a double tube sampler with the core obtained inside.

When especially crumbly ground is drilled or special care in the sample recovery is required, the triple tube samplers, like the one shown in Photograph 3.5, are used. These

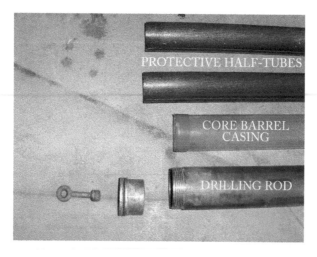

Photograph 3.5 Elements of a single tube sampler.

samplers accommodate a third tube, placed a little bit forward and with a cutting and retractable coring bit, inside which the sample or core is collected.

With the traditional drilling system, the full drill string, which includes the rods and the sample, has to be hoisted out of the borehole after every run, which produces significant down times.

To avoid this disadvantage, for boreholes of certain length (about 80 m) the *wire-line* system is often used. In this system, the drill string is constituted by a pipe with a diameter similar to that of the core barrel. Inside, a cable with a hook-shaped harpoon is used, which attaches to the core barrel head and is hoisted out of the borehole. In Photograph 3.6 a detail of this assembly can be observed.

In this way, it is not necessary to extract the whole drill string in each run and the drilling performances substantially increase with respect to conventional sampler performances.

All the elements used in drilling have standardized dimensions, as indicated in Table 3.9.

Photograph 3.6 Detail of the wire-line overshot assembly.

Table 3.9 Most common drilling diameters

		Bits				Casing pipe		
System	Type	Drilling diameter (mm)	Core diameter (mm)	Reaming Shell (mm)	Type	Outer diameter (mm)	Inner diameter (mm)	Weight (kg/m)
Metric	46.0	46.0	31.7	46.3	44.0	44.0	37.0	3.5
	56.0	56.0	41.7	56.3	54.0	54.0	47.0	4.4
	66.0	66.0	51.7	66.3	66.0	66.0	57.0	5.2
	76.0	76.0	61.7	76.3	76.0	76.0	67.0	6.3
	86.0	86.0	71.7	86.3	86.0	86.0	77.0	7.2
	101.0	101.0	83.7	101.3	101.0	101.0	88.0	10.5
	116.0	116.0	101.7	116.3	116.0	116.0	103.0	12.4
	131.0	131.0	116.7	131.3	131.0	131.0	118.0	138
	146.0	146.0	131.7	146.3	146.0	146.0	133.0	15.4
American (Wire-Line)	AQ	47.6	27.0	48.0	EX	46.0	38.1	4.1
	BQ	59.9	36.4	59.9	AX	57.2	48.4	5.6
	NQ	75.8	47.6	75.7	BX	73.0	60.3	10.4
	HQ	96.7	63.5	96.1	NX	88.9	76.2	12.8
	PQ	122.0	84.9	122.6	HX	120.3	103.1	17.4

3.5.2 Core orientation

The orientation of the joints present in a rock mass can vary with depth so, in many cases, the distribution obtained through the geomechanical stations may not represent the existing conditions at depth.

In these cases, it is necessary to obtain the actual joint orientation at the depth where the tunnel is being designed. Several techniques may be employed for this purpose, as described below.

3.5.2.1 Measures on the core

Of the two parameters which define the joint orientation, dip and dip direction, dip can be directly measured on the extracted core using a protractor, if the drill tilt at which the borehole was made is known and its deviations are assumed to be small, as shown in Photograph 3.7.

However, obtaining the joint dip direction from direct measurements taken from the borehole core is only possible if the core is oriented or if relative orientations are taken with respect to a discontinuity whose orientation is previously known from direct measurements.

This latter case is illustrated in Photograph 3.8, which shows a core box of rocks with stratification easily identifiable.

If the borehole has been drilled at moderate depth and there is an outcrop near where the stratification orientation has been directly measured, the core could be correctly oriented in relation to the stratification.

This makes it possible to relatively orientate any other joint and, through a simple geometric transformation, obtain its dip and dip direction.

3.5.2.2 Acoustic televiewer

The acoustic televiewer generates an image of the borehole walls from the amplitude and travel time of the acoustic waves reflected at the interface between the drilling fluid and the borehole walls.

The quality of the images recorded depends on the contrast between the characteristic impedances of the drilling fluid and the drilled ground. When the properties of both are

Photograph 3.7 Joint dip measure on the core.

Photograph 3.8 Cores with an stratification easily identifiable.

similar, there is little reflection and the image has a poor quality. There are several factors that influence the quality of the record obtained:

- Wall roughness
- Drilling fluid characteristics
- Oval profiling of the drilling

The images obtained are oriented through magnetometers and accelerometers, usually referencing them with respect to the magnetic North. The main application of this type of probe is to obtain the orientation, spacing and aperture of the structural discontinuities present in the rock mass.

Complementarily, with the same probe, the real diameter of the borehole and its orientation can be obtained, so it is possible to check if it has diverted from the planned direction. Figure 3.23 shows an image of a record obtained with this kind of probe and in Photograph 3.9 a detail of one of these devices can be observed.

3.5.2.3 Optical televiewer

Borehole logging with an optical televiewer allows imaging of the drilled ground walls, both in wet and dry boreholes.

The main drawback may arise when the borehole has water inside as for the optical recording the fluid should be transparent, with no material in suspension.

The main use of this geophysical method is to know accurately, through optical images, detailed information of the structure, orientation and geological characteristics of the material found when drilling. These devices give a 360 degrees image.

The equipment is lowered inside the borehole with a cable, which is connected to the surface. Inside the probe there is an optical color sensor (CCD camera) and an orientation sensor with a three-axis magnetometer and accelerometers. Moreover, it has a ring with LEDs to light the borehole walls.

The use of the data obtained with the camera, together with the data provided by the orientation sensors, makes it possible to obtain an oriented image of the borehole walls. Photograph 3.10 shows the appearance of the probe's active part.

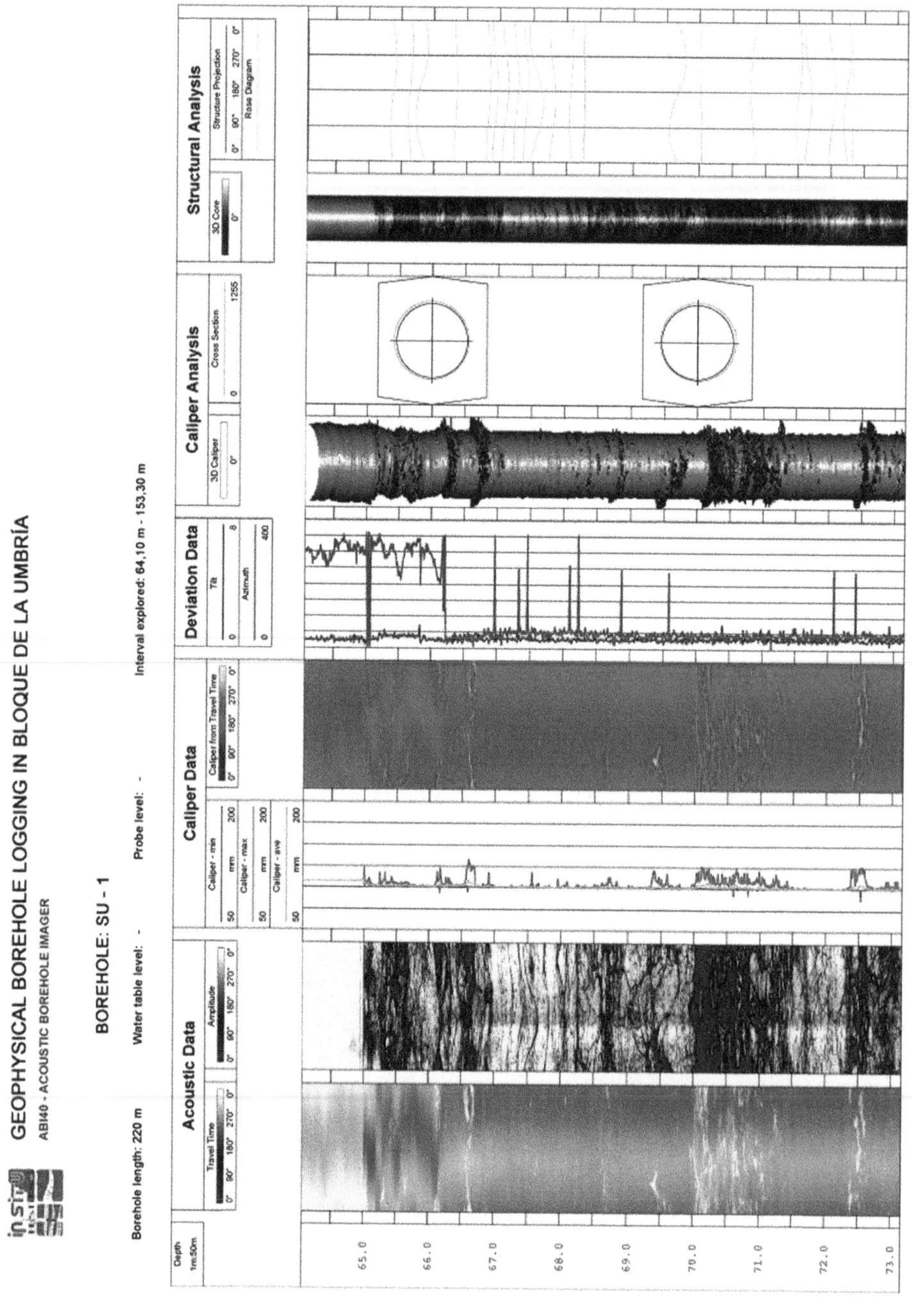

Figure 3.23 Geophysical borehole logging with acoustic televiewer.

Source: In Situ Testing, S.L.

Photograph 3.9 Detail of an acoustic televiewer probe.

Source: In Situ Testing, S.L.

Photograph 3.10 Probe OBI40-2G.

Source: Advanced Logit Technology, 2016.

Figure 3.24 shows the log of a borehole between 14 and 18 m deep; in the left column, a 360-degree image is shown, the next column indicates the orientation of the crossed joints and the two right columns show the three-dimensional view of the core at 0° and 180° which would correspond to the core.

3.5.3 Directional boreholes

The controlled directional drilling technique with core recovery, was developed in Norway in the late 1980s, later, in 2001, the wire-line technique was used for its implementation.

The execution of a directional borehole includes three main stages: planning, drilling and trajectory control.

The planning phase includes the use of a specific software where the path to be followed by the borehole, the checkpoints and the required tolerance are entered. From these parameters, the drilling path is planned, by setting the pitch and roll angles.

The roll angle controls the drilling direction, while the pitch angle controls the path curvature.

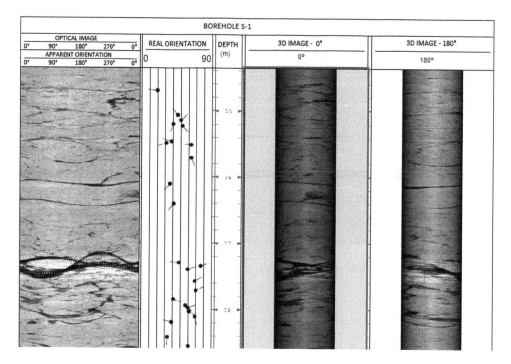

Figure 3.24 Log with optical televiewer of a borehole in limestone.

Table 3.10 Technical specifications in controlled directional drilling

Parameter	Specification
Maximum pitch angle	9° in 30 m (radius 180 m)
Maximum round trip length	3 m in stretches with controlled directional drilling 3 or 6 m in straight stretches
Core diameter	31.5 mm in stretches with controlled directional drilling 47 mm in straight stretches
Drilling diameter	76 mm

The drilling of the borehole straight stretches is carried out with a conventional wire-line system, while for curved stretches a controlled directional drilling system is used.

The path control is performed using an electronic device which controls the borehole azimuth and inclination at predetermined time intervals. The planned trajectory is compared with the measured one to establish if the drill is within the permitted tolerance.

Table 3.10 shows the technical specifications of a controlled directional drilling.

Until 2011, the longest borehole executed with controlled directional drilling with core recovery reached 1,250 m. This borehole was drilled in Hong Kong for the research of an underwater tunnel.

The drilling performances range between 9 and 15 m in each 12 h shift, although it must be noted that these performances are clearly influenced by the admissible trajectory tolerances.

The controlled directional drilling technique is very expensive and is often used under very special circumstances. Obviously, this high cost requires a prior high level of geological

Photograph 3.11 Directional drilling for the characterization of the Piora Syncline.

Source: Swiss Federal Archives, Alptransit Portal, 2016.

knowledge, to use it only in the investigation of critical areas along the tunnel alignment. This happened, for example, in the Gotthard Base Tunnel (Switzerland).

During the design phase of this tunnel, an area in Faido Section was identified, coinciding with the Piora Syncline, where there were suspicions about the existence of highly fractured dolomites and a water load exceeding 15 MPa.

For the investigation of this area, a 5.5 km long gallery with 5 m of diameter was excavated, using a Tunnel Boring Machine.

The gallery was 365 m above the tunnel crown elevation and six directional boreholes were made inside with lengths ranging from 200 m to 1,073 m. Photograph 3.11 shows one of these drillings.

Figure 3.25 shows a longitudinal profile with the geological distribution obtained with the directional boreholes. From this interpretation, the existence of a gypsum layer above the tunnel crown can be deduced which, due to its impermeability, interrupts the hydraulic load.

3.5.4 Layout of boreholes

Boreholes have to be placed in the appropriate spots, defined after the geological model is set up, to identify the areas of interest which, in general, will be the following locations:

- Portal areas
- Location of pits and air shafts
- Topographical depressions on the tunnel axis, as they normally reflect structural failures
- Fault zones or interface surfaces between different lithologies
- Potential aquifer horizons

For the right borehole layout, it will be necessary, on many occasions, to build specific access roads or to use special systems, such as the pontoon shown in Photograph 3.12.

In some cases, it is necessary to use more sophisticated resources to place the drilling probe, as happened in the drilling campaigns carried out to identify the ground where the Gibraltar Strait Tunnel was going to be excavated. In this case, it was necessary to drill

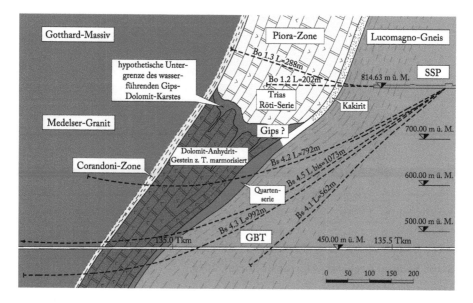

Figure 3.25 Scheme of the geotechnical investigation of the Piora Syncline, at the otthard Base Tunnel.
Source: Ehrbar et al., 2016.

Photograph 3.12 Borehole drilling from a pontoon.

boreholes up to 325 m in depth, crossing around 275 m of the water depth with strong sea currents. This was done with the help of the *Kingfisher* ship, which had the ability to withstand the sea currents and remain in a fixed position. Photograph 3.13 shows the *Kingfisher* and its drill rig.

Another important issue is the depth that boreholes should have, which in too many cases is planned to reach the tunnel alignment, but that is not always necessary.

The aim of boreholes is to provide the data necessary to establish a geological model of the ground where the tunnel is going to be designed and to obtain ground representative samples to be tested in the laboratory.

Photograph 3.13 The *Kingfisher* vessel used in the drilling of boreholes for the Gibraltar Strait Tunnel (Spain–Morocco).

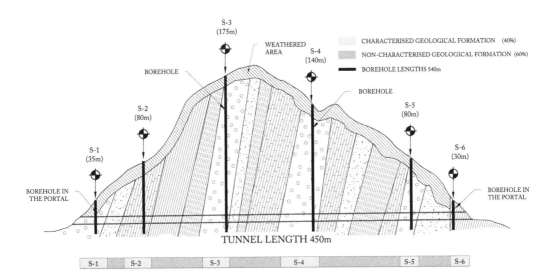

Figure 3.26 Example of a poor borehole layout.

Source: Dodds, 1982.

Boreholes drilled as far as the tunnel alignment will provide specific data about some points of the alignment, but may be of little value in defining the geological model if large aquifers are crossed. This can lead to serious problems during tunnel construction.

In Figure 3.26, reproduced from Dodds (1982), a vertical borehole campaign reaching the tunnel alignment is illustrated depicting an inefficient layout.

In this case, six boreholes have been planned to identify the ground where a 450 m long tunnel is going to be built, with a total borehole length of 540 m representing 120% of the tunnel length. Although the total borehole lengths in relation to the tunnel length are very high, there are grounds that are not identified by the boreholes.

Figure 3.27 illustrates an alternative campaign to the previous one, in which the cumulative borehole length is smaller, totaling 465 m, which represents 103% of the tunnel

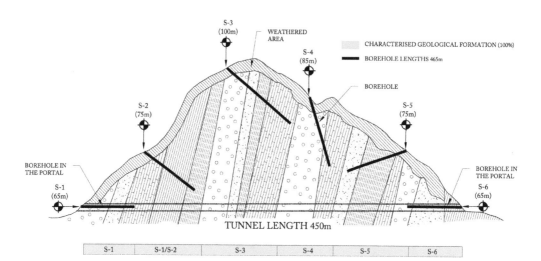

Figure 3.27 Example of an optimized borehole layout.

Source: Dodds, 1982.

length, but where all grounds that must be excavated have been previously identified by the boreholes.

3.5.5 Borehole logging and sampling

Borehole logging consists of the description of the recovered cores and their features, as well as the drilling process itself.

This work should be performed by a qualified professional with experience in this kind of task so that all the information that could be obtained from the borehole is extracted.

Wherever possible, the borehole logging should be done in situ and while the drilling progresses, as some grounds become quickly altered modifying their properties. In Photograph 3.14 an example of the site where a borehole is being drilled and the core boxes ready for its logging can be observed.

Photograph 3.14 Borehole drilling site.

As part of the borehole drilling process, the driller must fill a work order, in which all relevant information is reflected. As an example, the following data should be included:

- Project name
- Brand and model of the equipment used
- Identification of the drillers
- Borehole name
- Borehole x, y, z coordinates
- Borehole orientation and tilt
- Dates
- Performed round trips
- Simplified description of grounds traversed
- Drilling fluid
- Any fluid gains or losses occurring
- Drilling diameters, lining and type of bit used
- Round trip length
- Elevation and results of the tests carried out and of the samples taken
- Water table level at the beginning and at the end of each working day
- Percentage of the core recovered.

Cores have to be placed and kept in the adequate boxes, correctly labeled and numbered.

As an important part of the borehole logging process, color photographs of the core boxes should be taken, as illustrated in Photograph 3.15.

Figure 3.28 depicts an example of the borehole logging data.

3.5.6 Geophysical borehole logging

Boreholes allow access to the ground in depth and, therefore, provide an excellent opportunity to make geophysical measurements inside a borehole and across the boreholes, for which, usually, the techniques included in Table 3.11 are used.

The following sections describe the borehole logging techniques through the measurement of gamma radiation, spontaneous potential and full wave sonic probe, which are the ones most commonly used.

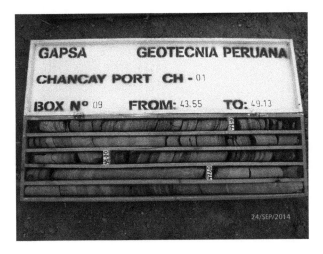

Photograph 3.15 Core box.

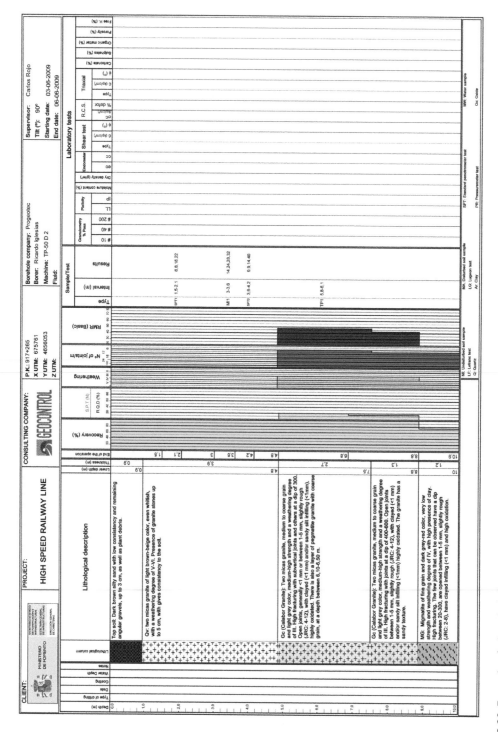

Figure 3.28 Example of a borehole logging.

Table 3.11 Geophysical borehole logging techniques

Methods	Probe
Electrical	Spontaneous potential
	Strength
	Normal resistivity
	Lateral resistivity
	Focused resistivity
	Fluid resistivity
	Induction
Seismic	Full wave sonic
Radioactive	Natural gamma
	Gamma gamma
	Neutron
	Spectrometry
Mechanical	Flowmetry
	Caliber
Others	Tilt
	Dip
	Temperature
	Gravity
	Magnetism
	Radar
	Optical televiewer
	Acoustic televiewer

3.5.6.1 Natural gamma radiation

Gamma rays are spontaneously emitted by some radioactive elements present in certain lithologies, as part of the nuclear decay process where the mass is transformed into energy.

The measure of these gamma rays is made using probes, like the one shown in Photograph 3.16 which measures the radioactive events emitted per unit of time.

It is important to consider that the obtained records are influenced by several factors, such as:

- Type of detector in the probe
- Speed at which the record is taken
- Drilling diameter
- Nature and density of the drilling fluid
- Casing presence or not

The use of this probe is of special interest in granite materials, as they contain a high concentration of radioactive isotopes. Uranium, thorium and potassium isotopes are also present in clay soils, so the technique can also be used in this type of soils. On the contrary, sandy soils, with low lime content, do not have significant concentrations of this kind of isotopes, so this technique is not useful in this kind of soil.

Borehole logging with natural gamma is useful for detecting the boundaries between lithological layers and identify horizons with high clay content.

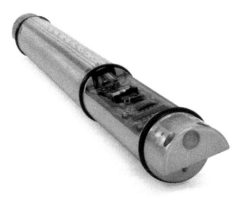

Photograph 3.16 Probe to measure the natural gamma radiation.

Source: CBG Corporation.

3.5.6.2 Probe to measure the spontaneous potential

The probe of spontaneous potential measures the existing potential difference between two electrodes, one fixed on the ground surface while the second one is placed inside the borehole.

The difference in the measured spontaneous potential is caused by the natural ionic imbalances, which exist in the interfaces solid-liquid and solid-solid with different permeability, so that these measures are useful to detect permeable ground levels.

In Figure 3.29 a scheme with the application of this technique can be seen. In clay materials the spontaneous potential value remains almost constant, so the obtained record is

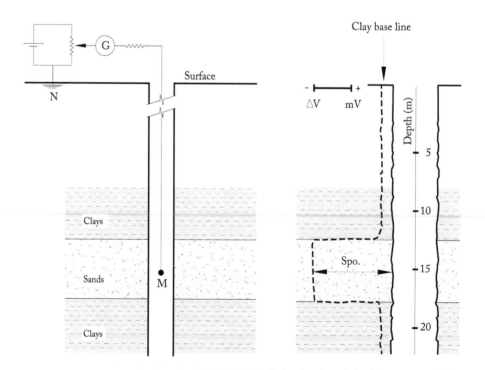

Figure 3.29 Working scheme of a spontaneous potential log.

Photograph 3.17 Full wave sonic tool.

Source: In Situ Testing, S.L.

practically a straight line called "clay base line" and in materials of higher permeability, this line undergoes a displacement which may be in the positive or negative direction depending on the water salinity.

The nature and salinity of the slurry have to be taken into account; they may influence the measurement.

This technique is especially useful in detecting ground levels with high permeability.

3.5.6.3 Velocity of compression and shear waves

The "full wave" logging tools make it possible to record the wave train traveling through the ground, after applying a mechanical pulse, which is composed of the compression waves, the fastest ones, shear waves being slower and called "Stoneley".

The full wave logging tool has a sender, which generates a pulse, and three receivers which detect the wave arrival times through different media. Photograph 3.17 shows a detail of this type of logging tool.

Wave logging tools allow determination of the velocity of compression (V_p) and shear waves (V_S), so, knowing the ground density (ρ), the Young modulus (E) and the Poisson coefficient (υ) can be calculated through the following expressions:

TESTIFICACIÓN GEOFÍSICA NUEVO TRAZADO VALLE DE LA UMBRÍA
2SAA-1000/F SÓNICO DE ONDA COMPLETA

SONDEO: SU - 8

Longitud sondeo: 307 m   Cota nivel freático: 13,50 m   Cota tubería: PVC   Tramo testificado: 22,22 m - 305,02 m

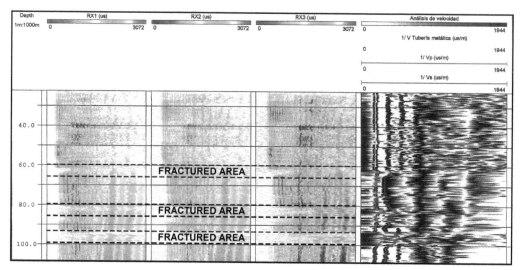

Figure 3.30 Record with a full wave sonic tool.

Source: In Situ Testing, S.L.

$$
\upsilon = \frac{\left(\dfrac{V_p}{V_s}\right)^2 - 2}{2 \cdot \left(\dfrac{V_p}{V_s}\right)^2 - 2}
$$

$$
E = 2 \cdot (\rho \cdot V_s^2) \cdot (1 + \upsilon)
$$

Figure 3.30 shows a borehole logging obtained using a full wave sonic tool. This borehole logging technique requires having the borehole full of water, which is difficult to achieve in permeable grounds.

3.6 TRIAL PITS AND ADITS

Trial pits, also known as adits, are shallow excavations made with conventional backhoe loaders, so that its depth hardly exceeds 5 m. Photograph 3.18 shows the excavation of a trial pit.

Due to the limited depth that can be reached with the trial pits, its relevance is focused on the identification of the upper ground layers, in the tunnel portal areas.

Figure 3.31 reproduces the information obtained with a trial pit.

Photograph 3.18 A trial pit.

3.7 FIELD AND IN SITU TESTS

Laboratory tests are carried out on small volume samples and, in some cases, on slightly altered samples, resulting in data representation being limited. Field and in situ tests reduce, to some extent, the scale effect existing between the ground properties determined at rock mass scale and the ones determined in the laboratory, because they involve a bigger ground volume. Thus, these tests are very relevant for the ground characterization.

The following field and in situ tests are most commonly used.

3.7.1 Ground penetration tests

Ground penetration tests are based on ramming a metallic piece in the ground, by hitting or applying a constant force, measuring the successive penetrating lengths achieved. Table 3.12 indicates some ground penetration tests and the regulation which govern them.

In the following sections, the penetration tests are listed by hitting, or other dynamic action, and at a constant penetration rate or static tests.

3.7.1.1 Dynamic penetration tests

Dynamic penetration tests consist of ramming a steel pipe into the ground with a steel cone placed at its end. The driving is achieved through the blow of a mass which is dropped from a certain height and rhythm, recording the blows needed to achieve an established penetration depth.

Depending on the mass weight and the drop height, there are several test variants. The most frequently used are the "super-heavy" test and the standard penetration test.

The Standard Penetration Test (SPT) is normally carried out in boreholes with the equipment shown in Photograph 3.19, in accordance with the following procedure:

- A borehole is drilled until the desired level and a sampling of normalized dimensions is placed at the excavation bottom, consisting of a footing, a split-spoon and the coupling to the drilling rod.

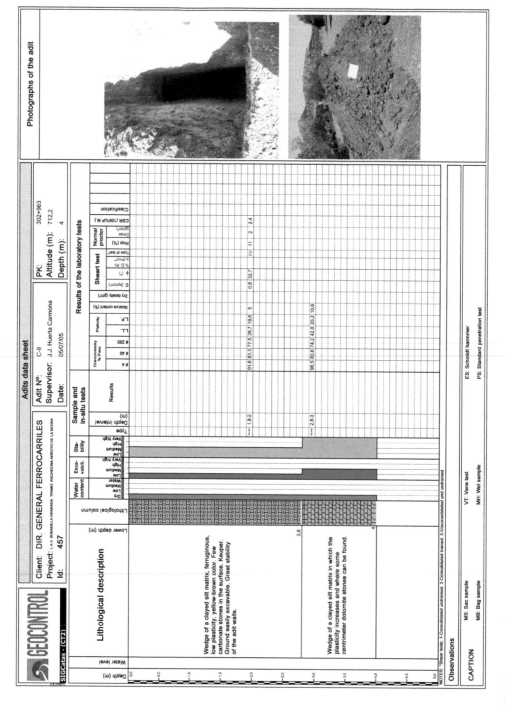

Figure 3.31 Information obtained with a trial pit.

Table 3.12 Most common ground penetration tests

Test	Abbreviation	Regulation

Investigation and Geotechnical Tests. In Situ Tests.
  Part 1: Penetration tests with the electric cone and
  the piezocone.
(ISO 22476-1:2012)
	CPT
	CPTU
		UNE-EN ISO 22476-1:2015

Investigation and Geotechnical Tests. In Situ Tests.
  Part 2: Dynamic penetration test.
(ISO 22476-2: 2005)
Investigation and Geotechnical Tests. In Situ Tests.
  Part 2: Dynamic penetration test.
Modification 1. (ISO 22476-2:2005/Amd 1:2011)
	DPL
	DPM
	DPH
	DPSH
		UNE-EN ISO 22476-2:2008
		UNE-EN ISO 22476-2:2008/A1:2014

Investigation and Geotechnical Tests. In Situ Tests. Part 3: Standard penetration test. (ISO 22476-3:2005)	SPT	UNE-EN ISO 22476-3:2006
Investigation and Geotechnical Tests. In Situ Tests. Part 12: Mechanical cone penetration test (CPTM). (ISO 22476-12:2009)	CPTM	UNE-EN ISO 22479-12:2010

Photograph 3.19 Dynamic penetration equipment.

- The sampling is driven 45 cm into the ground in three batches, counting the necessary hits to penetrate each 15 cm.
- The driving is done with a hammer with a mass of 63.5 kg falling through a distance of 76 cm, that theoretically corresponds to 0.5 kJ/hit.
- The first 15 cm are ignored and the sum of the hits in the remaining 30 cm constitutes the N30 parameter or the standard penetration resistance (NSPT).
- The test is discarded in cases where 50 or more hits are applied to advance through a 15 cm stretch, writing down the result as "reject".

It is a simple test and can be even done in soft or weathered rocks, for that case the regulation allows increasing the limit in the number of hits to 100, instead of the indicated 50 hits. Gravels complicate the test implementation and the resulting interpretation, and so does the presence of large size boulders. The test is mainly adequate for sands and loses its applicability in grounds with very coarse or very fine grading.

Table 3.13 Correlations of different parameters with results obtained from the SPT test

Parameter	Correlation	Source	Observations
Internal friction angle (ϕ)	$\phi = 27.1 + 0.3 \cdot N - 0.00054 \cdot N^2$	Wolff (1989)	—
	$\phi = \tan^{-1}\left[\dfrac{N}{12.2 + 20.3 \cdot \left(\dfrac{\sigma_0}{P_a}\right)}\right]^{0.34}$	Kulhawy and Mayne (1990)	σ_0': effective stress at the test depth P_a: reference pressure (1 MPa)
Modulus of deformation (E)	$\dfrac{E}{P_a} = 5 \cdot N$	Kulhawy and Mayne (1990)	Sands with silt and clay
	$\dfrac{E}{P_a} = 10 \cdot N$		Clean sands normally consolidated
	$\dfrac{E}{P_a} = 15 \cdot N$		Clean sands over-consolidated
Undrained cohesion (c_u)	$c_u\left(\dfrac{kN}{m^2}\right) \approx 6 \cdot N$	Terzaghi and Peck (1967)	Clay soils
Over-consolidation ratio (OCR)	$OCR = 0.193 \cdot \left(\dfrac{N}{\sigma_0}\right)$	Mayne and Kemper (1988)	σ_0': effective stress at the test elevation MPa

There are correlations of different geotechnical parameters with the NSPT value. Table 3.13 includes some of them which, in any case, have to be adopted with caution and always considering them as an estimate.

3.7.1.2 Static penetration tests

The static penetration test or Cone Penetration Test (CPT) consists of the penetration into the ground, at a constant rate of 2 cm/s, of a conical tip that allows measuring, by sensors, the cone penetration resistance (q_c) and the sleeve friction (f_s) as the cone penetrates. If the water pore pressure (u) is also measured then, the test is called CPTU.

The advantages of the CPTU test, when compared with the CPT, are that the CPTU distinguishes between drained, partially drained and undrained penetration; corrects the parameters obtained with the cone; evaluates the soil consolidation characteristics and enables evaluating the hydrostatic balance. Figure 3.32 shows the record obtained in a CPTU test carried out in sands up to a depth of 35 m.

3.7.2 Field tests to estimate the ground strength

The most commonly used in situ tests to determine the ground strength are the point load test, the Schmidt hammer test and the shear test, which are described in the following sections.

3.7.2.1 Point load test

The point load test consists of applying a compressive force between two conical tips on a rock sample until its failure.

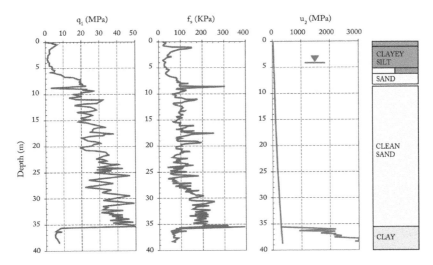

Figure 3.32 Record of a CPTU test.

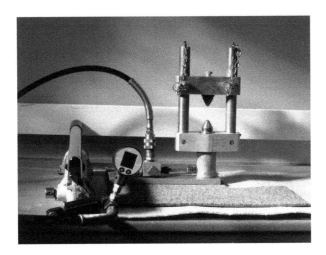

Photograph 3.20 Device for the point load test.

Source: Cepasa, S.A.

The device consists of a frame, a hydraulic cylinder, two conical tips, one driven by the cylinder and the other fixed to the frame, and of a manometer to measure the hydraulic pressure at failure, as illustrated in Photograph 3.20.

This test is governed by the following regulations:

- UNE 22950-5:1996. *Mechanical Properties of Rocks. Strength Determination Tests.* Part 5: resistance to a point load.
- ASTM D5731–08. *Standard Test Method for Determination of the Point Load Strength Index of Rock and Application to Rock Strength Classifications.*

In addition, this test is standardized by the IRSM (1985), in the publication *Suggested Method for Determining Point Load Strength.*

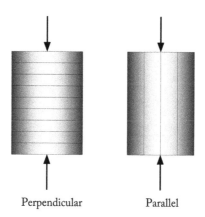

Perpendicular   Parallel

Figure 3.33 Load orientation with respect to the anisotropy.

It should be noted that there are some reservations involving the use of the point load test (which estimates but not determines the strength of a rock sample) and these were discussed in a detailed study by Bieniawski (1975).

In essence, the test is usually carried out on cylindrical cores and there are two types of tests depending on the direction of the applied load: axial and diametrical, as illustrated in Figure 3.33. However, blocks or irregular pieces of suitable size can also be tested.

To compare the results obtained from tests conducted on samples of different size and shape, the corrected point load index (I_{s50}) has been defined, which refers to the tests performed with rock cores of 50 mm in diameter and length greater than 25 mm. The I_{s50} is calculated by the expression:

$$Is_{50}\,(MPa) = \left(\frac{D_e}{50}\right)^{0.45}\frac{1.000P}{D_e^2}$$

where P is the force applied at failure (kN) and D_e is the equivalent sample diameter (mm), calculated by the expression:

$$D_e = \frac{4A}{\pi}$$

where A (mm²) = W (mm) × D (mm) is the minimum area of the cross section.

Figure 3.34 shows the specifications for W so that the test would be representative. A number of authors have correlated the I_{s50} with the uniaxial compressive strength value (Bieniawski, 1975). Recently, Nazir and Amín (2015) showed that the correlation depends on the type of rock tested.

3.7.2.2 Tests with the Schmidt hammer

The Schmidt hammer is a field test which involves a steel impact plunger being released by a spring, which in turn rebounds depending on the material strength on which the impact plunger is placed. Figure 3.35 illustrates the Schmidt hammer operation and Photograph 3.21, its implementation.

Initially, the Schmidt hammer was devised to detect the position of the steel reinforcement in the inspections carried out on reinforced concrete structures; but it has gradually

I.- Diametral test

II.- Axial test

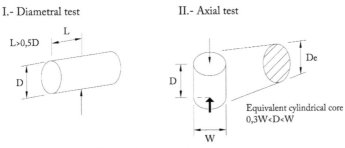

III.- Test with a parallelepiped rock sample

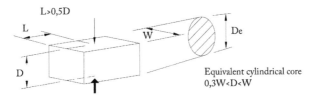

IV.- Test with an irregular rock sample

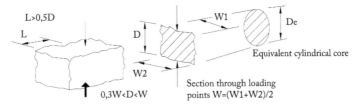

Figure 3.34 Samples specifications for point load tests.

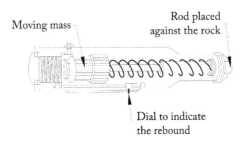

Figure 3.35 Schmidt hammer operation.

Source: Ulusay, 2014.

been adapted to estimate the rock uniaxial compressive strength of the rock material at the geomechanics stations.

This test is regulated by the ASTM D5873–14 norm *Standard Test Method for Determination of Rock Hardness by Rebound Hammer Method* and it is also standardized by the IRSM in *Suggested Method for Determination of the Schmidt Hammer Rebound Hardness: Revised Version*, Aydin, A. ISRM (Ulusay, 2009).

Photograph 3.21 Schmidt hammer implementation.

There are several correlations between the Schmidt rebound index (σ_{ci}) among which the following two are predominant (Kilic and Teymen, 2008; Aufmuth, 1973):

$$\sigma_{ci} = 0.0137\, R_L^{2.2721}$$

$$\sigma_{ci} = 0.33 \left(R_L \cdot \gamma \right)^{1.35}$$

where:

R_L = Rebound index obtained with the Schmidt hammer with an impact energy of 0.735 N·m and

γ = ground density.

It is also frequently used with a Miller chart, shown in Figure 3.36, in which the position of the hammer and rock density are taken into account.

The ISRM recommends taking 20 different readings at each test and suggests using the average of the ten readings with the highest values. Instead, the ASTM norm recommends taking ten readings excluding those that differ by more than seven points from the average.

3.7.2.3 In situ shear tests

In situ shear tests are used to determine the ground shear strength, of soft rocks and of the discontinuities in the rock masses.

These tests are normalized by the ASTM D4554 – 12 norm *Standard Test Method for In Situ Determination of Direct Shear Strength of Rock Discontinuities* and by the ISRM: *Suggested Methods for Determining Shear Strength*. Part 1: Suggested method for "in situ" determination of direct shear strength (1974).

To perform the test, two hydraulic cylinders are needed: one provides the stress normal to the shear plane, which remains constant during the test, and the other applies the shear stress.

For tests performed in soils or soft rocks, cubic blocks whose edge varies between 40 and 100 cm are cut in situ. In these cases, as illustrated in Figure 3.37, the reaction necessary to apply the normal stress is provided with a dead load.

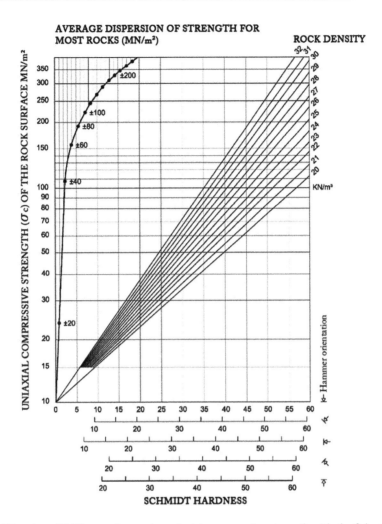

Figure 3.36 Miller chart (1965) to estimate the uniaxial compressive strength with the Schmidt hammer.

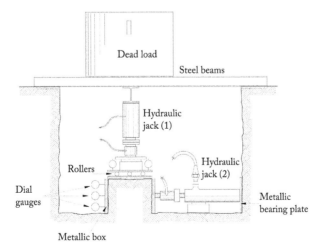

Figure 3.37 Device to perform an in situ shear test in soils or soft rocks.

Photograph 3.22 Shear test conducted on gravels.

Photograph 3.22 shows a shear test conducted on gravels.

In situ shear tests for the shear strength determination in a discontinuity are much more complex, because it is necessary that the shear occurs strictly in the joint plane to be tested, as illustrated in Figure 3.38.

From each shear test a curve can be obtained which represents the increase in shear stress as the displacement in the shear plane increases and to characterize the ground at least three tests have to be performed. Figure 3.39 shows the result of three shear tests conducted in gravels, with normal stresses of 0.5, 1.5 and 3 kg/cm².

3.7.3 Tests to determine the modulus of deformation

Knowledge of the modulus of deformation of the ground is essential if the ground displacements are to be accurately determined during the tunnel construction. The value of this parameter is very sensitive to the existing ground discontinuities, which is the reason why laboratory test results are strongly influenced by the scale effect. Among the most common tests done to determine the in situ modulus of deformation are the plate bearing test and the pressuremeter or dilatometer test.

3.7.3.1 Plate bearing tests

The plate bearing test consists of the application of increasing pressure to the ground through a steel plate, considered rigid, measuring the displacements produced as the pressure increases.

This test can be applied both in soils and rocks. The first case is standardized by the norms UNE 103808: 2006 *Load Test of Plate Soils* and ASTM D 1195 *Standard Test Method for Repetitive Static Plate Load Test of Soils*.

The load is applied in a vertical direction and in a staggered way using a smooth and rigid plate, whose dimensions are 30 cm × 30 cm although plates of 60 cm × 60 cm are also usual, in tests performed on soils. The test arrangement is illustrated in Figure 3.37.

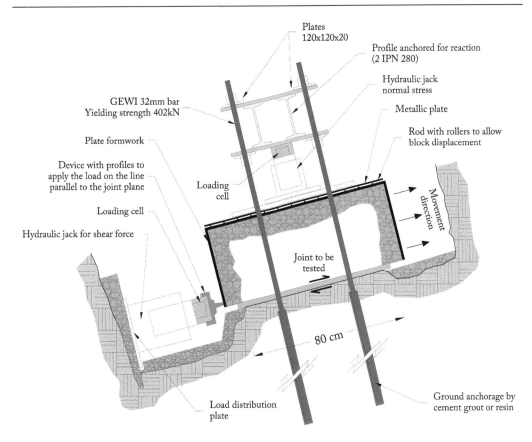

Figure 3.38 In situ shear test on a rock joint.

Source: In Situ Testing, S.L.

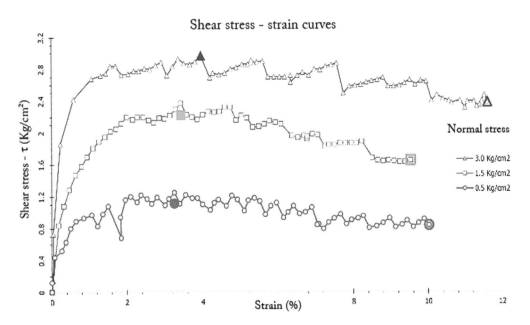

Figure 3.39 Shear tests conducted in gravels from Lima (Peru).

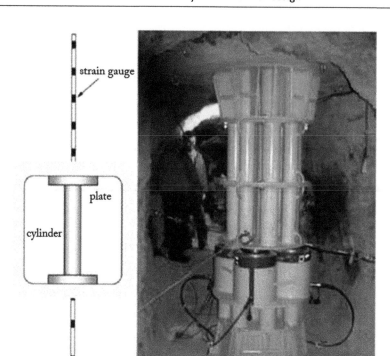

Photograph 3.23 Plate bearing test in the Bakhtiary dam (Iran).

Source: Agharazi et al., 2012.

The plate bearing test performed on rocks is regulated by the ASTM D 4394 *Standard Method for Determining the In Situ Test of Deformation of Rock Mass Using the Rigid Plate Loading Method.*

This method was extensively assessed and employed by Bieniawski (1979) featuring 57 in situ tests. Comparisons with the results obtained by dilatometers at the same sites were included. Rock mass quality was recorded using the RMR.

This test plate bearing is usually performed in previously excavated galleries, as shown in Photograph 3.23, which shows the equipment used to determine the rock deformability in the Bakhtiary Dam (Iran).

The plate bearing tests can also be performed by applying the pressure horizontally, obtaining for this case the horizontal modulus of deformation.

Figure 3.40 presents the scheme of a test with this arrangement and Photograph 3.24 shows the implementation of the test in gravels.

The plate bearing tests are usually performed in cycles in which the applied pressure increases, until it reaches the maximum value. Figure 3.41 shows a typical curve obtained in one of these tests. The modulus of deformation is often defined as the slope of the line passing through the points representing the 30% to 70% of the maximum stress, achieved in the last load cycle.

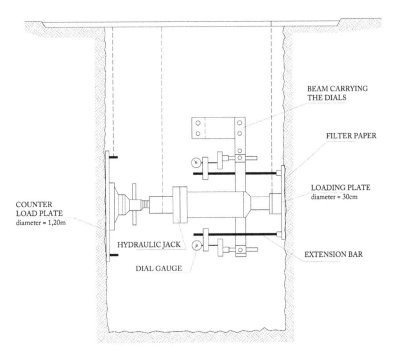

Figure 3.40 Test scheme with vertical plate in a shaft.

Photograph 3.24 Test with vertical plate in gravels.

3.7.3.2 Pressuremeter/Dilatometer test

The plate bearing tests performed on the ground surface are not very representative of the ground deformational behavior at depth and those carried out in shafts and galleries are extraordinarily expensive.

To avoid these drawbacks, for tests performed in soils, the Menard pressuremeter was developed with which the operation consisted of applying an increasing pressure in a rubber cell located inside a borehole; Figure 3.42 illustrates the Menard pressuremeter operation.

As the pressure increases inside the cell located inside the borehole the ground in contact with it is deformed which necessitate injecting more fluid in order to maintain the pressure.

Throughout the test, a curve of the changes in volume increase as a function of the fluid pressure is obtained, as shown in Figure 3.43. The straight part allows calculation of the modulus of deformation.

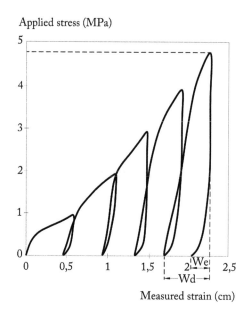

Figure 3.41 Typical curve obtained in a plate bearing test.

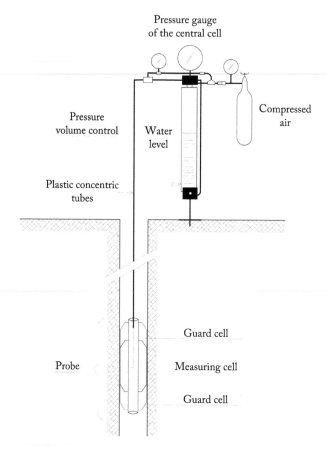

Figure 3.42 Menard pressuremeter operation.

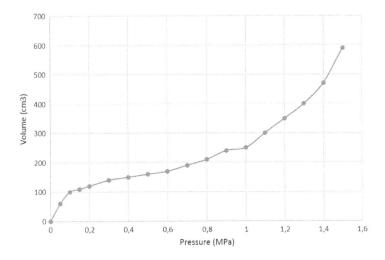

Figure 3.43 Typical curve obtained in the Menard pressuremeter test.

The Menard pressuremeter can only reach pressures of 10 MPa, therefore it is only appropriate for soils.

In the 1960s, the National Civil Engineering Laboratory in Lisbon (Portugal) developed a device similar to the Menard pressuremeter which was able to apply pressures up to 20 MPa and which was called a dilatometer.

This dilatometer did not measure the ground displacements during the test by the volume increase of the fluid injected into the cell, but by linear differential transformers, allowing measurements with an accuracy of microns.

Photograph 3.25 shows the sensors of a dilatometer.

Two decades after the dilatometer launch, OYO Corporation marketed a device that also provided pressures up to 20 MPa, but had a unique displacement measuring device that was much less complex and cheaper than the linear differential transformers.

This device is usually known as an OYO pressuremeter and can be used both in soils and rocks. Photograph 3.26 shows the appearance of this device.

Photograph 3.25 Detail of the sensors of a dilatometer.

Photograph 3.26 OYO pressuremeter.

Source: In Situ Testing S.L.

The pressuremeter test is regulated by the norms UNE-EN ISO 22476-5 *Flexible Dilatometer Test* (2014); UNE-EN ISO 22476-7 *Rigid Dilatometer Test* (2015) and ASTM D 4719-87 (1994).

Before conducting the test, the pressuremeter must be calibrated to correct the measured records. This is because during the test the borehole radius in which the test is done is not measured directly, only the inner radius of the pressuremeter casing. Therefore, the measured stiffness does not correspond to the ground stiffness but to the system formed by the casing and the ground.

To isolate the casing effect the following calibrations have to be done:

- Initial calibration to determine the pressure necessary to expand the casing
- Tube calibration to know the compression suffered by the casing when subjected to high pressures

The probe loading is done by hydraulic pumps, compressed air or gas. It is common and advisable to use nitrogen for this purpose, given its lower deformability.

Figure 3.44 shows the curve obtained in a perfect pressuremeter test, whose stages are as follows:

- **Casing fitting to the borehole:** R_0 is the radius at which the rubber casing is in full contact with the ground and P_0 is the pressure corresponding to such strain.

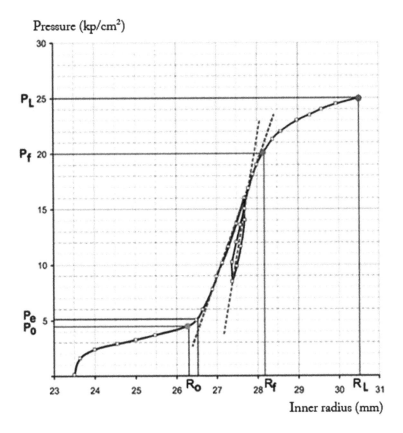

Figure 3.44 Curve obtained in a perfect pressuremeter test.

- **Elastic strain phase:** between R_0 and R_f.
- **Yield strain phase:** R_f corresponds to the radius at the beginning of the ground yielding and P_f is the yielding pressure at which the ground stops behaving elastically.
- **Ground failure:** P_L is the pressure at which the ground shears and does not support further pressure increments and is known as the limit pressure. R_L is the radius corresponding to the beginning of the ground shearing.

To obtain a complete pressuremeter curve, the pressure system has to be able to shear the ground and the rubber casing has to withstand the corresponding diameter increase without breaking.

The test is analyzed considering the hypothesis of the expansion of a cylindrical cavity in an elastoplastic infinite medium and, depending on the probe expansion capacity and the ground resistance, in ideal conditions it is possible to determine the following parameters:

- Shear modulus
- Yielding pressure
- Limit pressure

The ground modulus of deformation is calculated by the expression:

$$E = 2 \cdot G \cdot (1 + \upsilon)$$

where G is the shear modulus and υ is the Poisson coefficient.

Photograph 3.27 Permeability test at constant level.

3.7.4 Test to determine the permeability

The following sections describe the most common in situ test performed to determine the ground permeability.

3.7.4.1 Permeability measure in boreholes

The measure of the permeability in boreholes is carried out by filling them with water and controlling the flow transferred to the ground using two different methodologies:

- Tests at a constant level, where a previously known flow is introduced into the borehole to maintain the water level constant
- Tests at variable level, in which a water volume is introduced (or removed) from the borehole, leading to an instantaneous rise (or drop) of the borehole water level.

Photograph 3.27 shows the implementation of a permeability test at constant level.

Within this kind of tests, Lefranc and Gilg-Gavard tests must be highlighted, which differ in the formulation used for their interpretation. Both can be conducted at constant and variable levels.

Figures 3.45 and 3.46 include two separate schemes of tests performed at constant and variable level, respectively, as well as the formulation for their interpretation according to the methods mentioned.

3.7.4.2 Groutability estimation

Ground injection, with chemical products or with cement grout, is an option to keep in mind in the design of tunnel sections which have to be excavated in grounds with very poor quality, especially if the expected water inflow is large.

To evaluate the ground groutability, Lugeon tests are conducted by injecting pressurized water in a borehole, as illustrated in Figure 3.47.

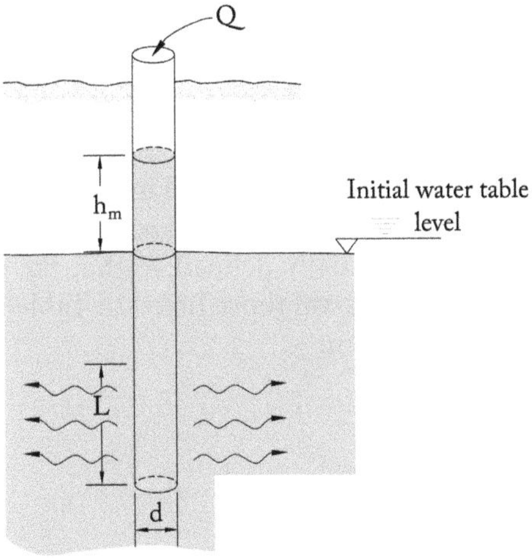

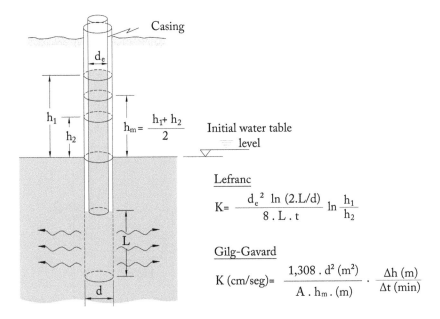

$$\underline{\text{Lefranc}}$$

$$K= \frac{Q}{C.h_m}$$

$$C= \frac{2.\pi.L}{\ln [L/d +\sqrt{(L/d)^2+1}]}$$

$$\underline{\text{Gilg-Gavard}}$$

$$K \text{ (cm/seg)}= \frac{Q \text{ (l/min)}}{600.A.h_m(m)}$$

A= (1,032 . L (m)+30 . d (m))   if L>6m

A= [(1,032 . L (m)+30 . d (m))] . [-0,014 . L² (m²)+0,178 . L(m)+0,481]   if L≤6m

Figure 3.45 Permeability tests at constant level.

$$\underline{\text{Lefranc}}$$

$$K= \frac{d_e{}^2 \ln (2.L/d)}{8.L.t} \ln \frac{h_1}{h_2}$$

$$\underline{\text{Gilg-Gavard}}$$

$$K \text{ (cm/seg)}= \frac{1,308 . d^2 \text{ (m}^2)}{A . h_m . \text{(m)}} . \frac{\Delta h \text{ (m)}}{\Delta t \text{ (min)}}$$

$$h_m = \frac{h_1+h_2}{2}$$

Figure 3.46 Permeability tests at variable level.

The stages to follow for the test implementation are as follows:

• **Insertion of the injection pipe into the borehole:** once the borehole is drilled, usually of 86 mm in diameter, a pipe is placed inside through which pressurized water will be injected. This pipe has a packer attached to it at the appropriate depth, depending on the rock section to be tested. The packer can be single, if the test is carried out at the borehole bottom, or double, if an upper and a lower packer is placed. The section to be tested usually ranges between 0.5 to 5.0 m in length.

- **Packing of the section to be tested:** once the packers have been placed at the testing depth, the testing section is isolated by a packer of rubber mechanism or an inflatable chamber.
- **Performing the grouting cycles:** usually successive loading and unloading steps of 0, 1, 2, 5 and 10 kg/cm² are applied, although sometimes the rock fractures sooner. For that reason, the maximum test pressure (P_{MAX}) is usually defined so that the confinement stress (σ_3) is not exceeded and the injection steps are those listed in Table 3.14.
- **Lugeon index calculation,** through the expression:

$$\text{Lugeon index} = \frac{q}{L} \times \frac{P_0}{P}$$

where:

q = grouted flow at the maximum pressure
L = length of the grouted borehole section
P_0 = reference pressure ($P_0 = 1$ MPa)
P = maximum pressure

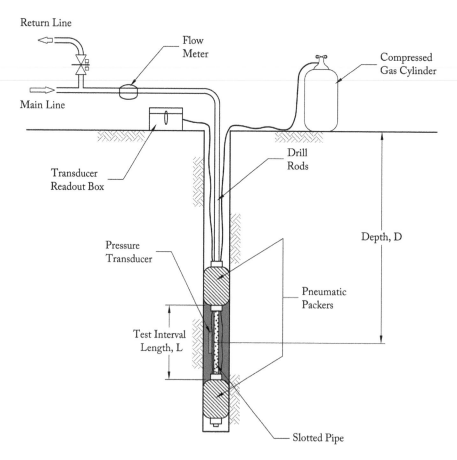

Figure 3.47 Lugeon test.

Table 3.14 Pressures applied in each injection step

Grouting stage	Grouting pressure	Grouting pressure value
First	Low	$0.5 \cdot P_{MAX}$
Second	Medium	$0.75 \cdot P_{MAX}$
Third	Maximum	$1.0 \cdot P_{MAX}$
Fourth	Medium	$0.75 \cdot P_{MAX}$
Fifth	Low	$0.5 \cdot P_{MAX}$

Table 3.15 Relationship between the Lugeon index and the grouting, permeability and the ground joint openings

Lugeon index	Grouting classification	Permeability range (cm/s)	Joint opening
<1	Very low	$<10^{-7}$	Very tight
1–5	Low	$10^{-7}–6\times10^{-7}$	Tight
5–15	Moderate	$6\times10^{-7}–2\times10^{-6}$	Few partly open
15–50	Medium	$2\times10^{-6}–6\times10^{-6}$	Some open
50–100	High	$6\times10^{-6}–10^{-5}$	Many open
>100	Very high	$>10^{-5}$	Closely spaced or voids

One Lugeon is equivalent to one liter of water grouted into one meter of borehole at a pressure of $P_0 = 1$ MPa, which roughly corresponds to a permeability of $k = 1.3 \times 10^{-7}$ m/s.

Table 3.15 relates the Lugeon index with the grouting, permeability and the joint openings existing in the ground.

3.7.4.3 Pumping tests

This type of tests consists in pumping a known water flow into a well drilled for the test, observing and recording the drop caused by the pump, both in the pumping well and in the observation piezometers placed around it, as shown in Figure 3.48.

Both the pumping well and the piezometers must be drilled using techniques that do not modify the natural permeability of the ground.

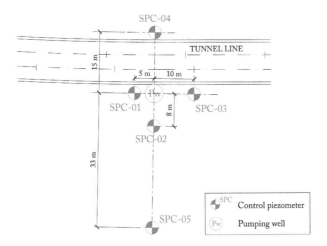

Figure 3.48 Pumping test layout.

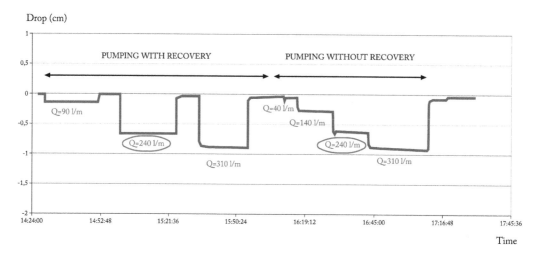

Figure 3.49 Flow and water level records to determine the optimal flow.

Pumping tests can be classified into two types:

- **Pumping or interference tests:** in which the drops produced in other wells or piezometers close to the pumping well are observed, inside which the water level changes are often measured.
- **Recovery test:** tests in which the water levels are measured after the termination of pumping in the well itself and/or in the observation wells and piezometers.

Frequently, in a first phase of the pumping test, a quick dynamic test is performed, with and without recovery, the aim being to determine the optimal flow at constant flow, as reproduced in Figure 3.49.

The pumping tests can be also divided into two classes:

- **At constant flow:** the usual duration of this test is 72 hours in unconfined aquifers and 24 hours in confined aquifers, although there is a wide range of practices depending on the specific site characteristics.
- **At variable flow:** in staggered pumps, in which increasing flows are pumped out and the corresponding water level drops are measured.

Figure 3.50 shows the water level drop distribution, over time, in a staggered pumping.

Although there are several analytical methods to interpret the pumping tests, nowadays the most practical tool consists in the use of a hydrogeological calculation program, as this type of program can take into account the characteristics of the ground involved in the pumping test and more representative results can be obtained.

3.8 WATER CHARACTERIZATION

While the ground investigation works are conducted it is essential to take samples of the water present in the ground, in order to analyze it chemically and evaluate its possible aggressiveness to concrete.

Table 3.16 reproduces the criteria allowed in the Spanish Concrete Instruction (EHE-08) to evaluate the water aggressiveness.

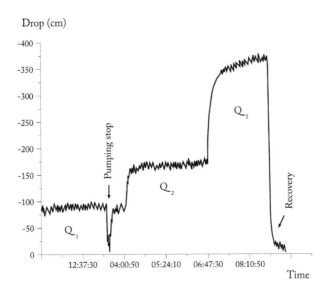

Figure 3.50 Staggered pumping test.

Table 3.16 Classification of the water chemical aggressiveness according to the EHE-08

Parameter	Type of attack to concrete		
	Soft (Qa)	Medium (Qb)	High (Qc)
PH	6.5–5.5	5.5–4.5	< 4.5
Aggressive CO_2 (mg CO_2/l)	15–40	40–100	> 100
Ammonium ion (mg NH_4^+/l)	15–30	30–60	> 60
Magnesium ion (mg Mg^{2+}/l)	300–1000	1000–3000	> 3000
Sulfate ion (mg SO_4^{2-}/l)	200–600	600–3000	> 3000
Dry residue	75–150	50–75	< 50

3.9 INTENSITY OF THE SITE EXPLORATION CAMPAIGN

It is clear that insufficient knowledge of the ground for tunnel design may lead to serious consequences, such as delays in the implementation deadlines and budget overruns.

Hoek and Palmieri presented in 1998, at the Congress of Geological Engineering in Vancouver (Canada), a study where they analyzed the budgetary deviations of an important set of underground works, correlating the ratio between the total accumulated borehole length to the tunnel length with the budgetary deviations.

Figure 3.51 shows the result of this study; the analysis of this figure reveals a surprising fact: for very low ratios of cumulative borehole length, below 0.1, the budgetary deviations range between 0% and 65%. On the other hand, not surprisingly, for very high ratios the

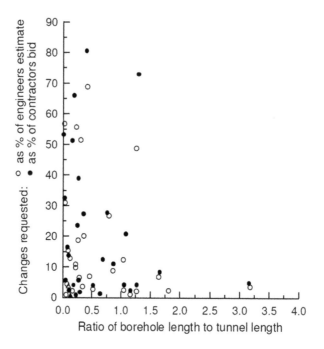

Figure 3.51 Comparison between the accumulated borehole length and the budgetary deviations in the design and during the works.

Source: Hoek and Palmieri, 1998.

budgetary deviations are less than 10%. From these facts two clear conclusions can be drawn:

1. Budgetary deviations depend on the tunnel construction difficulty, which explains why an easy tunnel can be designed and constructed without budgetary deviations, although only a few meters of boreholes were drilled during the design phase.
2. There is an optimum ratio of boreholes that ensures reasonable budgetary deviations in tunnels of normal difficulty.

On several occasions many authors have tried to establish criteria of the adequate cost of the geotechnical site investigation campaigns for tunnel design; as is shown in the following sections.

3.9.1 International criteria for the extent of site investigation campaigns

In 1980, in France, Panet and Verclós analyzed the construction overruns of 15 tunnels and they related them to the engineering and site investigation costs invested in the design phase, concluding that in the 15 tunnels analyzed, overruns occurred with respect to the initial budget; but, it also follows that, for a site investigation campaign and an engineering cost of around 7.4% of the construction value, the overrun is less than 10%.

In 1981, West and other authors published an extensive article in the *International Journal of Rock Mechanics*, entitled "Site Investigation for Tunnels" indicating:

Case histories of site investigations for tunnels in the U.K. show that the cost of the Stage II site investigation is generally less than about 3% of the cost of the works and may be as low as 0.5%. No hard and fast rules can be given about the cost of the site investigation because this will depend on both the complexity of the site geology and the complexity of the scheme.

In 1984, the US National Committee on Tunneling Technology analyzed the data from the site investigation campaign featuring costs for the design of 84 tunnels. The document, titled "Geotechnical Site Investigations for Underground Projects: Review of Practices and Recommendations", makes the following recommendation: "Expenditures for geotechnical site exploration should average 3% of the estimated Project cost".

Also from the United States, the Engineer Manual 1110-2-2901 *Engineering and Design. Tunnels and Shafts in Rock*, published in 1997 by the US Army Corps of Engineers, quantified the site exploration campaigns for tunnel design, establishing that the exploration intensity can be measured by three indices:

- Total cost of the site investigation campaign, referred to the tunnel construction cost
- Spacing between boreholes
- Accumulated borehole length, expressed as percentage of the tunnel length

Table 3.17 establishes the typical values of these indices.

For each of these three indices two separate correction factors are set as shown in Table 3.18, depending on some singularities.

In addition, the US Army Corps of Engineers points out that the borehole spacing should be increased as the tunnel depth increases.

Bjorn (2014) provided an interesting novelty to quantify the site investigation campaigns for tunnel design, in the article "Ground investigations for Norwegian tunnelling" in the No. 23 publication of the Norwegian Tunnelling Society.

Based on the work done by Palmstron et al. ((2003)), tunnel projects are classified into four groups, depending on the level of difficulty of the geology and the degree of safety required, as indicated in Table 3.19.

For each of these four groups, the tunnel length has been related to the cost of the site exploration works, expressed as a percentage of the excavation cost, as shown in Figure 3.52.

Table 3.17 Typical values of the indices to measure the site exploration campaigns

Indices	Normal cases	Extreme cases
Costs of boreholes and tests, expressed as a % of the construction cost	0.5/0.8	0.3/10
Spacing between boreholes (m)	150–300	15/1000
Accumulated borehole length, each 100 m of the tunnel total length (m)	15–25	5/1000

Table 3.18 Correction factors which multiply the typical indices

Singularities	Borehole cost as a % of the construction cost	Borehole spacing (m)	Borehole length in 100 m of tunnel
Easy geology	0.5	2/2.5	0.5
Complex geology	2/3	0.3/0.5	2/3
Tunnel in a rural area	0.5	2/2.5	0.5
Tunnel in densely urbanized area	2/4	0.3/0.4	2/5

Table 3.19 Classification of tunnel projects

	Geological difficulty of the project		
Safety required in the project	Low	Medium	High
Low	A	A	B
Medium	A	B	C
High	B	C	D

Source: Bjorn (2014).

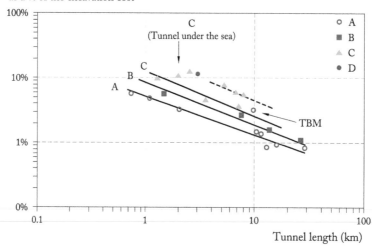

Figure 3.52 Site exploration costs as a function of the tunnel length and the project classification.

Source: Bjorn, 2014.

It should be emphasized that, in this figure, the tunnel length scale is logarithmic and the excavation costs only include blasting and loading of the excavated ground, multiplied by 1.25.

3.9.2 Specific recommendations

Tunnel design can be constrained by a number of particular project features, so no rigid rules, regarding the cost or scope of the geotechnical site exploration can or should be established.

Furthermore, the planning of an adequate geotechnical campaign should be a dynamic task, depending on the results obtained as it progresses, Oteo (2005).

Thus, it should be possible to increase, decrease or change the number or type of the initially planned investigations, if the preliminary results so indicate.

However, based on the criteria set out in the previous section and the experience gained in tunnel design, a set of guidelines or general recommendations can be given to plan the ground exploration campaigns.

In most cases, the best way to explore the ground in tunnel design is by using mechanical boreholes with core recovery.

As borehole costs are proportionally much higher than those from other exploration methods it is normal to establish the relative scope of the site exploration campaigns by the number of drilled boreholes or, more commonly, as the accumulated drilling total length expressed as a percentage of the tunnel length.

3.9.2.1 Urban tunnels

In the case of urban tunnels, it must be taken into account that they are almost always built at shallow depths and normally in soils.

It must also be kept in mind that, besides the usual lithological variations in these grounds, variations caused by human activity can be found which significantly alter the ground stress–strain behavior.

In urban areas, it is not possible to do mapping and the geophysical techniques may be limited by the influence of building foundations close to the tunnel alignment.

In these cases, the borehole spacing must range between 50 and 200 m, depending on the ground homogeneity and the access existing on the surface.

3.9.2.2 Tunnels under countryside areas

In site investigation campaigns for the design of tunnels in rural areas, the focus must also be on boreholes, although for long tunnels, the access to the places to be drilled can be very problematic and make some drillings unfeasible.

A relevant aspect about boreholes is related to the elevation that they must reach, as sometimes the technical specification documents require that the boreholes must be several meters below the tunnel level, which is not always necessary and, in some cases, can be an important conceptual error.

Accordingly, a characterization campaign with boreholes must be carefully defined adapting it to the existing geological model and in general, it will be preferable to drill more but shorter boreholes than few of great length, as Figures 3.26 and 3.27 reveal.

It is therefore interesting to consider the tunnel length, distinguishing the following three cases:

- Tunnels with length shorter than 500 m (L<0.5 km)
- Tunnels between 500 m and 5 km (0.5 km<L<5 km)
- Tunnels longer than 5 km (L>5 km)

Tunnels with lengths shorter than 500 m, located beneath non-built areas, do not often present serious problems in site investigation, as the ground to be investigated is easily accessible and it is possible to map the surface, do geophysical campaigns and set geomechanics stations.

In addition, horizontal boreholes are almost always drilled at both portals, the length of which can reach up to 150–200 m; such borehole length may cover a large proportion of the length of these shorter tunnels.

The cost of these site investigation works, typically represents between 0.5% and 1.5% of the tunnel construction cost.

In tunnels whose length is between 0.5 km and 5 km, it is possible to give criteria to estimate the cumulative borehole length to be drilled and the cost of the investigation campaign. For this it is useful to take the figure from the study carried out by Hoek and Palmieri and adjust the curve that crosses the mean values of the observed deviations, as shown in Figure 3.53.

From the adjustment shown in this figure it follows that, if the borehole cumulative length is equal to 30% of the tunnel length, the average budgetary deviations both in the design and construction phase, will be around 20%.

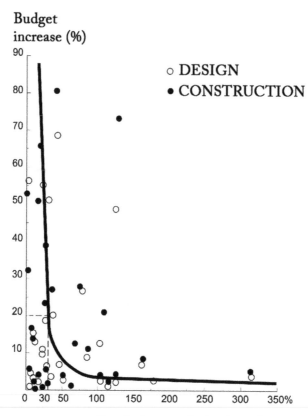

Figure 3.53 Adjustment of the budgetary deviations in the projects and construction works.
Source: Hoek and Palmieri, 1998.

To estimate the cost that the site investigation campaign to design a tunnel should have, whose length ranges between 0.5 km and 5 km, the following equation is proposed:

$$CC = \left(0.75 + \frac{L}{2}\right) \cdot \frac{CT}{100} \cdot DC$$

where:
-    CC   = Total cost of the site investigation campaign to be performed
-    L   = Tunnel length, expressed in km.
-    CT   = Estimated total cost for the tunnel construction
-    DC   = The factor that takes into account the tunnel construction difficulty, according to the following values:
        Normal construction difficulty: DC = 1
        Low construction difficulty: DC = 0.8
        High construction difficulty: DC = 1.4

According to what was presented in the previous section, for a tunnel 0.5 km long and normal constructive difficulty the total ground investigations cost would be 1% of the total tunnel cost. If the tunnel length changes to 4 km and the construction difficulty is normal, the total ground investigations cost becomes 2.75% of the total tunnel cost.

In tunnels with a length greater than 5 km, the site exploration campaign should be specifically studied in each case.

BIBLIOGRAPHY

Advanced Logic Technology, "OBI40-2G Tool Specifications", Luxemburg. 2016.

AENOR, Norma UNE 103808:2006: Ensayo de carga vertical de suelos mediante placa estática. 2006.

AENOR, Norma UNE-EN ISO 22476–5: Investigación y ensayos geotécnicos. Ensayos de campo. Parte 5. Ensayo con dilatómetro flexible. 2014.

AENOR, Norma UNE-EN ISO 22476–7: Investigación y ensayos geotécnicos. Ensayos de campo. Parte 7. Ensayo con dilatómetro rígido. 2015.

Agharazi, A., Tannant, D.D., Martin, C.D., "Characterizing rock mass deformation mechanisms during plate load tests at the Bakhtiary dam project", *International Journal of Rock Mechanics & Mining Sciences*. Vol. 49. January 2012, pp. 1–11.

ASTM, ASTM D1195, *Standard Test Method for Repetitive Static Plate Load Tests of Soils and Flexible Pavement Components, for Use in Evaluation and Design of Airport and Highway Pavements*. 2015.

ASTM, ASTM D4394, *Standard Test Method for Determining In Situ Modulus of Deformation of Rock Mass Using the Rigid Plate Loading Method*. 1998.

ASTM, ASTM D4719-87, *Standard Method for Pressuremeter Testing in Soils*. 1987.

Aufmuth, E.R., "A systematic determination of engineering criteria for rocks", *Bulletin of Engineering Geology and the Environment*. No. 11. 1973, pp. 235–245.

Barton, N., Choubey, V., "The shear strength of rock joints in theory and practice", *Rock Mechanics*. Vol. 10 Nos. 1–2. 1977, pp. 1–54.

Barton, N., Lien, R., Lunde, J., "Engineering classifications of rock masses for the design of tunnel support", *Rock Mechanics*. Vol. 6, No. 4. 1974, pp. 189–239.

Bieniawski, Z.T., "A comparison of rock deformability measurements different in situ methods", *Proceedings of the Rapid Excavation and Tunneling Conference*, Ch. 52. 1979, pp. 901–915.

Bieniawski, Z.T., *Engineering Rock Mass Classifications*. John Willey, New York. 1989.

Bieniawski, Z.T., "The point load test in geotechnical practice", *International Journal of Engineering Geology*, Vol. 9. No. 1. 1975, pp. 1–11.

Bieniawski, Z.T., Van Heerden, W., "The significance of in situ tests on large rock specimens", *International Journal of Rock Mechanics and Mining Sciences*. Vol. 120. No. 4. 1976, pp. 101–113.

Bjorn, N., *Ground Investigations for Norwegian Tunnelling*, Norwegian Tunnelling Society Publication. No. 23. 2014, pp. 19–33.

Boart Longyear, "Drilling Equipment Product Line". 2014.

Briaud, J.L., *The Pressuremeter*. Balkema, Rotterdam. 2005.

CBG Corporation, *CBG Corporation Product Catalog*. Austin, Texas. 2016.

Celada, B., "Criterios sobre el costo de las campañas de reconocimiento para el diseño de túneles", *Proceedings of the 4th Simposio Internacional sobre Túneles y Lumbreras en Suelos y Roca*. Ciudad de México, México. 2016.

Clarke, B.G., *Pressuremeters in Geotechnical Design*, Blackie Academic & Professional. London. 1995.

Deere, D.U., "Technical description of rock cores for engineering purposes", *Rock Mechanics and Engineering Geology*. Vol. 1. 1963, pp. 16–22.

Devicenzi, M., Frank, N., *Ensayos geotécnicos in situ. Su ejecución e interpretación*. Ingeotest, S.L. España. 2004.

Dodds, R.K., "Preliminary investigations", *Tunnel Engineering Handbook*. Eds. J.O. Bickel and T.R. Kuesel. Van Nostrand. New York. 1982, pp. 11–34.

Ehrbar, H., Gruber, L.R., Sala, A., *Tunnelling The Gotthard Base Tunnel*. Swiss Tunnelling Society. Esslingen, Germany. 2016.

El Tani, M., "Water inflows into tunnels", *Proceedings of the ITA World Tunnel Congress*. Rotterdam. 1999, pp. 61–70.

Federal Highway Administration, "Manual on Subsurface Investigations" (FHWA NHI-01-031). Washington, DC. 2001.

Federal Highway Administration, "Road Tunnel Manual" (FHWA-NHI-09-010). Washington, DC. 2009.

González de Vallejo, L.I., *Ingeniería Geológica*. Prentice Hall, Madrid, Spain. 2002.

Goodman, R.E., Moye, D.G., van Schalkwyk, A., Javandel, I., "Groundwater inflows during tunnel driving", *Engineering Geology*. Vol. 2. No. 1. 1965, pp. 39–56.

Heuer, R.E., "Estimating rock tunnel water inflow", *Proceedings of the Rapid Excavation and Tunelling Conference*. Society for Mining, Metallurgy and Exploration, Denver, CO. 2005, pp. 394–407.

Herrera, J., Castilla, J., "Utilización de Técnicas de Sondeos en Estudios Geotécnicos", *E.T.S.I. de Minas*. Madrid, Spain. 2012.

Hoek, E., Brown, E.T., "The Hoek–Brown failure criterion – a 1988 update", *15th Canadian Rock Mechanics Symposium*. Toronto, Canada. 1988.

Hoek, E., Palmieri, A., "Geotechnical risk on large civil engineering projects", *Proceedings of the International Association of Engineering Geology Congress*. Vancouver, Canada. 1998.

Instituto Geológico y Minero de España, "Mapa Geológico de España a Escala 1:50.000", *Magna*. Vol. 50. No. 2ª Serie. 1972–2003.

Kilic, A., Teymen, A., "Determination of mechanical properties of rocks using simple methods", *Bulletin of Engineering Geology and the Environment*. No. 67. 2008, pp. 237–244.

Kulhawy, F.H., Mayne, P.W., *Manual on Estimating Soil Properties for Foundation Design*. Electric Power Research Institute, Palo Alto, CA. 1990.

Littlejohn, G.S., "Ground: reducing the risk" *Proceedings of the Institute of Civil Engineering*. Vol. 102. No. 1. 1994, pp. 3–4.

López Jimeno, C. et al., *Manual de Túneles y Obras Subterráneas*. Madrid, Spain. 2011.

Mayne, P.W., Kemper, J.B., "Profiling OCR in Stiff Clays by CPT and SPT", *Geotechnical Testing Journal ASTM*. Vol. 11. No. 2. 1988, pp. 139–147.

Miller, R.P., "Engineering classification and index properties for intact rocks", Thesis. University of Illinois, Urbana, IL. 1965.

Nazir, R., Amin, M., "Prediction of unconfined compressive strength of rocks: A review paper", *Journal Teknologi (Sciences & Engineering)*. Vol. 74. 2015.

Oteo, C., "Geotecnia, Auscultación y Modelos Geomecánicos en los túneles ferroviarios", *Túnel de Guadarrama*. ADIF, Madrid, Spain. 2005.

Palmstrom, A., Nielsen, B., Pedersen, K.B., Grundt, L., "Proper extent of site investigations for underground excavations", Report R&D Project Tunnels for the citizens, Publicación No. 101, Directorate of Public Roads, Oslo, Norway. 2003.

Parker, H.W., "Geotechnical Investigations", *Tunnel Engineering Handbook*. Chapman & Hall, Boca Raton, FL. 1996.

Priest, S.D., Hudson, J.A., "Discontinuity spacings in rock", *International Journal of Rock Mechanics and Mining Sciences*. No. 13. 1976, pp. 135–148.

Rocscience Inc., *Software DIPS*. Rocscience, Toronto, Canada. 2015.

Rodríguez, A., Lozano, A., Pescador, S., "Testificación Geofísica de sondeos para estudios geotécnicos", *Ingeopres*. No. 131. Madrid, Spain. 2004, pp. 26–30.

Singh, M., Rao, K.S., "Empirical methods to estimate the strength of jointed rock masses", *Engineering Geology*. Vol. 77. Nos. 1–2. 2005, pp. 127–137.

Swiss Federal Archives SFA, The Alptransit Portal. 2016. www.alptransit-portal.ch.

Terzaghi, K., Peck R.B., *Soil Mechanics in Engineering Practice*, 2nd Edition. John Wiley, New York. 1967.

Ulusay, R., *The ISRM Suggested Methods for Rock Characterization, Testing and Monitoring: 2007-2014*. Springer Int. Publishing, Cham, Switzerland. 2015.

Ulusay, R., Hudson, J.A., *The Complete ISRM Suggested Methods for Rock Characterization, Testing and Monitoring: 1974–2006*. Ankara, Turkey. 2007.

US Army Corps of Engineers, *Tunnels and Shafts in Rock* (1110-2-2901). Department of the Army. Washington, DC. 1997.

US National Committee on Tunnelling Technology, *Geotechnical Site Investigations for Underground Projects: Review of Practice and Recommendations*. New York. 1984.

West, G., Carter, P.G., Dumbleton, M.J., Lake, L.M., "Site investigations for tunnels", *International Journal of Rock Mechanics and Mining Science & Geomechanics*. Vol. 5. 1981, pp. 345–367.

Wolff, T.F., "Pile capacity prediction using parameter funtions", *Proceedings of the Predicted and Observed Axial Behaviour of Piles, Results of a Pile Prediction Symposium*, ASCE Geotechnical Special Publication. No. 23. 1989, pp. 96–106.

Chapter 4

In situ state of stresses

Juan Manuel Hurtado Sola

Reasoning, observation and experience; the Science Holy Trinity

Robert G. Ingersoll

4.1 INTRODUCTION

As noted earlier, when a tunnel is excavated there is a distribution of the original in situ stresses leading to the induced stresses and, after this redistribution, the ground must withstand the tangential stress around the excavation while the tunnel support has to bear the radial stress.

To calculate the tangential and radial stresses around the tunnel it is essential to know the original, also known as virgin, state of stress.

The aim of this chapter is to analyze the factors, which influence the in situ stress state and describe the most common methods for its estimation and direct determination.

4.2 ORIGINAL IN SITU STATE OF STRESSES

If an infinitesimal element of the ground is oriented with respect to the axes placed on three pairs of perpendicular surfaces composing this element, the original in situ stress state is defined by the nine stress components as indicated in Figure 4.1.

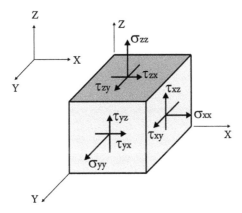

Figure 4.1 Stress components of a solid infinitesimal element.

133

If the rotational balance of the infinitesimal solid is kept, the shear stress acting on two adjacent sides has to be equal and the stress components are reduced to six:

$$\sigma_X; \sigma_Y; \sigma_Z$$

$$\tau_{XY}; \tau_{XZ}; \tau_{ZY}$$

If the solid principal directions are taken as axes of reference, in which there are only compressional components, the stress components are reduced to three: $\sigma_1; \sigma_2; \sigma_3$ but the three direction cosines orientating the principal axes must also be added as unknowns as illustrated in Figure 4.2.

Assuming the hypothesis that the gravity direction is one principal direction, the other two stress components would be on a horizontal plane. These components are often called σ_H and σ_h, so that the three stress components will be $\sigma_v; \sigma_H; \sigma_h$.

It is also often assumed that the stress σ_v corresponds to the ground weight acting in the observation site, so its value will be given by:

$$\sigma_v = \gamma \cdot H \tag{4.1}$$

where γ is the ground specific gravity and H is the depth of the point at which the stress state is going to be estimated.

Figure 4.3 shows the results from several ground vertical stress measures as a function of depth.

This figure represents the line $\sigma_v = \gamma \cdot H$ for a value of $\gamma = 0.027$ MN/m³, which fits relatively well with the collected measurements, and this validates the assumption that the vertical stress component is mainly caused by the ground weight.

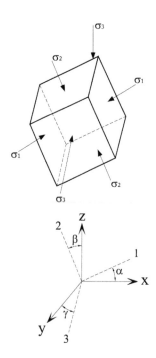

Figure 4.2 Stress components referred to the principal axes.

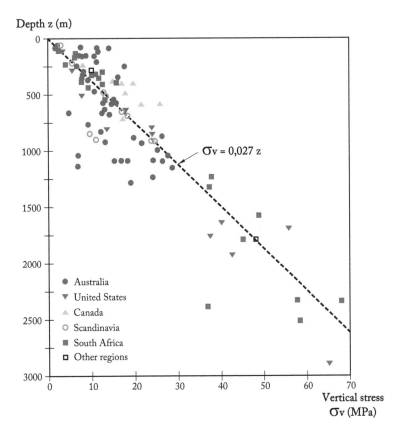

Figure 4.3 Vertical in situ stress variation with depth.

Source: Hoek and Brown, 1980.

With these considerations the stress components to be determined are reduced to σ_H and σ_h as well as one direction cosines of the OX and OY axes, which are placed on a horizontal plane.

Usually it is common to express σ_H and σ_h as a function of σ_v through the principal field stress ratios, K_{0H} and K_{0h}, and following that the value of the three principal stresses is calculated with the following expressions:

$$\sigma_v = \gamma \cdot H$$

$$\sigma_H = K_{0H} \cdot \gamma \cdot H \qquad (4.2)$$

$$\sigma_h = K_{0h} \cdot \gamma \cdot H$$

4.3 FACTORS WHICH INFLUENCE THE IN SITU STRESS STATE

Ideally, stresses below the earth surface must correspond to the lithostatic effect produced by the lithostatic weight. However, through geological history, grounds could have been exposed to tectonic effects, which substantially alter the ideal stress state.

Table 4.1 summarizes the tectonic actions and residual effects that modify the in situ stress state actions.

The following sections present the factors with a greater influence on the in situ stress state.

Table 4.1 Tectonic activities and residual effects which influence the in situ stress state

Tectonic	Active	Regional scale	Slab pull
			Ridge push
			Shear forces
			Suction
		Local scale	Bending
			Isostatic compensation
			Volcanism
			Heat flow
	Remnant		Folding
			Faulting
			Jointing
Residual			Diagenesis
			Metamorphism
			Metasomatism
			Magma cooling
			Changes in pore pressure

4.3.1 Tectonic effects

The tectonic movements of the earth's crust significantly change the original lithostatic arrangement, causing important changes in the virgin stress state. Figure 4.4 shows a scheme, by Zoback et al. (1989), of the most common tectonic effects induced by the movement of two tectonic plates.

One of the most impressive examples of tectonic movements is the Mountain of Seven Colors, Salta (Argentina), illustrated in Photograph 4.1.

Through geological history, the intensity and persistence of tectonic phenomena are very strong, resulting in major changes in the in situ stress state, producing schistosity, faults and areas with tensile and compressive stresses.

From the typology of the faults, present in the geotechnical mapping, it is possible to know the virgin stress state that has produced them.

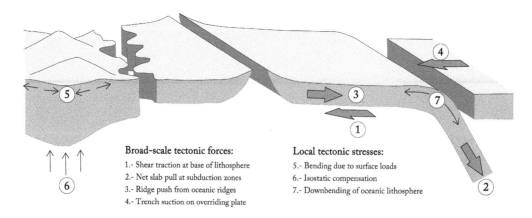

Broad-scale tectonic forces:
1.- Shear traction at base of lithosphere
2.- Net slab pull at subduction zones
3.- Ridge push from oceanic ridges
4.- Trench suction on overriding plate

Local tectonic stresses:
5.- Bending due to surface loads
6.- Isostatic compensation
7.- Downbending of oceanic lithosphere

Figure 4.4 General stresses due to the relative movement of two tectonic plates.

Source: Zoback et al., 1989.

Photograph 4.1 Mountain of Seven Colors, Salta (Argentina).

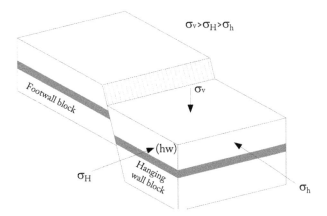

Figure 4.5 Stress state in a normal fault.
Source: Zoback et al., 1989.

Figure 4.5 shows the relative magnitudes of the principal stresses necessary to generate a normal fault characterized by the downwards displacement of the block that moves.

In this case the major principal stress is σ_v, leading to values of K_{0H} and K_{0h} smaller than unity.

Reverse faults and thrust faults are characterized by the upward displacement of the block that moves, a situation that leads to a stress state as the one represented in Figure 4.6.

In reverse faults σ_v is the smallest principal stress leading to values of K_{0H} and K_{0h} smaller than the unit.

Figure 4.7 shows the stress state necessary to produce a strike-slip fault, characterized by the horizontal displacement of the block that moves.

In this case σ_v is the intermediate principal stress and the major principal stress is σ_H, which acts in the direction of the displacement. This situation leads to principal field stress ratios close to unity.

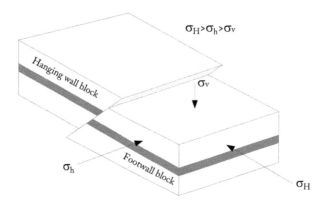

Figure 4.6 Stress state in a reverse fault.

Source: Zoback et al., 1989.

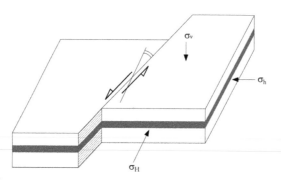

Figure 4.7 Stress state of a strike-slip fault.

Source: Zoback et al., 1989.

4.3.2 Effect of topography

In the ground close to the surface there cannot be a stress component perpendicular to it, which means that in a natural valley the vertical stress is much lower than the horizontal stress, as is illustrated in Figure 4.8 in these cases, the principal field stress ratios are much higher than unity.

A similar situation takes place in the hillside tunnels, as illustrated in Figure 4.9.

In this case, before the tunnel construction, the horizontal stresses are smaller than the ones existing at the same depth in a normally stratified field, which generates very low principal field stress ratios, which have a negative influence on the tunnel design.

4.3.3 Erosion and thaw effect

Along the geological history there have been widespread erosions in important areas of the ground surface, sometimes reaching up to hundreds of meters.

A similar effect takes place with the thaw effect in the areas which were under glaciation effects. Figure 4.10 illustrates this situation.

Widespread erosion leads to a decrease in the vertical stresses. However, the horizontal stresses are not dissipated at the same rate, which means an increase in the principal field stress ratios.

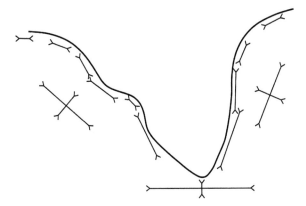

Figure 4.8 Topography influence on the stress state.

Source: Goodman, 1989.

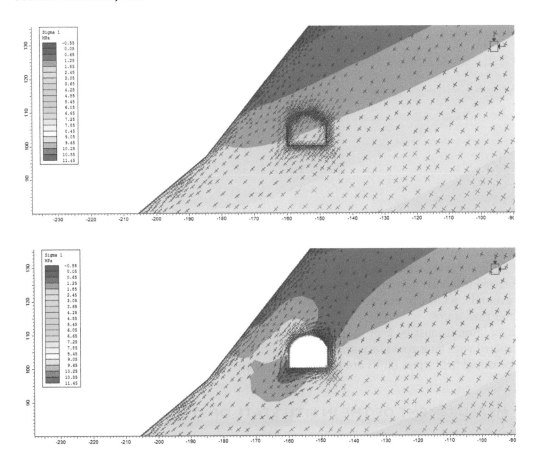

Figure 4.9 Stress state in a hillside tunnel.

4.3.4 Ground heterogeneity

Grounds have a differentiated stress–strain response depending on their mechanical characteristics and therefore, when two grounds with a strong contrast between their strengths are in contact, there will also be a contrast in the in situ stress distribution.

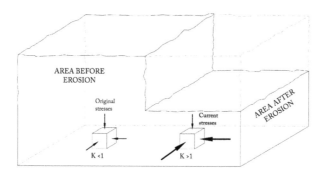

Figure 4.10 Erosion effect on the in situ stresses.

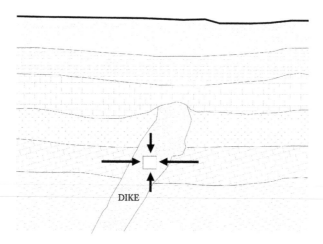

Figure 4.11 Intrusion into a stratified medium.

This situation is very common when there is an intrusion of a magmatic rock into a stratified medium, as illustrated in Figure 4.11.

Intrusions similar to the one presented in Figure 4.11 cause high horizontal stresses on the grounds, which, in practice, result in values of the principal field stress ratios that are greater than unity.

In addition it must be kept in mind that, in grounds with very poor quality, RMR < 20 associated with fault zones or with zones susceptible to presenting strong yielding, ICE < 15, the principal field stress ratios are usually close to unity.

4.4 MEASUREMENT OF THE IN SITU STATE OF STRESSES

The following sections describe the overcoring and the hydraulic fracturing techniques, which are the most commonly used techniques to measure the in situ stress state.

4.4.1 Overcoring

In situ stress measurement with the overcoring method consists of the installation of a strain gauge at the borehole end, so that the in situ stress state is not altered at the point tested, and measuring the strains induced in the ground by the drilling of a borehole concentric to the initial one using a larger coring bit.

Figures 4.12 and 4.13 illustrate the location of an exploration borehole and the overcoring process.

During the overcoring process the strains induced in the ground are measured which, afterward, are used to calculate the initial stress state.

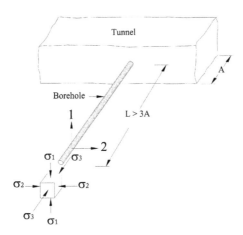

Figure 4.12 Borehole location to measure in situ stresses.

Source: Leeman, 1964.

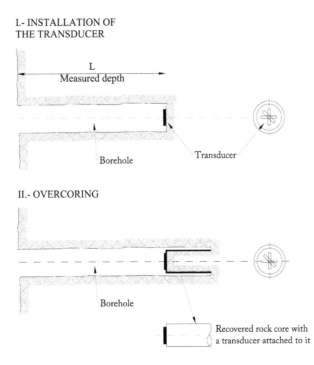

Figure 4.13 Transducer (strain gauges) installation and overcoring.

Source: Leeman, 1964.

Photograph 4.2 Doorstopper.

Source: Leeman, 1964.

When the overcoring technique was developed, the available strain gauges were only able to measure the strains on a plane. Photograph 4.2 reproduces the Doorstopper, developed in 1964 by the Rock Mechanics Division of the Council for Scientific and Industrial Research (CSIR), in South Africa, Leeman (1964).

To measure the strains, the Doorstopper had three strain gauges, as shown in Photograph 4.2, which enabled determination of the principal stresses and their orientation on the plane defined by the borehole end.

To determine the complete in situ stress state with the Doorstopper it was necessary to make measurements in three orthogonal boreholes, making this operation expensive and very difficult to perform.

To avoid these problems, the triaxial cell was developed by Leeman in 1969, also at the CSIR of South Africa and by the CSIRO (Australia), which enabled three strain measurements in three orthogonal axes, which solved the existing problems to measure the in situ stress state in a single borehole.

Photograph 4.3 shows a device with a triaxial cell to measure the in situ stresses, developed by Geosystems, in Australia.

Figure 4.14 shows the strains measured during the overcoring of a 55 m long borehole, inside which a triaxial cell was installed.

The measurements of the in situ stress state with the overcoring technique provide very accurate data about the in situ stress state, but the data is very local; moreover, overcoring measurements are based on assuming a homogeneous and isotropic ground, which is often inconsistent with the reality. Therefore, to measure the in situ stresses by overcoring it is necessary to carry out tests consisting of many measurements, which is expensive.

4.4.2 Hydraulic fracturing

The technique to measure in situ stresses by hydraulic fracturing was extensively studied and improved upon by Haimson (1978a,b) in the United States. This technique is also performed in boreholes, as in the overcoring process, but in this case no strains are measured because the fractures present in the ground open inside the borehole when pressurized water is injected in in the borehole.

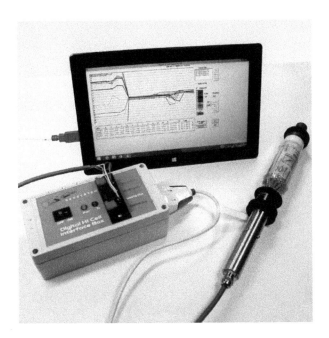

Photograph 4.3 Device to measure the in situ stress state with a triaxial cell.

Source: Geoystems, Australia.

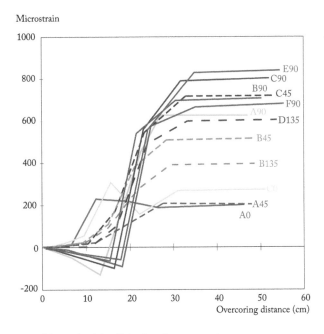

Figure 4.14 Strains measured in a triaxial cell during the overcoring process.

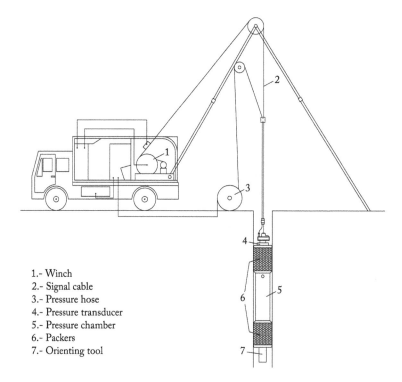

1.- Winch
2.- Signal cable
3.- Pressure hose
4.- Pressure transducer
5.- Pressure chamber
6.- Packers
7.- Orienting tool

Figure 4.15 Equipment necessary to perform the hydraulic fracturing test.

Figure 4.15 shows the device necessary to perform a hydraulic fracturing test. The development of this test involves the following stages:

  I. Inspection of the core recovered from the borehole, with the aim of identifying the soundest stretches close to the tested depth. This task could be complemented by the use of an acoustic televiewer probe, whose bases have been explained in Section 3.5.2.
  II. Installation of the hydro-fracturing probe in the selected section and its anchorage with an upper and a lower packer.
  III. Performance of a permeability test, by the instant pressurization of the section to be tested, recording for several minutes the pressure evolution. This test provides the permeability of the intact rock.
  IV. Increasing pressures are applied until the fractures appear and the breakdown pressure (P_c) is recorded.
  V. Several reopening cycles are performed to check the shut-in pressure (P_s) which is the pressure below which fractures close. It is important to fully relieve the pressure on the tested section between cycles. The pressure necessary to reopen the fractures (P_r) is recorded.
  VI. A new permeability test is performed, by pressure steps at constant flow, similar to the Lugeon test. This test provides permeability of the fractured rock mass.

Figure 4.16 shows a real record of one of these tests, indicating each of the stages described previously.

Once the test is completed it is necessary to have an oriented image of the fractures opened in the ground. This is achieved with an impression *packer* which is driven down to the area which is being investigated and provides an image as the one reproduced in Figure 4.17.

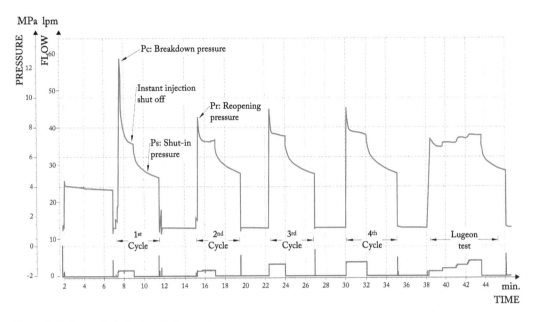

Figure 4.16 Record of a hydraulic fracturing test.

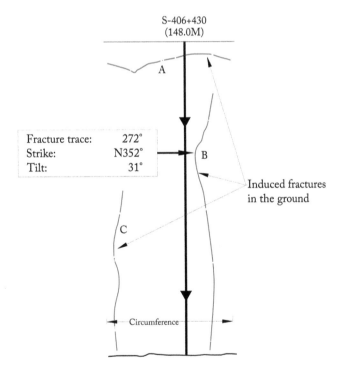

Figure 4.17 Fracture trace on the impression packer.

As an alternative to this impression system, a televiewer can be used to record and orientate the fractures.

The stress state in each tested depth is calculated by the following expressions:

$$\sigma_v = \gamma \cdot z$$

$$\sigma_h = P_s$$

$$\sigma_H = \sigma_t + 3 \cdot P_s - P_c - P_0, \quad \text{for the first pressure cycle.}$$

$$\sigma_H = 3 \cdot P_s - P_r - P_0, \quad \text{for the following cycles.}$$

where:
   γ = Average specific weight of the overlying rock
   P_s = Shut-in pressure
   P_c = Breakdown pressure
   P_r = Reopening pressure
   P_0 = Pore pressure at the investigated depth
   σ_t = Rock tensile strength
   σ_v = Vertical in situ stress
   σ_H and σ_h = Horizontal in situ stresses

Photograph 4.4 shows a snapshot of a hydraulic fracturing test.

The interpretation of the hydraulic fracturing tests is made by *assuming* that the vertical in situ stress component σ_v is one of the three directions. This is why this method does not

Photograph 4.4 Hydraulic fracturing test.

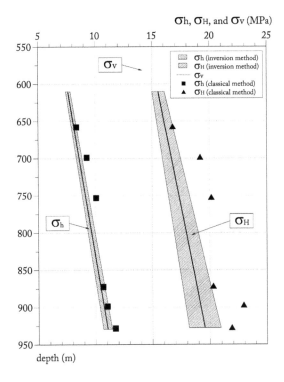

Figure 4.18 Results of a hydraulic fracturing test.

determine the complete state of in situ stresses. In essence, it measures the horizontal stress components, σ_h and σ_H. Figure 4.18 shows the results of a hydraulic fracturing test.

Nevertheless, the hydraulic fracturing method has two important advantages; each test affects a much larger ground volume than the overcoring tests and, in addition, in each borehole more than one test can be performed.

4.4.3 Measurements at great depth

The measurement of in situ stresses is especially relevant to tunnels of great length, deep mines and underground repositories for radioactive waste products.

A common feature of these projects is that measurements of the in situ stress state are performed at depths greater than 500 m; thus, it is necessary to modify the conventional hydro-fracturing measurement techniques.

4.4.3.1 Hydraulic fracturing tests at depths greater than 500 m

In order to perform hydraulic fracturing tests at great depth, two problems have to be solved; one is related to the greater hydraulic pressures needed to achieve hydraulic fracturing of the ground and the other is to do with the hydraulic isolation of the borehole section in which the test is performed.

Both problems are solved without great difficulty; this is because the highest hydraulic pressures are achieved by using more powerful booster pumps, which also require reinforcing the joints between the pipe sections, and the improvement of the hydraulic isolation of the testing cell is achieved by using four packers, as shown in Figure 4.19.

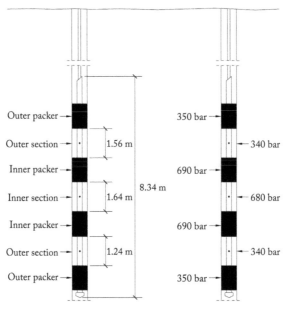

I.- Principle and key features   II.- Pressure capabilities

Figure 4.19 High pressure packers.

Source: Sjöberg, 2012.

Currently, with high pressure packers, it is possible to perform hydraulic fracturing tests at depths of around 1,000 m.

4.4.3.2 Overcoring at depths greater than 500 m

The main problem with overcoring is the interference between the drilling equipment and the wires that send the ground displacement data during overcoring, which makes being successful with this technique at depths greater than 100 m very difficult.

This problem has been solved by introducing, in the measurement unit, a data logger that stores all the measurements; which makes the classic signal transmission wires unnecessary.

Sjöber and Klasson (2003) developed the Borre probe; whose operating principle is illustrated in Figure 4.20.

The Borre probe is an improvement on the Doorstopper and the triaxial cell developed in South Africa. It features the use of soft cells and its measurement principle is based on the linear elasticity of a continuous, homogeneous and isotropic ground.

The stages of using the Borre probe are:

1. Advance 76 mm-diameter main borehole to measurement depth. Grind the hole bottom using the flattening tool.
2. Drill 36 mm-diameter pilot hole and recover core for appraisal. Flush the borehole to remove drill cuttings.
3. Prepare the Borre probe for measurement and apply glue to strain gauges. Insert the probe with the installation tool into hole.

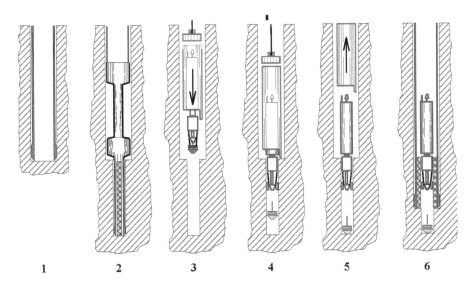

Figure 4.20 Operating principle of the Borre probe.

Source: Sjöberg, 2012.

4. Enter the tip of the probe with strain gauges into the pilot hole. The probe releases the installation tool, which also sets the compass, thus recording the installed probe orientation. Gauges are bonded to the pilot hole wall under pressure from the nose cone.
5. Allow glue to harden (usually overnight). Pull out installation tool and retrieve to surface. The probe is bonded in place.
6. Overcore the Borre probe and record strain data using the built-in data logger. Break the core off after completed overcoring and remove core barrel to surface.

Figure 4.21 shows the main elements of the Borre probe.

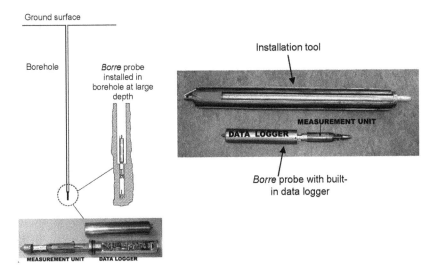

Figure 4.21 Main elements of the Borre probe.

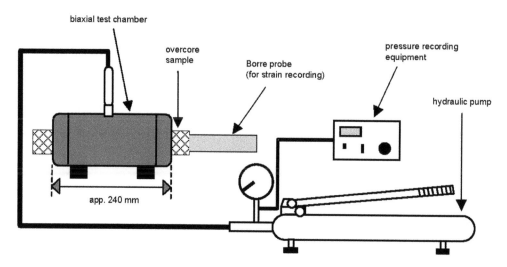

Figure 4.22 Schematic drawing of the biaxial load cell with pressure generator and recording equipment.

If it is possible to extract the rock sample from the overcored section, with the Borre probe attached to it, then it is possible to determine the elastic constants of the rock core. To do this, the overcore sample must be tested in a biaxial load cell in which the stress state corresponding to the measured strains is reproduced, using the device shown in Figure 4.22.

Since 1976, more than 1,000 measures of the virgin stress state have been performed with the Borre probe, mostly in Scandinavian and Middle Eastern countries, some of them at depths greater than 600 m.

4.5 ESTIMATION OF THE IN SITU STATE OF STRESS

Both the overcoring and the hydraulic fracturing tests are expensive; so for medium length tunnels they may not be affordable.

Accordingly, there are other approaches to *estimating* the in situ stress state, that can be used if necessary.

4.5.1 World Stress Map

The World Stress Map project (WSM) started under the auspices of the International Lithosphere Program (ILP) led by Mary Lou Zoback until 1992. Since 1995, the WSM project is led by the Heidelberg Academy of Sciences and Humanities, located at the Geophysical Institute in Kalsruhe University, Germany.

The main objective of this project is to collect global information about the in situ stress state in the earth's crust, with special emphasis on the orientation of the major horizontal stresses.

However, the database also contains the absolute measurements of the stress state in many points, with extensive information about them, such as depth, date of measurement and reference to the publication where it appears.

The data comes from a diverse set of tests performed in boreholes (hydraulic fracturing and overcoring), the estimation of the stress tensor from earthquake focal mechanisms and data from the geological observations.

To ensure the homogeneity of the different sources and to compare their results each data record is evaluated according to previously established criteria, assigning a representative value.

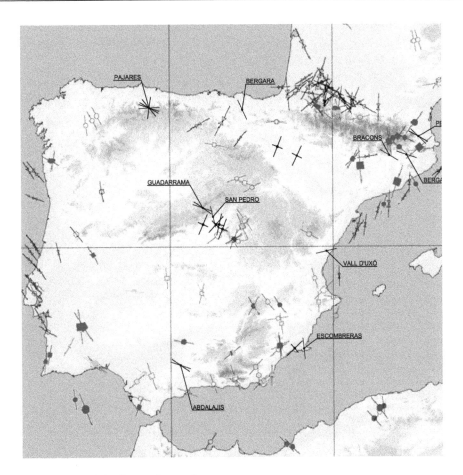

Figure 4.23 WSM map of the Iberian Peninsula.

Figure 4.23 shows an image of the Iberian Peninsula, which contains data from the WSM, indicating the in situ stress state measures which have been described before.

4.5.2 K₀ variation with depth

Figure 4.24 shows the variation of the principal field stress ratio (K_0) with depth, based on several measures of the in situ stress state, provided by Hoek and Brown (1980). In this figure the measures carried out by Geocontrol in Spain have been included, which are marked with a blue star.

From the data contained in Figure 4.24 it can be said that up to a depth of 500 m, the K_0 values are clearly higher than 1. For depths between 500 and 1000 m, the values range from 0.5 to 2.0.

However, it must be kept in mind that many of the in situ measures represented in Figure 4.24 have been made in unique places which do not represent most cases.

Sheorey (1994), based on the stress distribution of the measurements presented in Figure 4.24, has proposed the following adjustment:

$$K_0 = 0.25 + 7 \cdot E_h \cdot \left(0.001 + \frac{1}{h}\right) \tag{4.3}$$

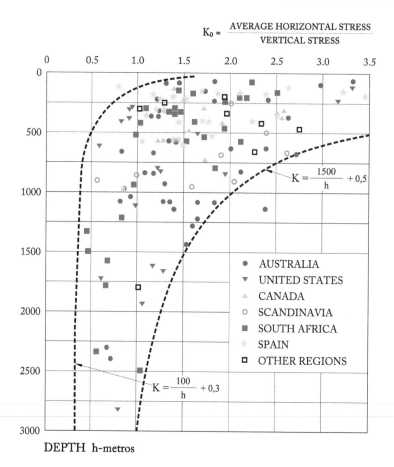

Figure 4.24 Changes of K_0 with depth.

Source: Hoek and Brown, 1980.

where E_h is the average modulus of deformation of the rock mass, measured in the horizontal direction and expressed in GPa, and h is the depth in meters.

Figure 4.25 shows this expression appearing for different values of the modulus of deformation.

4.5.3 General estimation of K_0

In general, the principal field stress ratio K_0 can be estimated considering the following factors:

• Lithostatic pressure
• Tectonic effects
• Erosion effects
• Effect of the topography

K_{0L}, in a normally stratified ground and exclusively exposed to lithostatic effects, can be calculated by the expression:

$$K_{0L} = \frac{\nu}{1 - \nu} \tag{4.4}$$

where ν is the Poisson coefficient of the ground.

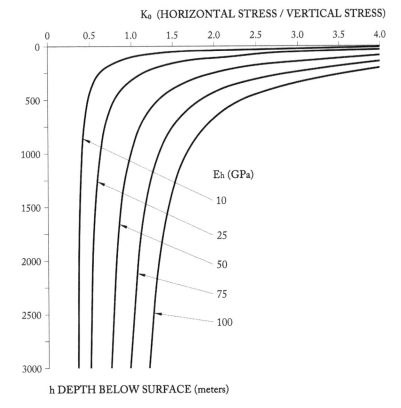

K_0 (HORIZONTAL STRESS / VERTICAL STRESS)

Eh (GPa)

10

25

50

75

100

h DEPTH BELOW SURFACE (meters)

Figure 4.25 K_0 variation with depth.

Source: Sheorey, 1994.

For a value of $\nu = 0.25$ the resulting K is $K_{0L} = 0.33$ and for $\nu = 0.35$ it is obtained that $K_{0L} = 0.54$, values that, in general, are quite low.

An estimation of K_0 could be made using the following expression:

$$K_0 = \frac{\nu}{1-\nu} + \Delta K_{0TE} + \Delta K_{0TO} + \Delta K_{0ER} \qquad (4.5)$$

where ΔK_{0TE}, ΔK_{0TO} and K_{0ER}, are the K_0 increments due to tectonic effects, the topography and the erosion which can be estimated considering what was covered in Sections 4.3.1., 4.3.2. and 4.3.3.

BIBLIOGRAPHY

Bielenstein, H.U., Barron, K., "In situ stresses", *Proceedings of the 7th Canadian Symposium Rock Mechanics.* Edmonton, Canada. 1972, pp. 3–12.

González de Vallejo, L.I., *Ingeniería Geológica.* Prentice Hall, Madrid, Spain. 2002.

Goodman, R., *Introduction to Rock Mechanics.* Second Edition, John Wiley & Sons, USA, 1989.

Haimson, B.C., "Deep hydrofracturing stress measurements", *Proceedings of the 19th U.S. Symposium On Rock Mechanics.* Reno, NV. 1978a, pp. 345–361.

Haimson, B.C., "The hydrofracturing stress measuring method", *International Journal of Rock Mechanics.* Vol. 15. 1978b, pp. 167–178.

Herget, G., *Stresses in Rock*. A. A. Balkema, Rotterdam. 1988.

Hijazo, T., *Estimación de las Tensiones Naturales y su Aplicación al Diseño de Túneles*. Thesis. Universidad Complutense de Madrid. 2010.

Hoek, E., Brown, E.T., *Underground Excavations in Rock*. Institution of Mining & Metallurgy, London. 1980.

Hoek, E., Kaiser, P.K., Bawden, E.F., *Support of Underground Excavations in Hard Rock*. Taylor & Francis, New York. 1995.

Leeman, E.R., "The measurement of stresses in rock", *Journal of the Southern African Institute of Mining and Metallurgy*. Vol. 65. No. 2. 1964, pp. 45–114.

Leeman, E.R., "The 'doorstopper' and triaxial rock stress measuring instruments developed by the CSIR", *Journal of the Southern African Institute of Mining and Metallurgy*. Vol. 69. No. 7. 1969, pp. 305–339.

Lindfors, U. *Forsmark site investigation. Overcoring rock stress measurements in borehole KFM07B*. Svensk Kärnbränslehantering AB, Stockholm, Sweden. 2007.

Ljunggren, C., Yanting, T., Janson, R., Chistiansson, R. "An overview of rock stress measurement methods", *International Journal of Rock Mechanics and Mining Sciences*. Vol. 40. Nos. 7–8. 2003, pp. 975–989.

Rodríguez, A., Peral, F., "Medida del Estado Tensional In Situ", *Ingeopres*. No. 75. 1999, pp. 18–22.

Rummel, F., *Rock Mechanics with Emphasis on Stress*. CRC Press, Boca Raton, FL. 2005.

Sheorey, R. "A theory for in situ stresses in isotropic and transversely isotropic rock", *International Journal of Rock Mechanics and Mining Sciences*. Vol. 31. 1994, pp. 23–34.

Sjöberg, J., *Stress Measurements: A Scandinavia Perspective*. Itasca Consultants AB, Stockholm, Sweden. 2012.

Sjöber, J., Klasson, H. "Stress measurements in deep boreholes using the Borre (SSPB) probe", *International Journal of Rock Mechanics and Mining Sciences*. Vol. 40. No. 7. 2003, pp. 1205–1223.

Zoback et al., "Global patterns of tectonic stress", *Nature*. Vol. 341. 1989, pp. 291–298.

Chapter 5

Laboratory tests

Benjamín Celada Tamames and Eduardo Ramón Velasco Triviño

Let us wait for what we want, but endure what takes place.

Marcus Tulius Cicero (106–43 BC)

5.1 INTRODUCTION

Samples obtained during site investigation have to be tested in the laboratory to determine their strength and other relevant properties.

Usually, the volume of the samples tested in the laboratory ranges between 150 and 1,600 cm³, and because of their small size, laboratory tests cannot fully take into account the effect of the discontinuities present in the ground in situ. Therefore, the results obtained in the laboratory yield more favorable data than those representing the real ground behavior.

Nevertheless, laboratory tests are very valuable, and sometimes the only means, for obtaining an insight into the complex behavior of rock under load. In fact, as early as the mid-1960s, laboratory tests led to establishing the detailed mechanism of brittle fracture of rock (Bieniawski, 1967).

According to Goodman (1989), 20 years later, the variability existing in the rock mass with respect to structures, textures and minerals, necessitates quantitative classification on the basis of some properties that are the easiest to measure and are known as index properties:

- **Porosity**: relative ratio between voids and solids.
- **Density**: adds information, indirectly, about the sample mineralogical composition.
- **Sonic velocity**: together with the petrographic description, enables evaluating the degree of strength of the specimen.
- **Permeability**: describes the pore interconnections among each other.
- **Durability**: indicates the specimen predisposition to deterioration.
- **Strength**: indicates the rock or soil competence against stresses when tested.

In soils, some of the above indexes lose their meaning and, for that reason, new ones must be determined:

- **Water content**: is the ratio between water and soil mass in a sample.
- **Grain specific gravity**: besides the test specimen density, the unit weight of the particles that constitute the ground must be determined.
- **Atterberg limits**: in fine grained soils, they represent the water content at which their mechanical behavior changes. These values are correlated with the strength,

permeability, compressibility and the swelling potential. It can be distinguished among the Liquid Limit (LL), the Plastic Limit (PL) and the Plasticity Index (PI), as a subtraction of the two previous ones.

- **Granulometry:** is the size distribution of the particles that constitute the sample. It is very useful to classify the soil.
- **Organic matter content:** in general, underground works are carried out below the soils affected by the organic matter content, but the presence of water in the ground may enhance the harmful effect of the organic matter.
- **Electrochemical classification:** gives information about the soil aggressiveness and its potential to attack steel and concrete.

The following sections present the most frequent laboratory tests performed on rock or soil samples.

5.2 UNIAXIAL COMPRESSION TESTS

In uniaxial compression tests, a sample is only loaded with an axial force which is increased until the sample fails.

Usually, conventional uniaxial compression tests end when the test specimen cannot withstand an increase in the applied axial force.

However, tests may also be carried out in the post-failure phase which, as shown in the next section, require the use of special equipment; therefore, these tests are much more expensive than conventional tests (Bieniawski et al, 1969).

In conventional tests it is common to measure only the evolution of the force applied during the test and the uniaxial compressive strength σ_c is calculated by the expression:

$$\sigma_c = \frac{F}{A}$$

where F is the maximum value of the applied force and A is the specimen cross section.

The use of this test is regulated by several standards, among which the following ones can be highlighted:

- UNE 22950-1:1990. *Mechanical Properties of Rocks. Strength Determination Tests. Part 1. Uniaxial compressive strength.*
- UNE 22950-3:1990. *Mechanical Properties of Rocks. Strength Determination Tests. Part 3: Modules of elasticity (Young) and Poisson's ratio determination.*
- ASTM D7012–14. *Standard Test Methods for Compressive Strength and Elastic Moduli of Intact Rock Core Specimens under Varying States of Stress and Temperatures.*

It is also advisable to follow the recommendations of the International Society for Rock Mechanics (ISRM).

- *Suggested Method for Determining the Uniaxial Compressive Strength and Deformability of Rock Materials.* ISRM (Ulusay and Hudson, 1979).
- UNE 103400:1993. *Simple Compression Rupture Test in Soil Test Specimens.*
- ASTM D 2166. *Standard Test Method for Unconfined Compressive Strength of Cohesive Soil* (on intact, remolded or reconstituted condition).

The presses that perform uniaxial compression tests on rocks must be able to apply forces larger than 1.000 kN, whereas to perform tests on soil samples it is enough if they can apply at least 50 kN.

Photograph 5.1 1.200 KN press for tests on rock specimens.

Source: Cepasa S.A.

In both cases, the presses must have a semi-spherical seating in one of the plates and the equipment must be complemented with an automatic system for data acquisition, as is shown in Photograph 5.1.

To perform a representative uniaxial compression test on rock, the following recommendations must be met:

- The samples should be cylindrical, with a slenderness of between 2.5 and 3.0.
- The diameter of the specimen should be larger than 54 mm, so that the largest grain of the rock is exceeded by ten times.
- The two bases of the test specimen must be parallel with a maximum deviation of 0.001 radians and with a roughness less than 0.02 mm.
- Samples should be tested in its in situ moisture state if possible, and the water content must be determined before the test.
- The load application rate must be constant and be kept within the range of 0.5 to 1.0 MPa/s.
- Batches of at least five test specimens from each lithotype should be tested.

Figure 5.1 shows the phases of rock failure under rock in uniaxial compression, including the actual test data, which includes the post-failure phase.

The diagram in Figure 5.2 summarizes the phases of a complete uniaxial compression test. In this curve the following zones can be distinguished:

- Zone A corresponds to the closure of the microfractures present in the specimen, when an axial force is applied.
- In zone B the behavior is elastic, which means that there is a linear correspondence between the stresses and the axial strains.
- In zone C the microfracture propagation begins and visible cracks appear on the specimen.
- In zone D the maximum strength is reached.
- In zone E the rock breaks into pieces that are kept together by friction until the residual strength is reached.

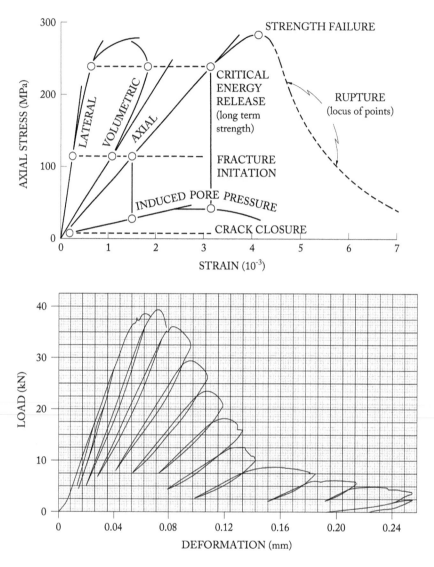

Figure 5.1 Representation of brittle fracture mechanism for rock (quartzite) in uniaxial compression.

Source: Bieniawski, 1967.

Figure 5.3 shows the result of a real uniaxial compression test in which the applied force and the axial and radial strains have been measured, which enables calculating the volumetric strain during the test. Following the fracture mechanism categories established by Bieniawski (1967), Goodman (1989) reported his results featuring measurements of the axial and radial strains recorded with strain gauges glued to the specimens or with strain transducers, placed between the press plates to measure the axial strain, or on the specimen itself if the radial strain is measured. Photograph 5.2 shows a transducer used to measure the radial strain of the specimen.

When transducers are used to measure the strains, the unit strains are calculated by the expressions:

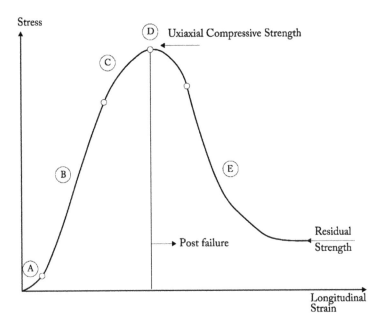

Figure 5.2 The phases of a uniaxial compression test on rock.

Source: Farmer, 1982.

$$\varepsilon_a = \frac{\Delta l_a}{L} \text{ and } \varepsilon_r = \frac{\Delta l_c}{\pi D}$$

where:

Δl_a = axial strain increment
Δl_c = circumferential strain increment
L   = initial specimen length
D   = diameter of the specimen

The volumetric strain of the specimen (ε_v) is calculated by the expression:

$$\varepsilon_v = \varepsilon_a + 2\varepsilon_r$$

The modulus of elasticity or Young's modulus (E) can be calculated, according to the ISRM, in either of the following two ways:

- **Tangent Young's modulus:** slope of the tangent to the curve (σ_1, ε_a) at the X-axis point $0.5 \cdot \sigma_c$
- **Secant Young's modulus:** slope of the secant between the origin and the X-axis point σ_c

The value of the Poisson coefficient υ is calculated by the ratio between the axial strain ε_a and the radial strain ε_r in the X-axis point $0.5 \cdot \sigma_c$.

Veermer and Borts (1984) proposed to model the volume change in a uniaxial compression process after failure, according to the model presented in Figure 5.4. In this Figure 5.4, υ is the Poisson coefficient and ψ is the angle of dilation.

Table 5.1 presents some typical values of the angle of dilation.

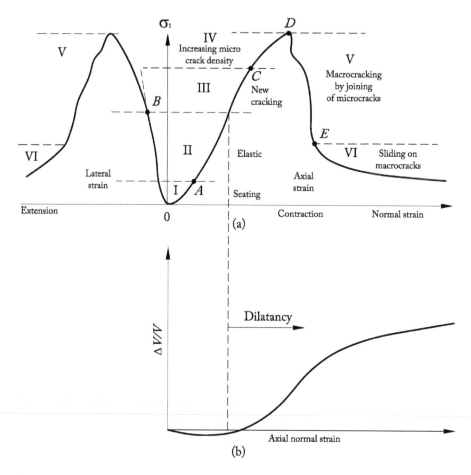

Figure 5.3 Curves of a uniaxial compression test on rock with measurements of the axial and radial strains.
Source: Goodman, 1989.

Photograph 5.2 Transducer to measure the radial strain of a specimen.

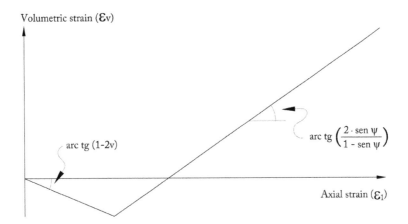

Figure 5.4 Veermer and Borts model for the dilatancy (1984).

Table 5.1 Typical values of the angle of dilation

Material	Angle of dilation (°)
Dense sand	15
Loose sand	< 10
Normally consolidated clay	0
Marble	12–20
Concrete	12

Table 5.2 Strength decrease in tests with saturated samples

Type of rock	Decrease in the uniaxial compressive strength in saturated samples
Hard	10–20%
Medium	30–40%
Soft	>70%

5.2.1 Influence of the water content

Water content variations in the samples have significant influence on the strength when compared to when a dry sample is tested, as it was noted in Section 1.3.4.2. Romana and Vásárhelyi (2007) quantify this reduction in Table 5.2.

Figure 5.5 shows the results of the uniaxial compression tests performed in carboniferous slates, Peng (1986), with dry samples, saturated at 48% and 100% by applying loads perpendicularly to the stratification.

In this figure it can be appreciated that while a dry sample can withstand an axial force up to 125 kN, a fully saturated sample can only withstand 35 kN, which means a strength reduction of 72%.

Figure 5.6 shows the uniaxial compressive strength variation in loam samples of the Lower Miocene as a function of their water content (Celada, 2011).

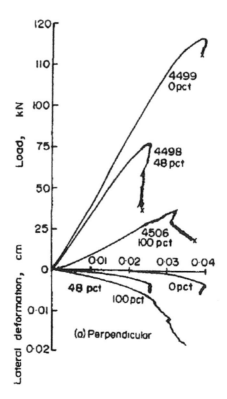

Figure 5.5 Influence of the degree of saturation in the compressive strength of carboniferous slates.

Source: Van Eeckhout and Peng, 1976.

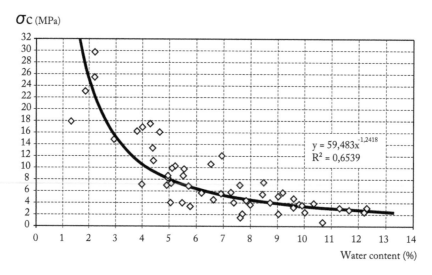

Figure 5.6 Uniaxial compressive strength variation in loam of the Lower Miocene depending on its water content.

Source: Celada, 2011.

In this case, samples with 2% of water content have an average strength of 24 MPa, but if the water content increases to 12% the uniaxial compressive strength decreases to 3 MPa, this means a strength decrease of 88%.

According to this, it is very important to determine the sample water content or the degree of saturation in each uniaxial compression test and to write this value in the test protocol.

These data will ease the analysis of the test results and will contribute to reducing the dispersion of the results obtained.

5.2.2 Influence of the sample size

Theoretically, laboratory tests are carried out on intact rock samples which should not contain the discontinuities present in situ.

From this point of view, there should not be large variations in the uniaxial compressive strength value when testing specimens with different diameters, but the reality is quite different, as shown in Figure 5.7 from Hoek and Brown (1980).

In this figure it can be easily appreciated that carrying out uniaxial compression tests with samples with a diameter smaller than 50 mm, implies a strength overestimation of between 10 and 30%.

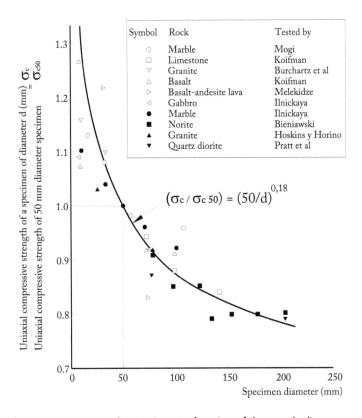

Figure 5.7 Uniaxial compressive strength variation as a function of the sample diameter.

Source: Hoek and Brown, 1980.

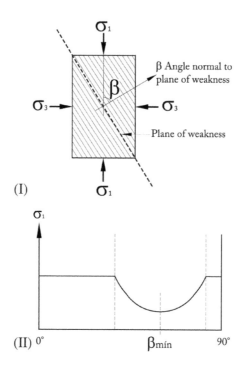

Figure 5.8 Influence of the schistosity on the uniaxial compressive strength. I. Schistosity orientation. II. Variation of the uniaxial compressive strength.

Source: Jaeger, 1960.

5.2.3 Influence of anisotropy

Jaeger (1960) analyzed the effect of anisotropy on samples at uniaxial compression tests and determined, as indicated in Figure 5.8, that there was an orientation of the schistosity, with respect to the load application direction, that gave a minimum strength whose value could be 40% of the strength without anisotropy.

The loading orientation to the minimum strength occurs for:

$$\beta_{min} = 45° + \left(\frac{\phi}{2}\right)$$

where β is the angle that the plane of weakness forms perpendicular with the direction of load application and ϕ is the internal friction angle of the tested rock.

5.2.4 Post-failure tests

When conventional presses are used, upon reaching the peak strength, the test specimen fails within a fraction of a second.

Therefore, to obtain the stress–strain graph at post-failure, as the one shown in Figure 5.9, it is first required that the stiffness of the press in which the test is going to be performed must be much higher to that of the specimen, otherwise, when the post-failure starts, the press will transmit the energy accumulated during the elastic strain and the rock specimen will literally explode.

Moreover, the press needs to have a servo-controlled system which enables reducing the force applied to the specimen at the same rate as the strain increases in the specimen.

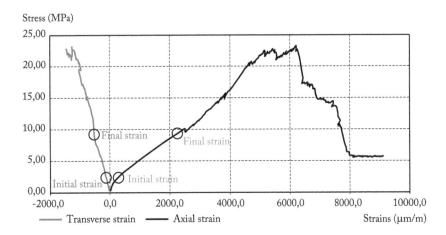

Figure 5.9 Graph obtained in a uniaxial compression test with post-failure.

Source: Cepasa, S.A.

Early post-failure tests were reported by Bieniawski et al, 1969 but today such tests are regulated in ISRM *Suggested Method for the Complete Stress-Strain Curve for Intact Rock in Uniaxial Compression* (Fairhust and Hudson, 1999).

Photograph 5.3 shows a 1,500 kN servo-controlled press to perform post-failure tests.

The uniaxial compression tests with post-failure are much more expensive than the conventional tests but are the only way to know the real ground behavior when the tangential stress around the tunnel exceeds the ground peak strength.

It is necessary to keep in mind that high stress concentrations can appear in the tunnel perimeter depending on the adopted geometry and on those places the ground will work at post-failure, although the remainder section can work in the elastic domain.

In these cases, to accurately calculate the ground displacements after the tunnel excavation, there is no choice but to know the post-failure soil behavior.

Photograph 5.3 1,500 kN servo-controlled press.

Source: Cepasa, S.A.

In this sense, in the flowchart of the active structural design, shown in Figure 2.22, it has been indicated that it is necessary to carry out post-failure tests when a ground intense yielding is expected.

5.3 TRIAXIAL COMPRESSION TESTS

The aim of using triaxial compression tests is to know the ground behavior in conditions similar to those in situ which, as indicated in Chapter 4, correspond to a triaxial stress state defined by σ_v, σ_H, σ_h.

To simplify the triaxial test implementation it is assumed that the horizontal stresses $\sigma_H = \sigma_h$; therefore, the test is performed by placing the specimen inside a steel chamber, where the stress σ_H is kept constant during the test. The specimen failure is achieved by progressively increasing the stress σ_v, as in a conventional uniaxial compression test. Figure 5.10 illustrates the stress distribution during a simplified triaxial test adopting the usual notation considering $\sigma_1 = \sigma_v$ and $\sigma_2 = \sigma_3 = \sigma_H$.

In 1968, Hoek and Franklin designed a triaxial cell which due to its simplicity and effectiveness has been internationally adopted to make these tests. Figure 5.11 shows a scheme of the Hoek–Franklin cell and Photograph 5.4 a shows a view of these cells to test specimens of several diameters.

The test to determine the triaxial compression strength is regulated by the following standards:

- UNE 22950-4:1992. *Mechanical Properties of Rocks. Strength Determination Tests. Part 4: Triaxial compression strength.*
- ASTM D7012-14. *Standard Test Methods for Compressive Strength and Elastic Moduli of Intact Rock Core Specimens under Varying States of Stress and Temperatures.* Method A.
- *Suggested Method for Determining the Strength of Rock Materials in Triaxial Compression.* ISRM (Ulusay and Hudson, 1983).
- UNE 103402:1998. *Determination of Strength Parameters of a Soil in Triaxial Set Apparatus.*
- ASTM D2850-15. *Standard Test Method for Unconsolidated-Undrained Triaxial Compression Test on Cohesive Soils.*

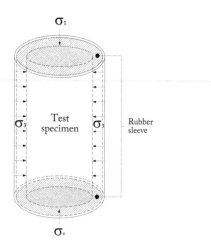

Figure 5.10 Stress scheme of a simplified triaxial test.

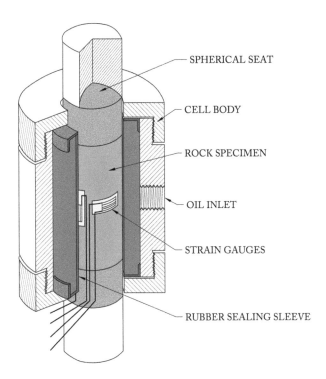

Figure 5.11 Hoek and Franklin triaxial cell.

Source: Brady and Brown, 1985.

Photograph 5.4 Triaxial cells of different diameters.

Source: Cepasa, S.A.

Figure 5.12 displays the results of the triaxial tests carried out by Brady and Brown (1985) where the different behaviors expected in post-failure as a function of the confining stress can be appreciated.

The first important remark is that, usually in rocks, the modulus of deformation of the samples tested is the same regardless of the confining stress used in the test.

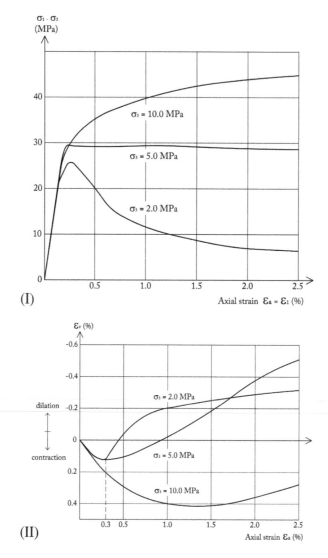

Figure 5.12 Distribution of the axial stresses and strains in triaxial tests. I. Axial stress distribution. II. Axial strain distribution.

Source: Brady and Brown, 1985.

If the confining stresses are very low, this leads to a mainly brittle behavior, with a residual strength due to the confining stress.

If the confining stresses are medium, the sample has a perfect elasto-plastic behavior until the peak strength is reached, which usually takes place for an axial strain larger than 0.3%.

For high confining stresses, the specimen strength and the axial strain increase progressively, maintaining this behavior for axial strains higher than 2.5%, which are about ten times larger than those corresponding to a peak strength with low confinements.

In this figure it can be also appreciated that, for medium to low confining stresses, the post-failure process becomes dilatant after axial strains of 0.3%, while for high confining stresses, the dilating effect cannot be observed until axial strains of 1.5%.

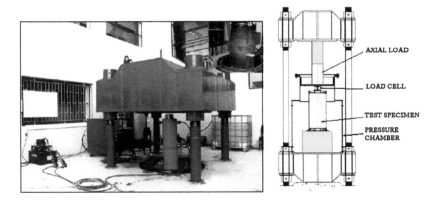

Photograph 5.5 Triaxial chamber of large diameter.

Source: Verdugo and Hoz, 2006.

Another very interesting example of triaxial compression tests are those performed at the Institute of Research and Material Testing (Investigación, Desarrollo e Innovación de Estructuras y Materiales, IDIEM in its Spanish acronym) in Chile, on the gravels from Santiago de Chile.

These gravels are soils of glacial origin with quite uniform grain size distribution, ranging from boulders of about 30 cm in diameter to clays, and which have extremely good stress–strain behavior, despite being a soil.

Verdugo and Hoz (2006) have carried out triaxial tests with specimens of 1 m in diameter and 2 m in height, remodeled with gravels of Santiago, for which the press shown in Photograph 5.5 was built.

This chamber can apply confining stresses up to 3 MPa and axial forces up to 20,000 kN.

Figure 5.13 reproduces the results of some triaxial tests carried out on the gravels from Santiago by Verdugo and Hoz (2006).

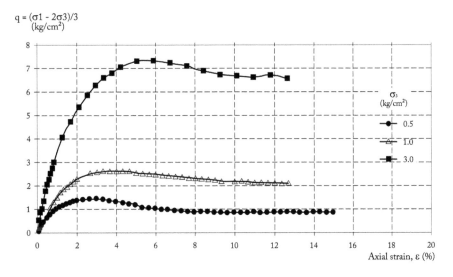

Figure 5.13 Results of triaxial tests carried out on the gravels from Santiago de Chile.

Source: Verdugo and Hoz, 2006.

In this figure it can be clearly appreciated that as the confining stress increases, so does the peak strength value and also the axial strain at which the peak strength is reached.

For the triaxial tests to be representative, the confining stress has to represent properly the conditions under which the ground is going to work when the tunnel is constructed.

As a general criterion to select σ_3 in tunnel design, one can take $\sigma_3 = 0.5 \cdot \gamma \cdot H$, an approach derived from the recommendations from Hoek et al. (2002), where γ is the ground specific gravity and H is the depth at which the tunnel is located.

5.4 TENSILE TESTS

Rock tensile strength is very small, as it is often between 5 and 10% of its uniaxial compressive strength, which creates many problems with making direct tensile tests on rock samples, as its preparation is very difficult.

To avoid these drawbacks, the rock tensile strength is determined indirectly, by applying a diametrical compression to the rock specimen which generates a tensile state in the center of the specimen as illustrated in Figure 5.14. In the central area of the specimen a biaxial stress state occurs, whose vertical compressive stress is three times higher than the generated horizontal tension.

This test, known as the Brazilian test, is regulated by the following standards:

- UNE 22950-2:1990. *Mechanical Properties of Rocks. Strength Determination Tests. Part 2: tensile strength. Indirect determination (Brazilian test).*
- ASTM D 3967-08. *Standard Test Method for Splitting Tensile Strength of Intact Rock Core Specimens.*
- *Suggested Method for Determining the Indirect Tensile Strength by the Brazil test.* ISRM (Ulusay and Hudson, 1978).

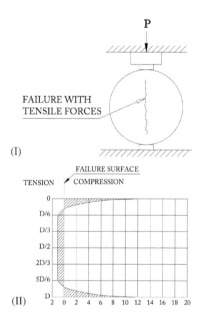

Figure 5.14 Diametrical compression test. I. Test scheme. II. Stress distribution.

The indirect tensile strength is calculated by the expression:

$$\sigma_t = \frac{2P}{\pi DL}$$

where D is the specimen diameter and L is its length. The suggested rate for the load application is 200 N/s, so that the failure occurs within 15–30 s.

The ISRM recommends performing the test by applying the load through steel jaws, as shown in Photograph 5.6. It also shows the ideal failure in this type of test, constituted by a vertical fracture line developed between the two loading points.

During the test it is necessary to stop the load application as soon as the first tensile crack is generated in the center of the specimen, otherwise the specimen will be subjected to a compression effect that will increase, unrealistically, the tensile strength.

The Brazilian test tends to overestimate the tensile strength, since the failure is not produced in the weakest plane of the specimen, but in another, previously determined by the point of application of the load.

Perras and Diedierich (2014) indicate the following relationships between the Direct Tensile Strength (DTS) and the Brazilian Tensile Strength (BTS) for different types of rock:

- Sedimentary rocks: DTS = 0.68 BTS
- Igneous rocks: DTS = 0.86 BTS
- Metamorphic rocks: DTS = 0.93 BTS

As in most tests, in the Brazilian Test, as the specimen diameter increases, the value of the indirect tensile strength decreases, as shown in Figure 5.15, Rocco et al. (1996). Note that the horizontal axis is in logarithmic scale.

5.5 SHEAR TESTS

For tunnel design, the most typical shear tests are performed on preexisting discontinuities in the ground, since shear tests on the matrix are very unusual as the information provided can be much more easily obtained from uniaxial compression and indirect tensile tests.

Photograph 5.6 Specimen failure in a Brazilian Test. I. Before the test. II. After the test.

Source: Cepasa, S.A.

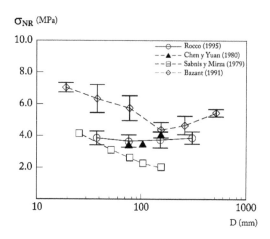

Figure 5.15 Indirect tensile strength variation with the specimen diameter.

Source: Rocco et al., 1996.

The following sections describe the performance of shear tests on rock discontinuities and on the sample matrix in the case of soils and soft rocks.

5.5.1 Shear tests on discontinuities

To determine the shear strength of a discontinuity it is necessary to take a piece of rock including the discontinuity to be tested.

To carry out the test, a normal force is applied to the discontinuity, which is kept constant during the test and another force tangential to the discontinuity, which is progressively increased.

The ISRM has published the document *Suggested Method for Laboratory Determination of the Shear Strength of Rock Joints: Revised Version*, about shear tests on discontinuities.

In order to carry out this test, the Hoek shear box has become very popular, which is shown in Figure 5.16. In this figure the position of the two hydraulic cylinders in charge of applying the normal and tangential forces during the test can be appreciated, although the most important thing is the position of the discontinuity tested, in order to ensure that the tangential force is parallel to the plane of the discontinuity.

To do this, each of the two rock fragments containing the discontinuity is placed in two half-cubes, which are filled with cement mortar. Once the mortar is set, the half-cubes are taken out of the molds and placed in the shear box.

Photograph 5.7 shows two half-cubes containing two fragments of a discontinuity, already prepared to start the test. In Photograph 5.8 the Hoek shear box is shown, prepared to make a shear test and in the Photograph 5.9 a detail of the half-cube installed in the shear box can be appreciated.

From each shear test, performed at a constant normal stress, a curve relating the displacement of the discontinuity (δ) with the shear stress (τ) is obtained.

This curve has two characteristic values of the shear stress; one, the maximum achieved during the test (τ_{max}) and another, the residual one at the end of the test (τ_{res}), due to the friction existing between the two planes of the discontinuity.

To determine the shear strength of a discontinuity it is necessary to perform at least three tests, each one at a different normal stress.

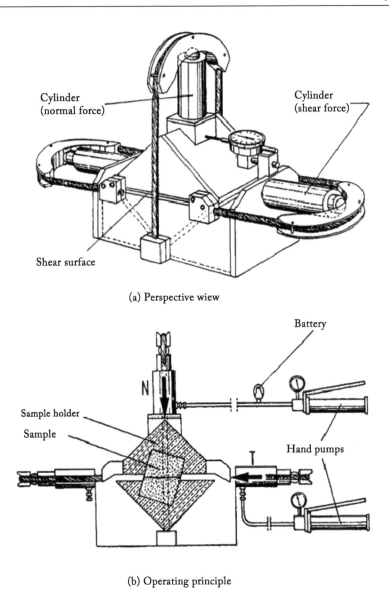

(a) Perspective wiew

(b) Operating principle

Figure 5.16 Scheme of the Hoek shear box.

The value of the normal stress to be applied in the shear tests has to be related to the stress range at which the discontinuity will actually work; these values usually depend on the excavation depth.

The results from the shear tests, under peak and residual conditions, are usually represented in a MohrCoulomb diagram, see left part of Figure 5.17, for four tests performed on the same discontinuity.

The Hoek shear box has the limitation of the small size of the sample, so that the displacement of the sliding block during the test can only be of few centimeters.

To avoid this disadvantage, tests with blocks much larger than those used in the Hoek shear box can be performed, but the sample preparation is much more difficult and the

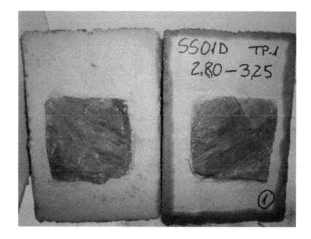

Photograph 5.7 Half-cubes containing a discontinuity.

Source: Cepasa, S.A.

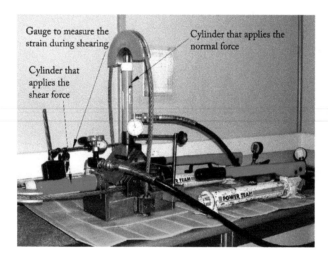

Photograph 5.8 Hoek shear box ready to do a test.

Source: Cepasa, S.A.

Photograph 5.9 Half-cubes installed in the shear box.

Source: Cepasa, S.A.

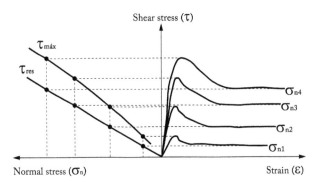

Figure 5.17 Display of the results from four shear tests performed on samples of the same discontinuity.

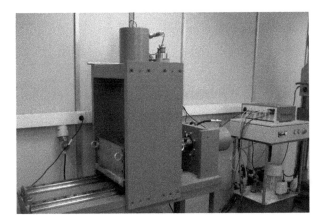

Photograph 5.10 Shear machine to test samples of 400×400 mm.

Source: Cepasa, S.A.

cost of the tests is much higher. Photograph 5.10 shows a shear machine that allows testing blocks with joints, with dimensions of 400×400 mm.

5.5.2 Shear tests on soil samples

Conceptually, direct shear tests on soil samples are similar to shear tests on discontinuities; but as soils have lower shear strength than the joints present in the rocks, it is possible to perform shear tests on the soil matrix. For that purpose, machines as the one shown in Photograph 5.11 are used.

Direct shear tests on soils are regulated by the following standards:

- UNE 103401:1998. *Determination of the Shear Strength of a Soil with the Direct Shear Box.*
- ASTM D3080M-11. *Standard Test Method for Direct Shear Test of Soils Under Consolidated Drained Conditions.*
- ASTM D6467-13. *Standard Test Method for Torsional Ring Shear Test to Determine Drained Residual Shear Strength of Cohesive Soils.*

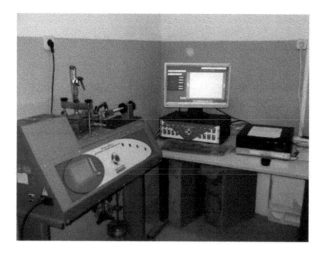

Photograph 5.11 Device to perform shear tests on soils.

Source: Cepasa, S.A.

5.6 TESTS RELATED TO EXCAVABILITY

In the following sections three tests related to the ground excavability are described: the Cerchar abrasivity index, the drilling rate index and the cutter life index.

5.6.1 Cerchar Abrasivity Index

In the 1960s, the *Centre d'Études et Recherches des Charbonnages de France* (CERCHAR) developed the Cerchar Abrasivity Index, which is used all over the world and known by its acronym, CAI.

The CAI is measured as the wear, in tenths of mm, of the steel pin after 10 mm of travel across the rock surface under a force of 70 N.

The CAI test is regulated by the following standards:

- NF P-94-430-1 (2000). *Détermination du Pouvoir Abrasif d'une Roche.* Partie 1: essai de rayure avec une pointe.
- ASTM D7625–10 (2010). *Standard Test Method for Laboratory Determination of Abrasiveness of Rock Using the CERCHAR Method.*
- *Suggested Method for Determining the Abrasivity of Rock by the CERCHAR Abrasivity Test* (Alber M. et al.) ISRM (Ulusay, 2014).

Photograph 5.12 shows the device used to determine the CAI and Table 5.3 indicates the criteria used to estimate the abrasivity depending on the CAI.

5.6.2 Drilling Rate Index

The Drilling Rate Index (DRI) was developed in Trondheim University, Norway, by Lien in 1961. It is a test to estimate the ground drillability by measuring the penetration rate of a drill bit into a sample and its brittleness, according to the document NTNUs 13A-98 *Drillability. Test Methods & SINTEF Standards.*

Photograph 5.12 Device used to determine the CAI.

Source: Cepasa, S.A.

Table 5.3 Ground abrasivity estimation depending on the CAI

CAI value	Abrasivity estimation
0.1–0.4	Extremely low
0.4–0.9	Very low
0.9–1.9	Low
1.9–2.9	Medium
2.9–3.9	High
3.9–49	Very high
> 4.9	Extremely high

The measure of the drilling rate is quantified by the index SJ, which is the penetration, expressed in tenths of a millimeter, achieved with a 8.5 mm drill bit under a weight of 20 kg after 200 revolutions. To perform this test the device shown in Figure 5.18 is used.

The brittleness should be determined with a sample of the same rock in which the SJ was obtained. To do this, a steel mass of 14 kg will be dropped 20 times on 0.5 kg of crushed rock.

The grain size distribution of the crushed rock has to be between 16.0 and 11.2 mm, while the brittleness is measured by the loss of weight after the rock sample, at the end of the test, passed through a sieve of 11.2 mm.

The loss of weight, referred to the 0.5 kg of the initial sample weight, defines the S_{20} index. Figure 5.19 illustrates the device to determine the S_{20} Index.

Finally, the DRI is determined with the abacus reproduced in Figure 5.20.

Table 5.4 presents the criteria to estimate the rock drillability from the DRI.

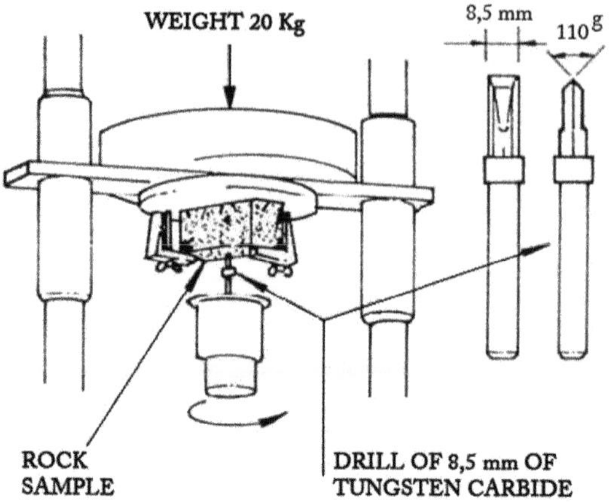

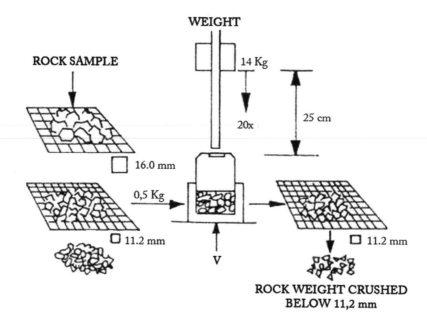

Figure 5.18 Device to determine the SJ Index.

Figure 5.19 Device to determine the S$_{20}$ Index.

5.6.3 Cutter Life Index

The cutter Life Index (CLI) has been developed by the Norwegian Institute of Technology (Norway) to estimate the lifetime of the Tunnel Boring Machine (TBM) cutters, Nilsen et al. (2007).

The CLI is calculated by the expression:

$$CLI = 13.84 \left(\frac{SJ}{AVS} \right)^{0.3847}$$

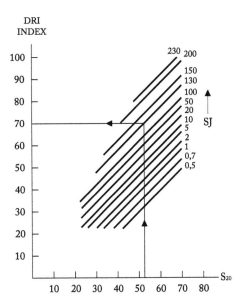

Figure 5.20 Graph for determining the DRI from the indexes S_{20} and SJ.

Table 5.4 Drillability estimation depending on the DRI

DRI value	Drillability estimation
< 21	Extremely low
21–28	Very low
28–37	Low
37–49	Medium
49–65	High
65–86	Very high
> 86	Extremely high

Where SJ is the drilling index defined in Section 5.6.2. and AVS is an index that measures the cutter wear produced by a given rock.

The AVS is determined with the device shown in Figure 5.21 according to the following procedure:

- The rock fragments to be tested are placed on a disk rotating at 20 rpm, with a diameter smaller than 1 mm and a flow rate of 80 g/m.
- On top of the rock fragments a steel piece made with the same steel as that of the cutter to test is placed, loaded with a weight of 10 kg.
- Once the steel piece is placed on the rock, the disk rotates at 20 revolutions.
- The steel piece is removed and weighed, the loss of weight expressed in milligrams obtained being the value of AVS.

Table 5.5 indicates the ground abrasivity as a function of the AVS.

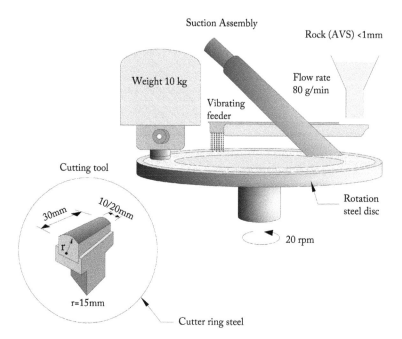

Figure 5.21 Procedure to determine the AVS.

Table 5.5 Abrasivity as a function of the AVS

AVS	Abrasivity
< 1	Extremely low
2–4	Very low
4–13	Low
13–25	Medium
25–35	High
35–44	Very high
> 44	Extremely high

5.7 NONCONVENTIONAL TESTS

The following sections present some tests which are only of interest in some specific types of rocks and with which attempts of quantification of the following phenomena are made:

- Alterability in presence of water
- Swelling pressures in presence of water
- Creep due to insufficient confinement
- Cyclic loads due to earthquakes

5.7.1 Alterability tests

Rocks of sedimentary origin and fine grading, mainly the argilites and limolites, tend to be mechanically altered in the presence of water, their degradation can be estimated using the slake–durability test (SDT). In the RMR_{14} the authors have proposed the ethylene glycol immersion test, as an easy way to evaluate the rock alterability.

In the following sections both tests are presented.

5.7.1.1 Slake durability test

This test is performed by undergoing ten rock fragments, each of them loaded with a weight of between 40 g and 60 g, to two drying cycles at 105°, and then they are placed in a drum semi-submerged in water that rotates 200 revolutions in 10 minutes. Finally, the fragments are weighed again and the percentage of weight lost with respect to the initial weight (I_D) is calculated.

This test is regulated by the document ASTM D4644-08. *Standard Test Method for Slake Durability of Shales and Similar Weak Rocks.*

Photograph 5.13 shows a device used to perform the SDT test and Table 5.6, the alterability estimation through the SDT test.

5.7.1.2 Immersion in ethylene glycol

Broch (1996) identified the degradation produced by the water to basalt samples, at the Hydroelectric Project in Lesotho, by immersing the samples in an ethylene glycol solution, which is the antifreeze fluid used in radiators of motor vehicles.

Afterward, this method was reviewed by Paige-Green (2008) and, later, it was evolved by Carter et al. (2010).

Photograph 5.13 Device to perform the SDT test.

Source: Cepasa, S.A.

Table 5.6 Rock durability estimation in presence of water from the SDT test

Rock durability in presence of water	In 1 cycle of 10 minutes	In 2 cycles of 10 minutes
Very high	< 1	< 2
High	1–2	2–5
Medium	2–15	5–40
Low	15–40	40–70
Very low	>40	> 70

Photograph 5.14 Appearance of the rock samples immersed in ethylene glycol.

Source: Carter et al., 2010.

Table 5.7 Piaggio alterability criterion (2015)

Sample state after 30 days immersed in ethylene glycol	Alterability degree
No disintegration or change in weight is observed.	0
Small surface cracks.	1
The samples are broken into two or three parts. It is not possible to weigh all the samples again.	2
Samples are disintegrated and cannot be weighed.	3

The procedure developed by Carter et al. (2010) consists in selecting 40 rock samples, with an approximate diameter of 1 inch, and immersing them into a container with ethylene glycol for 30 days. The samples should not have cracks, should be weighted before being immersed and should be observed in the following periods: 1 hour, 1 day, 5 days, 10 days, 15 days, 20 days, and 30 days. Photograph 5.14 shows the appearance of one of these samples during the test. Piaggio (2015) has classified the ethylene glycol immersion test results into five categories, according to the criteria presented in Table 5.7.

5.7.2 Swelling tests

Some rocks of clay nature have the capacity to absorb water molecules in their structure generating significant swelling pressures, but anhydrites moisture in the presence of water becomes gypsum, leading to much higher pressures.

The following sections present some tests to estimate the swelling pressures generated in these two processes.

5.7.2.1 Swelling of rocks and clay soils

The following tests are normalized to evaluate the swelling of clay soils:

- UNE 103600:1996. *Determination of Expansivity in a Soil in the Lambe Apparatus.*
- UNE 103601:1996. *Test for Free Swelling of Soils in Oedometer Device.*
- UNE 103602:1996. *Test Method for One-Dimensional Swell Pressure of a Soil in Consolidometer.*
- ISRM. *Suggested Methods for Laboratory Testing of Swelling Rocks.*

Photograph 5.15 Device to perform the Lambe tests.

Source: Cepasa, S.A.

These tests can be also be performed if the rock samples are crushed into sizes smaller than 0.1 mm, although the results obtained with them only provide a first indication about the sample behavior in the presence of water.

Photograph 5.15 shows a device to perform the Lambe tests and in Figure 5.22 the result from a Lambe test can be observed.

The tests to determine the swelling pressure and the free swelling are performed in a box for oedometer tests, as the one shown in Figure 5.23, by placing the sample between two porous plates.

The swelling pressure test is performed by constraining the sample so that no vertical displacement could take place during its moistening and that the pressure generated could be measured.

The free swelling test does not constrain the vertical displacement during hydration so that this parameter can be measured.

5.7.2.2 Swelling of the anhydrites

Calcium sulfate, SO_4Ca, is naturally present as a sedimentary rock called anhydrite. In the presence of water, the anhydrite and moisture change into gypsum which is less dense than the anhydrite according to the following table and equation:

	SO_4Ca	+	H_2O	⇔	$SO_4Ca \cdot 2H_2O$
	Anhydrite	+	Water	⇔	Gypsum
Mass (g)	136.14	+	36	=	172.14
Density (g/cm³)	2.96		1		2.32
Volume (cm₃)	45.99		36		74.2

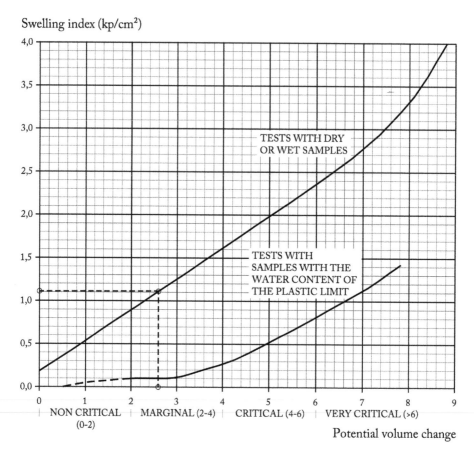

Figure 5.22 Result of a Lambe test.

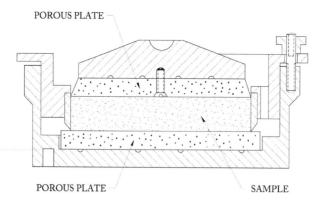

Figure 5.23 Box to perform swelling pressure tests and free swelling tests.

Accordingly, the transformation of anhydrite into gypsum results in the following volume increase:

$$\frac{AV}{V} = \frac{74.2 - 45.99}{45.99} \times 100 = 61\%$$

The transformation of anhydrite into gypsum can be stopped at a temperature of 20°C by applying a pressure of 1.6 MPa. This transformation is reversible in such a way that, if a pressure of 80 MPa is applied at 58°C, the gypsum changes back to anhydrite.

Pure anhydrite does not usually appear at depths less than 60 m, because at that depth the pressure due to the soil weight is around 1.6 MPa, which is the pressure that inhibits the moistening of gypsum.

Accordingly, when a tunnel in anhydrite must be excavated at depths over 60 m and with water intake from the ground, the anhydrite will be moistened and its volume increase will generate swelling pressures which usually range between 2 and 7 MPa.

This is an additional overload that must be withstood by the tunnel support, so if it is not taken into account in the calculations of the support, the tunnel support will be seriously damaged. The ISMR has proposed the following procedures to determine the swelling of the anhydrite:

- Time-dependent swelling of a radially confined sample submerged in water
- Axial and radial free swelling (without confinement)

Another option outside of the ISRM procedures is the Huder–Amberg radially confined swelling test. Figure 5.24 presents the results obtained by Wittke (1999) from a Huder–Amberg test.

In this figure stages 1, 2 and 3 correspond to the initial process, which includes two load cycles, and stage 4 corresponds to the sample moisture, in which there is not a stress increase but a unit strain.

Finally, stage 5 corresponds to the swelling process, at which a swelling pressure of about 700 kN/m^2 (0.7 MPa) is reached.

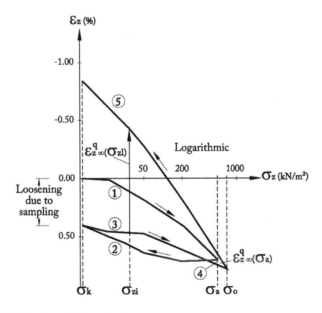

Figure 5.24 Results obtained in a Huder–Amberg test.

Source: Wittke, 1999.

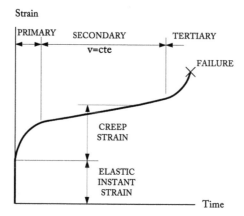

Figure 5.25 Standard curve of the strain phases of a sample with creep behavior at a constant load.

5.7.3 Creep tests

The strains that some rocks undergo over time when subjected to a constant load is referred to as creep. Figure 5.25 shows the typical strain distribution of the creep phenomena.

When the sample is initially loaded, it undergoes an instantaneous shortening, according to Hooke's law and from that moment three creep phases can be observed:

- **Primary creep**, usually lasts a few weeks and is characterized by a decreasing strain rate.
- **Secondary creep**, usually lasts several years and is characterized by a constant strain rate which usually ranges between 20 and 100 microns/day.
- **Tertiary creep**, usually lasts a few months and is characterized by an increasing strain rate which leads to the excavation collapse.

In rocks which are likely to have a creep behavior, this phenomenon does not appear until they are loaded with a stress equal or greater than 40% of their simple compressive strength.

Creep tests are regulated by the following norms:

- ASTM D7070-08. *Standard Test Methods for Creep of Core Under Constant Stress and Temperature.*
- ISRM. *Suggested Methods for Determining the Creep Characteristics of Rock.*

5.7.4 Dynamic tests with cyclic loads

In situations where the underground work has to be constructed in soils and could be subjected to dynamic loads, such as those due to earthquakes, the tests aimed at determining the soil dynamic characteristics become more important, through specialized tests such as the resonant column, the torque test, cyclic simple shear or dynamic triaxial.

Table 5.8 shows the most common tests, the applicable standards and the parameters obtained.

Figure 5.26 shows the curves that relate the shearing distortion with the damping and shear modulus, obtained with resonant column tests, performed in three types of soils in Santiago de Chile by Verdugo and Hoz (2006).

Table 5.8 Most common dynamic tests on soils

Test	Regulation	Paramaters obtained
Resonant Column	ASTM D 4015-15. Standard Test Methods for Modulus and Damping of Soils by Fixed-Base Resonant Column Devices	Shear modulus and damping.
Cyclic simple shear test	Without specific regulations	Change in the shear modulus and damping with the variation of the number of cycles.
Dynamic triaxial test	ASTM D 3999M-11e1. Standard Test Methods for the Determination of the Modulus and Damping Properties of Soils Using the Cyclic Triaxial Apparatus	Change in the shear modulus and damping with the variation of the number of cycles. Change in the pore pressure. Fatigue tests.

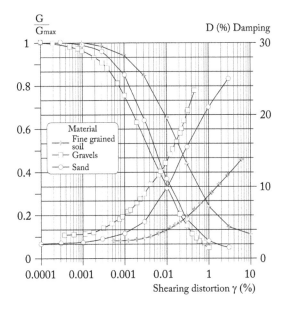

Figure 5.26 Evolution of the damping and shear modulus as a function of shear strain, for three types of soils in Santiago de Chile.

Source: Verdugo and Hoz, 2006.

In rocks, the degradation of the shear modulus with the strains has less importance than in soils and the dynamic tests are generally limited to the rock dynamic strength determination at uniaxial compression and tension. The main utility of these tests is the blasting design, impact and explosion-resistant designs, rockburst phenomena in fragile rocks subject to very high stress levels that, usually, are only found in tunnels and very deep mines.

BIBLIOGRAPHY

Brady, B.H.G., Brown, E.T., *Rock Mechanics for Underground Mining*. George Allen & Unwin Ltd, London. 1985.

Bieniawski, Z.T., "Mechanism of brtittle fracture of rock", *International Journal of Rock Mechanics and Mining Sciences*. Vol. 4. No. 4. 1967, pp. 395–423.

Bieniawski, Z.T., Denkhaus, H.G., Vogler, U., "Failure of fractured rock", *International Journal of Rock Mechanics and Mining Sciences.* Vol. 6. No. 3. 1969, pp. 323–341.

Broch, E., "Rock engineering projects outside Scandinavia", *Kilpailukykyinen kalliorakentaminen seminaari.* Suomen rakennusinsinöörien liitto, Helsinki, Finland. 1996, pp. 7–25.

Carter, T.G. et al., "Tunnelling issues of Chilean Tertiary volcaniclastic rocks", *Proceedings of the MIR 2010. XIII Ciclo di conferenze di Meccanica ed Ingegneria delle Rocce.* Torino. 2010, pp. 215–236.

Celada, B., "Caracterización de rocas sensibles al agua", *Ingeopress.* No. 209. Madrid, Spain. 2011.

Farmer, I., *Engineering Behaviour of Rocks.* Chapman and Hall, London. 1982.

Franklin, J.A., Hoek, E., "Developments in triaxial testing equipments", *Rock Mechanics*, No. 2. 1970, pp. 223–228.

Goodman, R.E., *Introduction to Rock Mechanics.* 2nd Edition. John Wiley & Sons, New York. 1989.

Hoek, E., Brown, E.T., *Underground Excavations in Rock.* The Institution of Mining and Metallurgy, London. 1980.

Hoek, E., Carranza-Torres, C., Corkum, B., "Hoek–Brown Failure Criterion". *Proceedings of NARMSTAC 2002, Mining Innovation and Technology.* Toronto, Canada. 2002, pp. 267–273.

Jaeger, J.C., "Shear Fracture of anisotropic rocks", *Geological Magazine*, No. 97. 1960, pp. 65–72.

Nilsen, B., Dahl, F., Holzhäuser, J., Raleigh, P., "New test methodology for estimating the Abrasiveness of soils for TBM tunnelling", *Proceedings of the Rapid Excavation and Tunnelling Conference (RETC).* Toronto, Canada. 2007.

Paige-Green, P., "A revised ethylene glycol test for assessing the durability of basic crystalline materials for road aggregate", *Proceedings of the 33rd International Geological Congress.* Oslo, Norway. 2008.

Piaggio, G., "Swelling rocks characterization: Lessons from the Andean Region", *ITA World Tunnel Congress.* Dubrovnik, Croatia. 2015.

Peng, S.S., *Coal Mine Ground Control.* 2nd Edition. Wiley-Interscience, New York. 1986.

Perras, M.A., Diedierich, M.S., "A Review of the Tensile Strength of Rock: Concepts and Testing", *Geotechnical and Geological Engineering*, Vol. 32. No. 2. 2014, pp. 525–546.

Rocco, C., Guinea, G.V., Planas, J., Elices, M., "Ensayo Brasileño: Efecto Tamaño y Mecanismos de Rotura", *Anales de Mecánica de la Fractura.* Madrid, Spain. 1996.

Romana, M., Vásárhelyi, B., "A discussion on the decrease of unconfined compressive strength between saturated and dry rock samples", *11th Congress of the International Society for Rock Mechanics.* Lisbon, Portugal. 2007.

Ulusay, R., *The ISRM Suggested Methods for Rock Characterization, Testing and Monitoring: 2007–2014.* Springer Int. Publishing, Cham, Switzerland. 2015.

Ulusay, R., Hudson, J.A., *The Complete ISRM Suggested Methods for Rock Characterization, Testing and Monitoring: 1974–2006*, Ankara, Turkey. 2007.

Van Eeckhout, E.M., Peng, S.S., "The effect of humidity on the compliances of coal mine shales", *International Journal of Rock Mechanics and Mining Sciences & Geomechanics Abstracts*, Vol. 13. No. 2. 1976, pp. 61–67.

Verdugo, R., Hoz, K., "Caracterización geomecánica de suelos granulares gruesos", *Revista Internacional de Desastres Naturales, Accidentes e Infraestructura Civil.* Vol. 6. No. 2. 2006, pp. 199–214.

Wittke, W., *Rock Mechanics: Theory and Applications with Case Histories.* Springer-Verlag, New York. 1999.

Chapter 6

Engineering classifications of rock masses

Z.T. Bieniawski von Preinl, Benjamín Celada Tamames, and Isidoro Tardáguila Vicente

> The things which create problems are not those we do not know; but those we think we know for certain.
>
> Sir Winston Churchill

6.1 INTRODUCTION

Rock masses are complex materials which present the engineering designer with unique problems. They are deeply heterogeneous because their origin is associated with mineralogical processes produced by sedimentary, tectonic and intrusive phenomena.

To design a tunnel it is essential to know the properties of the rock mass to be excavated, but too often this knowledge may be inaccurate due to difficulties in exploring rock masses at depth. Rock mass classifications have emerged as powerful design aids in civil, mining and geological engineering going back to 1948.

The origin of the science of classification goes back to Greek civilization and this field of endeavor forms an important aspect of most sciences, with similar principles and procedures having been developed independently in many disciplines.

Taxonomy is the formal name of the science of classification and it deals with theoretical aspects of classification, including its basis, principles, procedures and rules. A distinction should be made between *classification* and *identification*; classification is defined as the arrangement of objects into groups on the basis of their relationship, whereas identification means the allocation or assignment of additional unidentified objects to the correct class, previously established.

Classifications have played an indispensable role in engineering for centuries. For example, the leading classification society for shipping, Lloyd's Register of London, was established in 1760 and nowadays rigid standards are specified for the design, construction and maintenance before any ship in the world, having more than 100 tons of deadweight, can be insured.

Due to all of these reasons, rock mass classification is considered a basic activity of tunnel engineering and this chapter presents the most commonly used engineering classifications, the estimation of the rock mass stress–strain parameters based upon them and the criteria to classify rock masses from the point of view of their excavability.

6.2 ENGINNERING CLASSIFICATIONS IN TUNNELING

In tunnel engineering, the first rock mass classification was proposed in the United States by Terzaghi (1946), to estimate the load, which should be withstood by the steel arches used at the time as support elements in railway tunnels.

Table 6.1 contains the nine ground categories considered by Terzaghi to evaluate the ground pressure on tunnels with a width between 5 and 10 m. In this table, B and H correspond to the width and height of a tunnel as indicated previously in Figure 2.5.

The classification proposed by Terzaghi was very valuable at the time, but as it is a methodology based on loads imposed on the support elements, it has become obsolete today.

Twelve years later, Lauffer (1985) developed a classification with the objective of estimating the tunnel support which introduced the concept of "stand-up time" entailing great physical significance but one very difficult to determine accurately.

Another remarkable milestone in the field was the introduction of the Rock Quality Designation index (RQD) or "modified core recovery", by Deere in 1963.

The RQD is calculated as the ratio between the length of sound core pieces longer than 10 cm recovered from a borehole.

Using the RQD, rock core is classified into five categories, according to the criteria presented in Table 6.2. The RQD is very easy to calculate when cores are available, but when it has to be estimated at an outcrop or at the tunnel face, the process is much more complicated. RQD is still useful today, mostly as part of mass classifications such as the Q index.

Note that the RQD does not include the effects of joint conditions (tightness and infilling) nor their orientations.

In the 1970s two engineering rock mass classifications were proposed, which received acceptance worldwide and are still in use today: the Rock Mass Rating (RMR) by Bieniawski (1973) and the Q index by Barton at al. (1974). This is so because they include the most significant geo-engineering parameters characterizing rock mass formations. Moreover, proposed independently, and structured differently, they provide useful crosschecks for one another.

Table 6.1 Ground categories proposed by Terzaghi, 1946

No.	Ground description	Ground height that generates the load (m)	Pressure on steel arches (kp/cm²)	
			B=H=5 m	B=H=10 m
I	Strong and without joints	0	0	0
2	Strong and stratified	0–0.5 B	0–0.6	0–1.3
3	Massive with few joints	0–025 B	0–0.3	0–0.6
4	Stratified with few joints	0.25 B–0.35 (B+H)	0.3–0.9	0.6–1.8
5	Stratified with many joints	(035 a 1.1) (B+H)	0.9–2.9	1.8–2.9
6	Fully fractured but not altered	1.1 (B+H)	2.9	5.7
7	Rocks with creep at a moderate depth	(1.1 a 2.1) (B+H)	2.9–5.5	5.7–10.9
8	Rocks with creep at a great depth	(2.1 a 4.5) (B+H)	5.5–11.7	10.9–23.4
9	Rocks that generate swelling pressures	Up to 80 m for any value of (B+H)	Up to 20.8	Up to 20.8

Table 6.2 Ground classification depending on the RQD

RDQ (%)	Ground quality
0–25	Very poor
25–50	Poor
50–75	Fair
75–90	Good
90–100	Very good

After the development of the RMR system and the Q index, there was a large proliferation of other classifications or modifications, which claimed to provide alternative approaches, but over the years, the most widely used engineering rock mass classifications, worldwide, are still the RMR and Q.

Accordingly, in the following sections these two rock mass classifications proposed independently by Bieniawski and by Barton are described in detail.

6.2.1 RMR classification system

This engineering classification method features the RMR that was presented by Z. T. Bieniawski in 1973 and was slightly modified by him in 1989. The RMR_{89} is calculated by the sum of the ratings of the following five parameters:

1. Uniaxial compressive strength of the intact rock, rated from 0 to 15 points
2. RQD, rated from 0 to 20 points
3. Spacing of the discontinuities, rated from 0 to 20 points
4. Condition of the main discontinuities, rated from 0 to 30 points
5. Presence of water in the ground, rated from 0 to 15 points

In addition, the parameter orientation of discontinuities is indirectly included as an adjustment to the total sum of the rated parameters, in accordance with Table 6.4 that follows later.

It should also be noted that the RQD parameter was included originally for historical reasons because it was covered extensively in tunneling case histories at the time, featuring tunnel support by rock bolts. More recently, RQD was replaced by the fracture frequency parameter, which is correlated to the RQD, but which is more easily and more reliably determined from both rock cores as well as from surveys of surface outcrops and tunnel faces.

The evaluation of these input parameters can be done using the criteria contained in Table 6.3 or, preferably to avoid the interpolations necessary to apply these criteria, the best procedure is to use one of the three graphs shown in Figure 6.1.

The overall RMR value calculated with the above criteria is the sum of the ratings of the parameters, which is called the basic RMR and it is represented as $RMRb_{89}$.

When the relative position of the tunnel axis with respect to the orientation of the main joint set is considered, the behavior of the tunnel excavation at the face changes; as it is well known that it is easier to excavate the same ground when performed perpendicular to the dip direction of the discontinuities than when excavated in the parallel direction.

To take into account this effect, the RMR_b has to be adjusted according to the criteria indicated in Table 6.4.

When the excavation is done perpendicularly to the dip strike of the layers, two extreme situations occur; one favorable, in which the layers dip toward the excavation face and cannot slide when the support is placed and another when the layers dip toward the excavated tunnel and can slide easily at the tunnel face, creating an unfavorable situation. Figure 6.2 clarifies these two situations.

The RMR, calculated with the above criteria, varies between 0 and 100 rating points and allows classifying the rock masses into five classes, as indicated in Table 6.5.

The RMR in the tunnel face does not have an exact value as it depends on the judgment of the professional who is determining it. As a practical approach, it is recommended that two professionals, an engineer and a geologist with enough education and experience, should determine RMR values for the same tunnel face in the range of RMR $\pm$ 4%. This means, if the real value of the RMR of the tunnel face is 56 points, the expected range of the ratings will be 54 to 58 points.

Table 6.3 Criteria to calculate the RMR$_{89}$

Parameter			Range of values						
1	Strength of intact rock material	Point load strength index	> 10 MPa	4–10 MPa	2–4 MPa	1–2 MPa	For this low range – uniaxial compressive test is preferred		
		Uniaxial compressive strength	> 250 MPa	100–250 MPa	50–100 MPa	25–50 MPa	5–25 MPa	1–5 MPa	< 1 MPa
	Rating		15	12	7	4	2	1	0
2	Drill core Quality RQD		90%–100%	75%–90%	50%–75%	25%–50%	< 25%		
	Rating		20	17	13	8	5		
3	Spacing of discontinuities		> 2 m	0.6–2 m	200–600 mm	60–200 mm	< 60 mm		
	Rating		20	15	10	8	5		
4	Condition of discontinuities (See E)		Very rough surfaces Not continuous No separation Unweathered wall rock	Slightly rough surfaces Separation < 1 mm Slightly weathered walls	Slightly rough surfaces Separation < 1 mm Highly weathered walls	Slicken sided surfaces or Couge < 5 mm thick or Separation 1–5 mm continuous	Soft gouge > 5 mm thick or separation > 5 mm continuous		
	Rating		30	25	20	10	0		
5	Ground water	Inflow per 10 m tunnel length (l/m)	None	< 10	10–25	25–125	> 125		
		(Joint water press)/(mayor principal σ)	0	< 0.1	0.1–0.2	0.2–0.5	> 0.5		
		General conditions	Completely dry	Damp	Wet	Dripping	Flowing		
	Rating		15	10	7	4	0		

The individual ratings shown for the five input parameters are their average values (not the minimum values) for each range of the parameters.

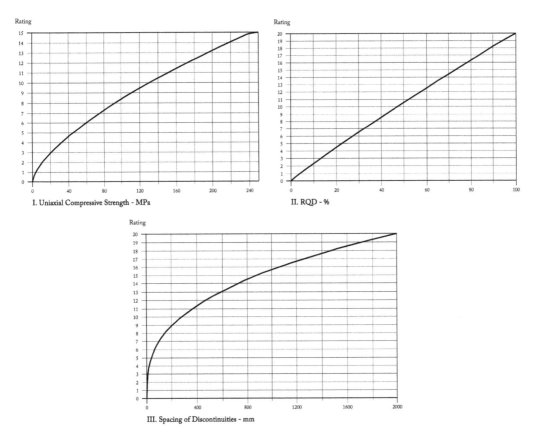

Figure 6.1 Graphs to evaluate the RMR input parameters: uniaxial compressive strength, RQD and the spacing of the discontinuities.

Table 6.4 Adjustment of the basic RMR as a function of the tunnel axis orientation with respect to the orientation of the main joint set

Strike perpendicular to the tunnel axis				Strike parallel to the tunnel axis		
Excavation with dip		Excavation against dip				
Dip 45–90	Dip 20–45	Dip 45–90	Dip 20–45	Dip 45–90	Dip 20–45	Dip 0°–20° any direction
Very favorable	Favorable	Fair	Unfavorable	Very unfavorable	Fair	Fair
0	−2	−5	−10	−12	−5	−5

The methodology to calculate the RMR has had two important changes over the years of long experience, which are presented in the following sections.

6.2.1.1 Modification to rating the RQD and the spacing of the discontinuities combined

To calculate the RMR_{89} it is necessary to know the RQD, which creates a significant problem when evaluating the RMR at the tunnel face or at an outcrop, where the RQD

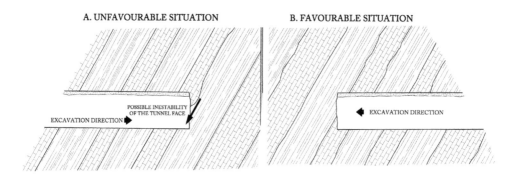

A. UNFAVOURABLE SITUATION

B. FAVOURABLE SITUATION

POSSIBLE INESTABILITY
OF THE TUNNEL FACE

EXCAVATION DIRECTION

EXCAVATION DIRECTION

Figure 6.2 Situations created when excavating perpendicularly to the dip strike of the layers.

Table 6.5 Rock mass classification using the RMR

RMR value	Class	Ground Classification
81–100	1	Very good
61–80	2	Good
41–60	3	Fair
21–40	4	Poor
0–20	5	Very poor

cannot be directly calculated and this parameter has to be estimated through any of the available correlations.

On the other hand, to calculate the RMR_{89} the spacing of the discontinuities has to be also evaluated, which is a parameter related to the RQD.

Figure 6.3 shows the correlation between the RQD and the spacing of the discontinuities, established by Bieniawski (1973), which shows the mean correlation between the density of the discontinuities or the number of discontinuities per meter of ground which corresponds to value ranges of the spacing of the discontinuities and the RQD.

In 2000, Geocontrol used the information from Figure 6.3 to group the RQD and the spacing of the discontinuities replacing them, in the RMR calculation, by the number of discontinuities present in a meter of ground.

Since both the RQD and the spacing of the discontinuities were rated with a maximum of 20 points each, the number of discontinuities per meter was rated with a maximum of 40 points, using the criteria contained in Table 6.6.

Subsequently, Lawson (2013) proposed the graph shown in Figure 6.4 to evaluate this parameter, with which a greater accuracy is obtained.

6.2.1.2 Update of the RMR in 2014

After almost four decades of using the RMR, some issues were identified that could lead to improved tunnel engineering practice:

- It is very rare to find grounds with RMR > 90.
- The favorable effect of the mechanical excavation was not included in the original RMR_{89}.

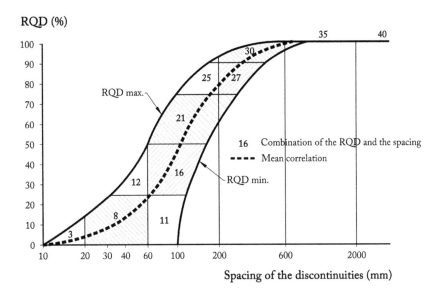

Figure 6.3 Correlation between the RQD and the Spacing.

Source: Bieniawski, 1973.

- The criteria to calculate the RMR_{89} tended to concentrate on rock masses in the RMR range from 40 to 60 points.
- The loss of strength of some rocks subjected to the water effect was not taken into account.

Between 2012 and 2014 Geocontrol carried out an industrial research project to improve the above issues, using information relating to RMR_{89} featuring mapping of 2,298 tunnel faces (Celada et al., 2014). By comparison, the original RMR was developed on the basis of 351 tunnel case histories, as listed by Bieniawski (1989).

As a result of this work, a new version of the RMR, called RMR_{14}, was developed, which retains the original RMR structure but introduces important improvements. The RMR_{14} is calculated by the expression:

$$RMR_{14} = (RMR_{14b} + F_0) \cdot F_e \cdot F_s \tag{6.1}$$

where:

RMR_{14b} = RMR_{14} of the rock mass, without the adjustment due to the tunnel axis orientation with respect to the discontinuities

F_0 = Factor that considers the effect of the tunnel axis orientation with respect to the discontinuities

F_e = Factor that considers the ground behavior improvement when excavating with TBMs

F_s = Factor that characterizes ground yielding effect at the tunnel face

The RMR_{14} is calculated by evaluating five parameters, which include the uniaxial compressive strength of the intact rock, the number of discontinuities per meter and the presence of water. These three parameters have the same ratings as in the RMR_{89} and are evaluated in the same way.

Table 6.6 Rating of the number of discontinuities per meter at the tunnel face

RMR (2+3)	No. of joints per meter (RQD and spacing between discontinuities)																			
Joints per meter	0	0.5	1	2	3	4	5	6	7	8	9	10	11	12	13	14	15	16	17	18
Rating	40	37	34	31	29	28	27	26	25	23	22	22	21	20	19	18	17	16	16	15
Joints per meter	19	20	21	22	23	24	25	26	27	28	29	30	31	32	33	34	35	36	37	38
Rating	15	14	14	12	12	11	11	10	10	10	9	9	9	8	8	8	8	8	7	7
Joints per meter	39	40	41	42	43	44	45	46	47	48	49	50								
Rating	7	7	7	7	5	5	5	4	4	4	4	3								

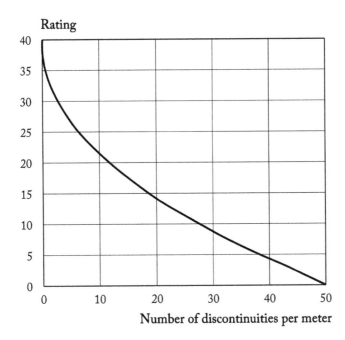

Figure 6.4 Continuous rating of the number of discontinuities per meter.

The other two parameters are new and represent the condition of the joints present in the ground and the alterability of the ground in the presence of water.

The effect of water on rock mass strength has a considerable importance in tropical countries, as evidenced by De Oliveira (2007).

The condition of the joints present in the rock mass is rated between 0 and 20 and is evaluated with the criteria shown in Table 6.7.

The water effect on the weathering of rocks is valued between 0 and 10 points, depending on the behavior observed during 24 hours when a rock sample is immersed in a solution of ethylene glycol, which is the liquid used in the radiators of motor vehicles.

This test is a variant of the one presented in Section 5.7.1.2 in which the time of observation has been reduced, in order to obtain approximate results in 24 hours.

In order to perform the weathering test quickly, it is recommended to immerse five rock fragments in a commercial solution of ethylene glycol, with their in situ moisture and with a side length of about 3 cm.

Table 6.7 Criteria to evaluate the strength of the discontinuities

Discontinuity length	< 1 m	1–3 m	3–10 m	> 10 m
Rating	5	4	2	0
Roughness	Very rough	Rough	Smooth	Sliding surfaces
Rating	5	3	1	0
Infilling	Hard		Soft	
	< 5 mm	> 5 mm	< 5 mm	> 5 mm
Rating	5	2	2	0
Weathering	Unweathered	Slightly weathered	Highly weathered	Decomposed
Rating	5	3	1	0

Table 6.8 Criteria to evaluate the strength of the discontinuities

Behavior after immersing 5 rock fragments in ethylene glycol.	Rating
The rock disintegrates in a few minutes.	0
The rock partially disintegrates after 8 hours.	1
The rock is superficially softened after 16 hours.	4
The rock is not altered after 24 hours.	10

Visually, the actual weathering of the samples at 1, 8, 16 and 24 hours can be observed and the results are evaluated with the criteria presented in Table 6.8.

Table 6.9 presents the latest form to calculate the RMR_{14b}.

Referring to Equation 6.1 given earlier, the tunnel orientation factor F_0 is the same that was used in the RMR_{89} and, therefore, is determined according to the content in Table 6.4.

The TBM effect factor F_e represents the improvement in rock mass behavior when using Tunnel Boring Machines and is evaluated using the graph shown in Figure 6.5 – a recent development. When an excavation is not made with TBMs, $F_e = 1$. Previously, another approach was proposed by Alber (1993).

The factor ground yielding effect at face F_s is only used when determining the RMR at the tunnel face, as in other cases its value is 1.

This factor takes into account the movements produced at the tunnel face which modify the joints appearance and the rock mass at the tunnel face seems to have a lower RMR than the one which would correspond to the ground before excavating.

This correction is based on the ICE value at the tunnel face, calculated as presented in Section 1.3.3.

The evaluation of the F_s factor is performed using the graph presented in Figure 6.6.

6.2.1.3 Correlation between the RMR_{89} and the RMR_{14}

Once the development of RMR_{14} was completed, it was tested using the mapping data of 2,298 tunnel faces, which constituted the database for correlating the RMR_{89} with the RMR_{14}.

Figure 6.7 shows the results of the correlation, which can be expressed by the equation:

$$RMR_{14} = 1.1 \cdot RMR_{89} + 2 \qquad (6.2)$$

6.2.2 Q system

In 1974, N. Barton, R. Lien and J. Lunde published their paper "Engineering classification of rock masses for the design of tunnel support" introducing the Q system. Developed after analyzing about 200 tunnel case records, permanent support was correlated with the Q index.

Unlike the RMR, which aims to classify rock masses and estimate their properties, the Q index focused on the selection of tunnel support, as described in the next section.

6.2.2.1 Structure of the Q system

The original paper introducing the Q index was divided into three parts:

- Part I: Estimation of the rock mass quality
- Part II: Estimation of the pressure on the tunnel support
- Part III: Selection of the tunnel support based on precedent cases

Table 6.9 Form to calculate the RMR$_{14b}$

1. Compressive strength of the intact rock				

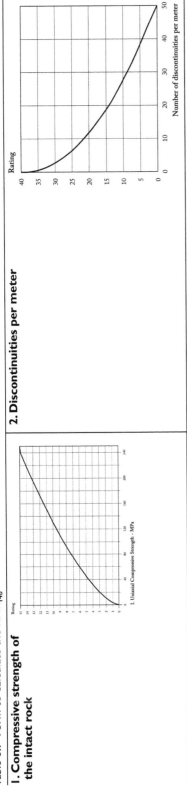

2. Discontinuities per meter				

3. State of the joints

Discontinuity length	< 1 m	1–3 m	3–10 m	> 10 m
Rating	5	4	2	0
Roughness	Very rough	Rough	Smooth	Sliding surfaces
Rating	5	3	1	0
Filling	Hard			Soft
Rating	< 5 mm	> 5 mm	< 5 mm	> 5 mm
	5	2	2	0
Weathering	Unweathered	Slightly weathered	Highly weathered	Decomposed
Rating	5	3	1	0

4. Water presence

State	Dry	Slightly wet	Wet	Water dropping
Rating	15	10	7	4

5. Alterability because of the water

Effect after immersing 5 rock fragments in ethylene glycol.	The rock disintegrates in a few minutes	The rock partially disintegrates after 8 hours	The rock is softened after 16 hours	The rock is not altered after 24 hours
Rating	0	1	3	10

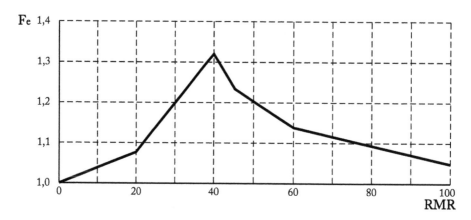

Figure 6.5 Evaluation of the F_e factor.

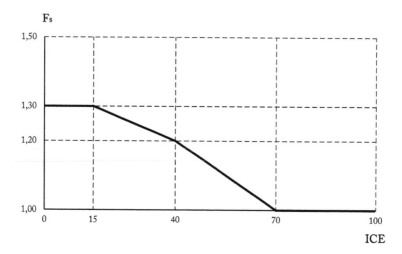

Figure 6.6 Evaluation of the F_s factor.

Applications of the Q index for estimating the quality of rock masses maintain their validity to this day, so this aspect is presented next at length.

It should be noted that, when the Q concept was introduced, selection of tunnel support offered a choice of 38 categories of recommended support measures, whose application required consideration of meticulous details.

The choice of the tunnel support from 38 recommended possibilities and, specifically, the fact of taking into account a great number of observations necessary for implementation, made so many recommendations impractical.

Eventually, ten years later, Grimstad and Barton (1994) synthesized the initial 38 recommended supports in nine reinforcement categories represented in Figure 6.8. The ESR factor (Excavation Support Ratio), included in Figure 6.8, is described in Section 2.4.3.2. It is related to the use for which the excavation is intended and the degree of safety demanded.

Over the years, the empirical design of the tunnel support has lost importance, compared to the design methodology based on the stress–strain analyses and, consequently, the selection of the tunnel support using the Q index should only be used in the early stages of a tunnel design project.

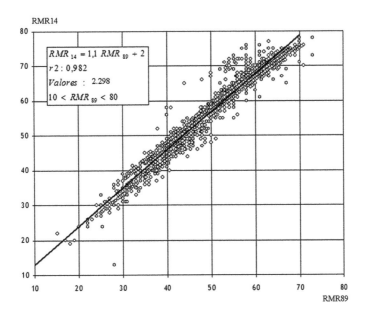

Figure 6.7 Correlation between the RMR$_{89}$ and the RMR$_{14}$.

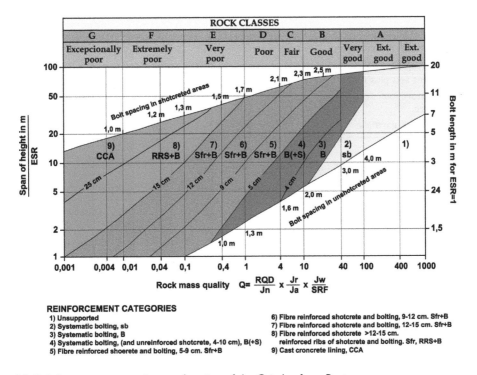

REINFORCEMENT CATEGORIES
1) Unsupported
2) Systematic bolting, sb
3) Systematic bolting, B
4) Systematic bolting, (and unreinforced shotcrete, 4-10 cm), B(+S)
5) Fibre reinforced shoerete and bolting, 5-9 cm. Sfr+B

6) Fibre reinforced shotcrete and bolting, 9-12 cm. Sfr+B
7) Fibre reinforced shotcrete and bolting, 12-15 cm. Sfr+B
8) Fibre reinforced shotcrete >12-15 cm.
   reinforced ribs of shotcrete and bolting. Sfr, RRS+B
9) Cast croncrete lining, CCA

Figure 6.8 Reinforcement categories as a function of the Q index from Barton.

6.2.2.2 Estimation of the rock mass quality through the Q index

The Q index is calculated by the following expression:

$$Q = \left(\frac{RQD}{J_n}\right) \cdot \left(\frac{J_r}{J_a}\right) \cdot \left(\frac{J_w}{SRF}\right) \tag{6.3}$$

where:

RQD = Rock Quality Designation (Deere, 1963)
J_n = Related to the joint sets in the ground
J_r = Associated with the roughness of the joints
J_a = Related to the degree of alteration along the joints
J_w = Reduction factor due to the presence of water in the joints
SRF = Stress reduction factor due to the level of in situ stress field

Tables 6.10 through 6.11 6.12 6.13 6.14 show, respectively, the descriptions, ratings and observations referred to as J_n, J_r, J_a, J_w and SRF.

Table 6.10 Description, rating and observations associated with J_n

Description		Rating	Observations
A.	Massive, no or few joints	0.5–1.0	Note:
B.	One joint set	2	(i) For intersections use
C.	One joint set plus random	3	$(3.0 \times J_n)$
D.	Two joint sets	4	(ii) For portals use $(2.0 \times J_n)$
E.	Two joint sets plus random	6	
F.	Three joint sets	9	
G.	Three joint sets plus random	12	
H.	Four or more joint sets, random, heavily jointed, "sugar cube", etc.	15	
J.	Crushed rock, earthlike	20	

Source: Barton et al. (1974).

Table 6.11 Description, rating and observations associated with J_r

Description		Rating	Observations
	(a) Rock wall contact and		Note:
	(b) Rock wall contact before 10 cm shear		(i) Add 1.0 if the mean spacing of the relevant joint set is greater than 3 m
A.	Discontinuous joints	4	
B.	Rough or irregular, undulating	3	(ii) $J_r = 0.5$ can be used for planar slickensided joints having lineations, provided the lineations are favourably orientated
C.	Smooth, undulating	2	
D.	Slickensided, undulating	1.5	
E.	Rough or irregular, planar	1.5	
F.	Smooth, planar	1.0	
G.	Slickensided, planar	0.5	
	(c) No rock wall contact when sheared		
H.	Zone containing clay minerals thick enough to prevent rock wall contact	1.0 (nominal)	
J.	Sandy, gravelly or crushed zone thick enough to prevent rock wall contact	1.0 (nominal)	

Source: Barton et al. (1974).

Table 6.12 Description, rating and observations associated with J_a

Description		Rating	Observations
	(a) *Rock wall contact*		
A.	Tightly healed, hard, non-softening, impermeable filling i.e., quartz or epidote	0.75	
B.	Unaltered joint walls, surface staining only	1.0	
C.	Slightly altered joint walls. Non-softening mineral coatings, sandy particles, clay-free disintegrated rock etc.	2.0	
D.	Silty-, or sandy-clay coatings, small clay-fraction (non-softening)	3.0	
E.	Softening or low friction clay mineral coatings, i.e., kaolinite, mica. Also chlorite, talc, gypsum and graphite etc., and small quantities of swelling clays. (Discontinuous coatings, 1–2 mm or less in thickness)	4.0 (8°–16°)	Note: (i) Values of $(\varphi)_r$ are intended as an approximate guide to the mincralogical properties of the alteration products, if present
	(b) *Rock wall contact before 10 cms shear*		
F.	Sandy particles, clay-free disintegrated rock etc.	4.0 (25°–30°)	
G.	Strongly over-consolidated, non-softening clay mineral fillings (Continuous, <5 mm in thickness)	6.0 (16°–24°)	
H.	Medium or low over-consolidation, softening, clay mineral fillings. (Continuous, <5 mm in thickness)	8.0 (12°–16°)	
J.	Swelling clay fillings, i.e., mont-morillonite (Continuous, <5 mm in thickness). Value of J_a depends on percent of swelling clay-size particles, and access to water etc.	8.0–12.0 (6°–12°)	
	(c) *No rock wall contact when sheared*		
K,L,M.	Zones or bands of disintegrated or crushed rock and clay (see G, H, J for description of clay condition)	6.0, 8.0 (6°–24°) or 8.0–12.0	
N.	Zones or bands of silty- or sandy clay, small clay fraction (non-softening)	5.0	
O,P,R.	Thick, continuous zones or bands of clay (see G, H, J for description of clay condition)	10.0, 13.0 (6°–24°) or 13.0–20.0	

Source: Barton et al. (1974).

The correct use of the previous rating criteria is not easy, as the information included in some of the tables to define them is difficult to interpret correctly.

In order to make the use of these tables easier, Barton and Bieniawski (2008) proposed the Figure 6.9 to clarify the rating of the parameters J_r and J_a.

However, the use of the Stress Reduction Factor (SRF) is not easy, as it requires advanced knowledge of the in situ stresses and specific tests, such as in the case of ground subjected to squeezing and swelling.

Based on the experience from Norway and China, Grimstad and Barton (1994) recommend assuming the values shown in Table 6.15. for the SRF. Q takes values between 0.001 and 1,000 which allow classifying rock mass quality in the nine categories shown in Table 6.16.

6.2.3 Correlations between the RMR and Q

Both the RMR and the Q aim at evaluating the quality of rock masses and, therefore, one can expect a correlation between both systems, if each correctly characterizes rock mass behavior.

Table 6.13 Description, rating and observations associated with J_w

Description		Rating		Observations
A.	Dry excavations or minor inflow i.e., <5 l/min. locally	1.0	<1	Note: (i) Factors C to F are crude estimates. Increase J_w if drainage measures are installed (ii) Special problems caused by ice formation are not considered
B.	Medium inflow or pressure occasional outwash of joint fillings	0.66	1.0–2.5	
C.	Large inflow or high pressure in competent rock with unfilled joints	0.5	2.5–10.0	
D.	Large inflow or high pressure, considerable outwash of joint fillings	0.33	2.5–10.0	
E.	Exceptionally high inflow or water pressure at blasting, decaying with time	0.2–0.1	>10.0	
F.	Exceptionally high inflow or water pressure continuing without noticeable decay	0.1–0.05	>10.0	

Source: Barton et al. (1974).

Bieniawski and Van Heerden (1975) presented such a correlation shown in Figure 6.10, where the following relationship between RMR and Q is evident:

$$RMR = 9 \cdot \ln Q + 44 \qquad (6.4)$$

Subsequently, Barton et al. (1985) proposed the following expression as the correlation between the RMR and Q:

$$RMR = 15 \cdot \log Q + 50 \qquad (6.5)$$

Figure 6.11 shows the correlations proposed by Bieniawski and Barton. From the content in Figure 6.11 it can be observed that both correlations are reasonable for grounds within the interval $50 < RMR < 80$, i.e., for rock masses with good to fair quality.

However, in rock masses with RMR smaller than 50 rating points the difference between both correlations is clearly excessive.

To correlate the RMR_{14} with Q, the RMR_{14} values for each of the 101 cases represented in Figure 6.10 have been used after ignoring the cases considered less representative.

As a result, the values presented in Figure 6.12 have been plotted together with a correlation derived from this data.

The equation that correlates Q with the RMR_{14} is:

$$RMR_{14} = 8.12 \cdot \ln Q + 49.8 \qquad (6.6)$$

Here, the correlation coefficient is $r^2 = 0.68$ which is quite low, but it is consistent with the fact that the correlation between RMR and Q does exist.

Table 6.14 Description, rating and observations associated with SRF

Description	Rating	Observations
(a) *Weakness zones intersecting excavation, which may cause loosening of rock mass when tunnel is excavated*		Note:
A. Multiple occurrences of weakness zones containing clay or chemically disintegrated rock, very loose surrounding rock (any depth)	10.0	(i) Reduce these values of SRF by 25–50% if the relevant shear zones only influence but do not intersect the excavation
B. Single weakness zones containing clay, or chemically disintegrated rock (depth of excavation $\leq 50\,m$)	5.0	
C. Single weakness zones containing clay, or chemically disintegrated rock (depth of excavation > 50 m)	2.5	
D. Multiple shear zones in competent rock (clay free), loose surrounding rock (any depth)	7.5	
E. Single shear zones in competent rock (clay free) (depth of excavation $\leq 50\,m$)	5.0	
F. Single shear zones in competent rock (clay free) (depth of excavation > 50 m)	2.5	
G. Loose open joints, heavily jointed or "sugar cube" etc. (any depth)	5.0	
(b) *Competent rock, rock stress problems* σ_c/σ_1 σ_t/σ_1		(ii) For strongly anisotropic stress field (if measured): when $5 \leq \sigma_1/\sigma_3 \leq 10$, reduce σ_c and σ_t to 0.8 σ_c and 0.8 σ_t; when $\sigma_1/\sigma_3 > 10$, reduce σ_c and σ_t to 0.6 σ_c and 0.6 σ_t where: $\sigma_c =$ unconfined compression strength, $\sigma_t =$ tensile strength (point load), σ_1 and $\sigma_3 =$ major and minor principal stresses
H. Low stress, near surface $>200 >13$	2.5	
J. Medium stress 200–10 13–0.66	1.0	
K. High stress, very right structure (Usually favorable to stability, may be unfavorable to wall stability) 10–5 0.66–0.33	0.5–2.0	
L. Mild rock burst (massive rock) 5–25 0.33–0.16	5–10	
M. Heavy rock burst (massive rock) <2.5 <0.16	10–20	
(c) *Squeezing rock; plastic flow of incompetent rock under the influence of high rock pressures*		(iii) Few case records available where depth of crown below surface is less than span width. Suggest SRF increase from 2.5 to 5 for such cases (see H)
N. Mild squeezing rock pressure	5–10	
O. Heavy squeezing rock pressure	10–20	
(d) *Swelling rock; chemical swelling activity depending on presence of water*		
P. Mild swelling rock pressure	5–10	
R. Heavy swelling rock pressure	10–15	

Source: Barton et al. (1974).

6.2.4 Criteria for the proper use of the engineering classifications of rock masses

In 2008 Barton and Bieniawski published their paper "Setting the record straight about RMR and Q", which specified ten recommendations for their proper use:

I. Ensure that the classification parameters are quantified (measured, not just described), from standardized tests for each geologically designated structural region employing boreholes, exploration adits and surface mapping plus seismic refraction for interpolation between the inevitably limited numbers of boreholes.

II. Follow the established procedures for classifying the rock mass by RMR and Q and determine their typical ranges and the average values.

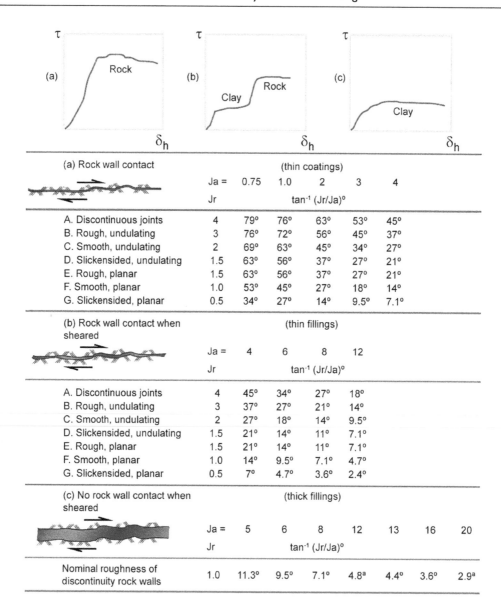

(a) Rock wall contact		(thin coatings)				
	Ja =	0.75	1.0	2	3	4
	Jr		tan⁻¹ (Jr/Ja)°			
A. Discontinuous joints	4	79°	76°	63°	53°	45°
B. Rough, undulating	3	76°	72°	56°	45°	37°
C. Smooth, undulating	2	69°	63°	45°	34°	27°
D. Slickensided, undulating	1.5	63°	56°	37°	27°	21°
E. Rough, planar	1.5	63°	56°	37°	27°	21°
F. Smooth, planar	1.0	53°	45°	27°	18°	14°
G. Slickensided, planar	0.5	34°	27°	14°	9.5°	7.1°

(b) Rock wall contact when sheared		(thin fillings)			
	Ja =	4	6	8	12
	Jr		tan⁻¹ (Jr/Ja)°		
A. Discontinuous joints	4	45°	34°	27°	18°
B. Rough, undulating	3	37°	27°	21°	14°
C. Smooth, undulating	2	27°	18°	14°	9.5°
D. Slickensided, undulating	1.5	21°	14°	11°	7.1°
E. Rough, planar	1.5	21°	14°	11°	7.1°
F. Smooth, planar	1.0	14°	9.5°	7.1°	4.7°
G. Slickensided, planar	0.5	7°	4.7°	3.6°	2.4°

(c) No rock wall contact when sheared		(thick fillings)					
	Ja =	5	6	8	12	13	16	20
	Jr		tan⁻¹ (Jr/Ja)°				
Nominal roughness of discontinuity rock walls	1.0	11.3°	9.5°	7.1°	4.8ᵃ	4.4°	3.6°	2.9ᵃ

Figure 6.9 Remarks about the rating of J_a and J_r.

Source: Barton et al., 1974.

III. Use both systems and then check with at least two of the published correlations of Bieniawski and Barton.

IV. Estimate the support requirements through Figure 2.17. The permanent support can be estimated with the graph included in this figure, but only if the support elements are of good quality.

V. Estimate the stand-up time from the graph in Figure 6.13 and the rock mass modulus with the correlations presented in Figure 6.14, and updated in Figure 6.18 for preliminary modeling purposes. A depth-dependent modulus of deformation may be needed if the depth of the tunnel is significant.

VI. Perform representative numerical modeling and check if enough information is available.

Table 6.15 Values of SRF recommended by Grimstad and Barton, 1994

Stress level	Relative ground strength		SRF
	σ_{cl}/σ_1	$\sigma_{cl}/\sigma_\theta$	
Low stress level. Near surface excavations with open joints in the ground.	> 200	< 0.01	2.5
Medium stress level. Favorable stress conditions.	200–10	0.01–0.3	1
High stress level with closed joints. Favorable to stability except for the excavation walls.	10–5	0.3–0.5	0.5–2
The excavations in massive rocks show slabbing problems after 1 hour.	5–3	0.5–0.65	5–50
The excavations in massive rocks show slabbing problems and rockburst after a few minutes.	3–2	0.65–1	50–200
Heavy rockbursts in massive rocks.	< 2	> 1	200–400

σ_{cl} = uniaxial compressive strength of the intact rock

σ_1 = higher main stress

σ_θ = maximum tangential stress

Table 6.16 Rock mass quality classification depending on the Q value

Q value of the ground	Ground classification
0.001–0.01	Exceptionally poor
001–0.1	Extremely poor
0.1–1	Very poor
1–4	Poor
4–10	Fair
10–40	Good
40–100	Very good
100–400	Extremely good
400–1,000	Exceptionally good

VII. If there is not enough information available for the ground characterization, request further geological exploration and parameter testing.

VIII. Consider the construction process, and in case of expecting the use of a TBM, estimate the advance rates, using the Q_{TBM} and Rock Mass Excavability (RME) methods.

IX. Ensure that all the rock mass characterization information is included in the Geotechnical Baseline Report, which discusses design procedures, assumptions and specifications.

X. During tunnel construction, determine the RMR and Q at the tunnel faces so that comparisons can be made of expected and encountered conditions, leading to design verification or appropriate changes.

6.3 ESTIMATION OF THE STRESS–STRAIN PARAMETERS OF THE ROCK MASS

The inherent purpose of the engineering classifications of rock masses is to classify them according to their quality and enable estimation of the parameters which represent their stress–strain behavior.

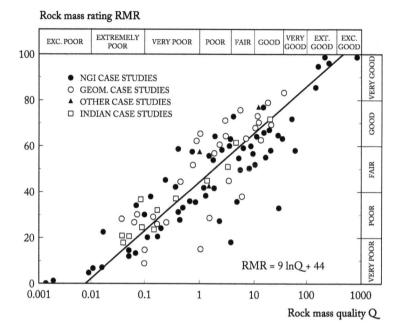

Figure 6.10 Correlation between Q and the RMR.

Source: Bieniawski and Van Heerden, 1975.

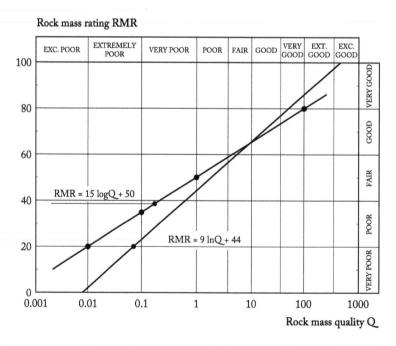

Figure 6.11 Representation of the correlations proposed by Barton and Bieniawski (2008).

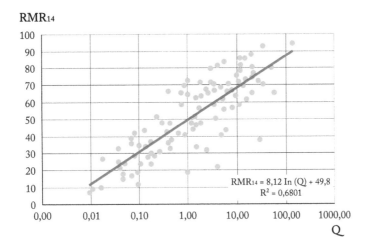

RMR₁₄

$$RMR_{14} = 8,12 \ln (Q) + 49,8$$
$$R^2 = 0,6801$$

Q

Figure 6.12 Correlation between Q and RMR₁₄.

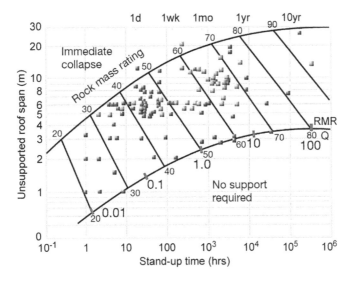

Figure 6.13 Graph for estimating the stand-up time of tunnels.

Source: Bieniawski, 1989.

The results obtained from the rock mass classifications which provide an estimate of the stress–strain parameters must be used with caution and should be only used in the preliminary design phases. In final tunnel design studies, it is essential to use the parameters obtained through specific tests, carried out in the laboratory and directly in situ.

The following sections describe the way to estimate the stress–strain parameters of rock masses using the RMR.

6.3.1 Determination of the uniaxial compressive strength

The uniaxial compressive strength of a rock mass is a parameter which has high relevance in tunnel design.

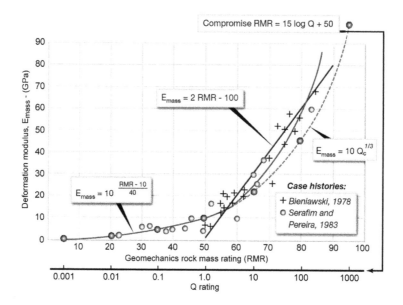

Figure 6.14 Correlations to estimate the modulus of deformation.

Source: Barton and Bieniawski, 2008.

However, this parameter is very difficult to measure directly, due to the scale effect produced by the discontinuities in the ground.

Between 1966 and 1973, the South African Council for Scientific and Industrial Research (CSIR) carried out an important program of large scale uniaxial compression tests in situ. An extensive series of 66 tests were carried out on coal specimens of different shapes cut in situ, whose dimensions reached two meters. Photograph 6.1 depicts one of the specimens during its test.

Decades later, Aydan and Dalgic (1998) provided data from about 20 shear tests performed in situ, on specimens whose RMRs ranged from 5 to 80 rating points. Figure 6.15 presents the results of these tests.

The expression that better correlates these results is:

$$\sigma_M = \sigma_c \cdot e^{\frac{RMR-100}{24}}\qquad(6.7)$$

where σ_c is the uniaxial compressive strength of the intact rock.

When the RMR_{14} is used, the above expression becomes:

$$\sigma_M = \sigma_c \cdot e^{\frac{RMR-100}{17}}\qquad(6.8)$$

6.3.2 Determination of the modulus of deformation of rock masses

The modulus of deformation of rock masses is an essential parameter to calculating the ground displacements when a tunnel is excavated.

Photograph 6.1 Coal specimen 2 m cube cut in situ and tested at uniaxial compression.

Source: Bieniawski and Van Heerden, 1975.

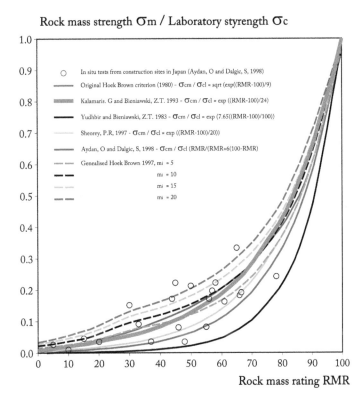

Figure 6.15 Correlation between the RMR and the rock mass compressive strength.

The modulus of eformation (E) is related with the ground stress (σ) by the well-known Hooke's law:

$$\sigma = \varepsilon \cdot E \qquad\qquad (6.9)$$

where ε is the ground strain.

Deere (1963) proposed a chart, shown in Figure 6.16, which shows the values of the modulus of deformation (σ/ε) from uniaxial compression tests performed in several lithologies.

Knowing the ground uniaxial compressive strength, this value can be correlated with the modulus of feformation, using the following graph.

Palmström and Singh (2001) presented the correlation between the RMR and the modulus of deformation shown in Figure 6.17, where the greatest amount of data is grouped in the RMR range between 55 and 75 rating points. However, there is a very wide variation in this interval of the modulus of deformation for a given RMR value.

On the other hand, for RMR values smaller than 55 rating points the grouping of the values of the modulus of deformation are much better.

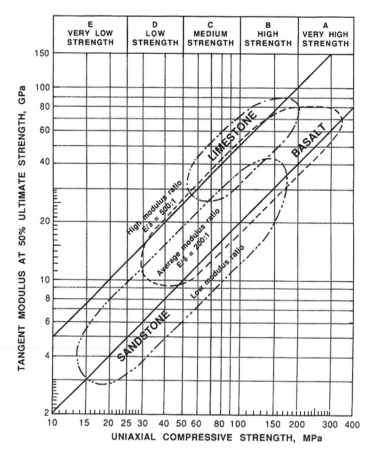

Figure 6.16 Correlation between the uniaxial compressive strength and the modulus of deformation in several lithologies.

Source: Deere, 1963.

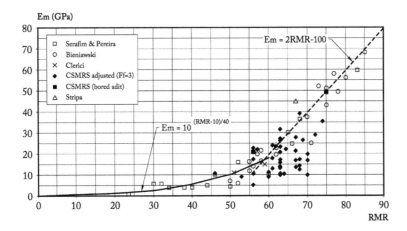

Figure 6.17 Correlation between the RMR and the modulus of deformation.

These findings led Lowson and Bieniawski (2013) to propose the following new correlations:
(a) For RMR < 55

$$E_m = 10^{\frac{RMR-10}{40}}$$ (6.10)

(b) For RMR > 55

$$E_m = 14 + (E_i - 14)\left[1 - \left(\frac{100 - RMR}{44}\right)^{\frac{RMR}{70}}\right]$$ (6.11)

where E_i is the modulus of deformation of the intact rock, obtained in the laboratory, expressed in GPa as well as the E_m.

Figure 6.18 shows the correlations between the RMR and E_m proposed by Lowson and Bieniawski.

When the RMR_{14} is used, Expressions 6.10 and 6.11 become:
(a) For RMR_{14} < 64

$$E_m = 10^{\frac{RMR_{14}-13}{44}}$$ (6.12)

(b) For RMR_{14} > 64

$$E_m = 14 + (E_i - 14)\left[1 - \left(\frac{100 - RMR_{14}}{40}\right)^{\frac{RMR_{14}}{145}}\right]$$ (6.13)

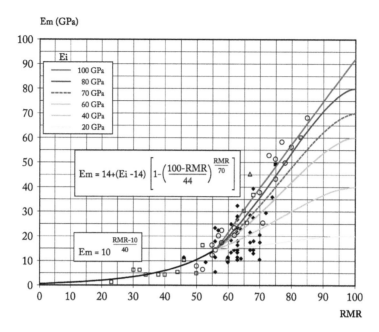

Figure 6.18 Correlations between the RMR and E_m.

Source: Lowson and Bieniawski, 2013.

6.4 ROCK MASS EXCAVABILITY USING TUNNEL BORING MACHINES

Using TBMs in Spain during tunnel construction has achieved major successes on a world scale, such as the three extensions of Madrid Subway, constructed between 1994 and 2007, where steady performances of around 800 m/month have been achieved and the construction of the Guadarrama Tunnel, consisting of two 28 km long tubes, which was completed with average performances of 550 m/month. The most spectacular results were achieved in the construction of La Cabrera Tunnel, on the high-speed Madrid–Levante railway line, and the Sorbas Tunnel, on the high-speed Murcia–Almería railway line because, in both cases, steady performances of more than 1,000 m/month were achieved.

However, in tunnels constructed with TBMs, there have also been some important failures, all of them caused by a poor fitting of the TBMs characteristics to the real rock mass behavior.

At the beginning of the 21st century, two important attempts have been made to develop procedures for classifying RME when using Tunnel Boring Machines. These were the systems featuring the RME index and the Q_{TBM}.

6.4.1 Rock Mass Excavability index

The Rock Mass Excavability index was developed using the data obtained during the construction of the Guadarrama, San Pedro and Abdalajis Tunnels in Spain, as well as those from the Katzenberg Tunnel, Germany, and the Gilge Gibe hydroelectric complex, Ethiopia.

The RME was first presented at the ITA Congress in Korea, by Bieniawski et al. (2006) and its aim is to classify rock masses from the point of view of their excavability, and to forecast the advance rates which can be achieved by the TBMs.

Rating

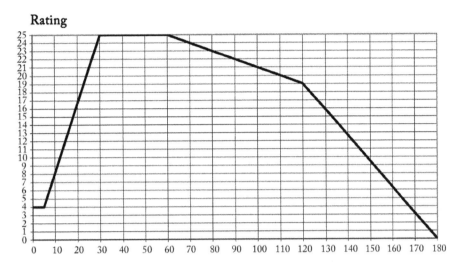

Figure 6.19 Rating of the uniaxial compressive strength of the intact rock.

The RME varies between 0 and 100 points, has a structure similar to the RMR and is calculated by rating the following five parameters:

1. Uniaxial compressive strength of the intact rock, rated between 0 and 25 points
2. Drilling Rate Index, rated between 0 and 15 points
3. Influence of the discontinuities at the tunnel face, rated between 0 and 30 points
4. Yielding of the tunnel face, rated between 0 and 25 points
5. Water inflow at the tunnel face, rated between 0 and 5 points

The uniaxial compressive strength of the ground is rated using the graph shown in Figure 6.19.

It should be kept in mind that the optimal grounds to be excavated by TBMs are those with a uniaxial compressive strength of the intact rock, between 30 and 60 MPa. Below these values problems in the steering of the TBM can take place, which lowers the advance rates. Above 60 MPa, the wear of the cutters begins to affect the performance negatively.

The DRI is a drillability test which has been developed by the University of Trondheim (Norway) and has been used by Bruland (2014) to forecast the advance rates of TBMs. Table 6.17 shows some typical DRI values.

The rating of the DRI for each rock is done using the graph shown in Figure 6.20.

The evaluation of the effect of the discontinuities present in the tunnel face, rated from 0 to 30 rating points, is performed as indicated in Table 6.18.

Figure 6.21 shows the criteria taking into account the tunnel face homogeneity, between 0 and 10 points, which is one of the factors that define the effect of the discontinuities on the TBM performance.

To apply the criteria from Figure 6.21 the following rock strength classification is identified:

Medium strong rocks	$\sigma_{ci} < 45$ Mpa
Strong rocks	45 MPa $< \sigma_{ci} < 120$ Mpa
Very strong rocks	$\sigma_{ci} > 120$ MPa

The evaluation of the number of joints at the tunnel face is done with the graph shown in Figure 6.22.

Table 6.17 DRI typical values

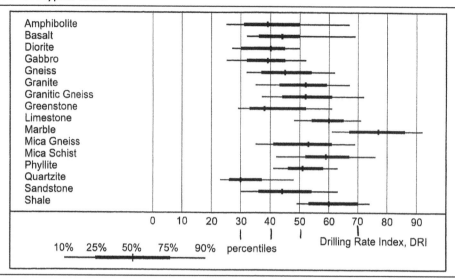

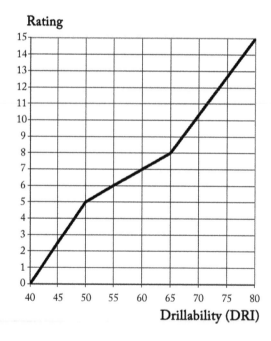

Figure 6.20 Rating of the DRI.

A highly fractured tunnel face does not provide good excavation performance because instability problems can appear which will produce overbreak and slow down the advance rates.

Tunnel faces without cracks are also not a positive condition because the lack of cracks makes it tougher to excavate the ground by a TBM, so the performance will also decrease.

The yielding of the tunnel face is evaluated using the ICE and the graph included in Figure 6.23. The ICE is defined in Section 1.3.3 as Índice de Comportamiento Elástico in Spanish; the index of elastic behavior.

Table 6.18 Rating of the effect of discontinuities at the tunnel face

Homogeneity		No. of joints per meter					Orientation with respect to the tunnel axis		
Homogeneous	Mixed	0–4	4–8	8–15	15–30	>30	Perpendicular	Oblique	Parallel
Rating for 10 the middle value	0	2	7	15	10	0	5	3	0

I.- Tunnel face consisting of hard and very hard rocks

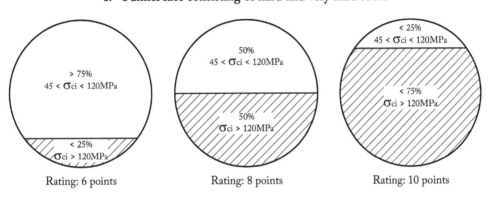

II.- Tunnel face consisting of hard and semi-hard rocks

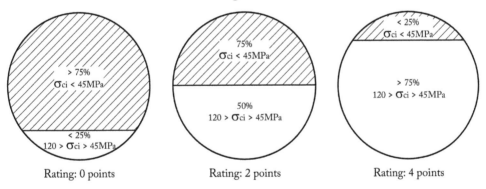

Figure 6.21 Rating of the tunnel face homogeneity.

The highest rating of the ICE from the tunnel face corresponds to a fully elastic state, ICE > 130, while the lowest rating, associated with the possibility of the TBM being immobilized, takes place when ICE = 15.

The water inflow at tunnel face when excavating with TBMs is not a major problem when the flow rates are moderate; but if they exceed 20 l/s, difficulties arise with the loading of the excavated ground and the performance decreases.

The presence of water at the tunnel face is evaluated using the graph in Figure 6.24.

The RME allows evaluating the excavability of rock masses classified into four groups, according to the criteria shown in Table 6.19.

Chapter 7 is dedicated to using the RME for predicting TBM advance rates.

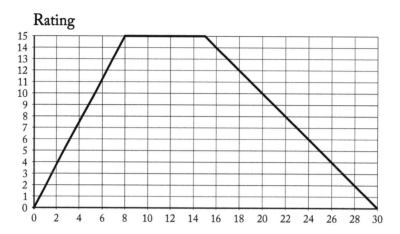

Figure 6.22 Evaluation of the number of joints at the tunnel face.

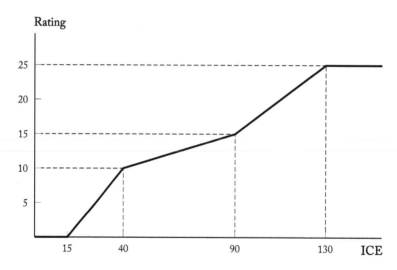

Figure 6.23 Evaluation of the ICE from the tunnel face.

6.4.2 Q_{TBM}

In 1999 Barton proposed a modification of the well-known Q index with the aim of being able to predict TBM performances; the new index was called Q_{TBM} and is calculated by the expression:

$$Q_{TMB} = \frac{RQD_o}{J_n} \cdot \frac{J_r}{J_a} \cdot \frac{J_w}{SRF} \cdot \frac{SIGMA}{F^{10}/20^9} \cdot \frac{20}{CLI} \cdot \frac{q}{20} \cdot \frac{\sigma_\theta}{5} \tag{6.14}$$

where:

J_n, J_r, J_a, J_w, SRF are the original parameters defining the Q.
RQD_o = RQD oriented along the tunnel axis.
F = Average thrust per cutter (Tons).
SIGMA = Estimated rock mass strength (MPa).
CLI = Cutter Life Index.
q = Quartz content in the ground (%).
σ_θ = Bi-axial stress in the tunnel face.

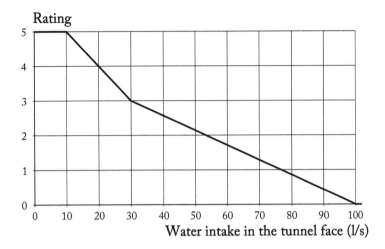

Figure 6.24 Evaluation of the water flow at the tunnel face.

Table 6.19 Ground excavability

Class	RME	Excavability
1	Excellent	80 < RME < 100
2	Very good	60 < RME < 79
3	Good	40 < RME < 59
4	Bad	RME < 40

The CLI was presented in Section 5.6.3 and has the typical values shown in Table 6.20.

Q_{TBM} is related to the Penetration Rate (PR), defined as the advance rate achieved by the TBM in revolving the cutterhead, by the graph shown in Figure 6.25.

As for the RME, the application of the Q_{TBM} to predict TBM advance rates is discussed in Chapter 7.

6.4.3 Specific energy of excavation

The concept of specific energy (SE) is quite simple as it refers to the energy consumed per unit volume excavated and is measured in MJ/m³.

Originally applied by the petroleum industry when drilling large diameter vertical boreholes, it was used by Teale (1965), who established the following relationship:

$$SED = \frac{F}{A} + \frac{2\Pi \cdot \omega \cdot T}{A \cdot u} \qquad (6.15)$$

where:

SED = Specific energy in drilling (MJ/m³)
F = Thrust of the cutterhead on the ground (kN)
A = Borehole area (m²)
ω = Rotation speed of the cutterhead (rev/sec)
T = Torque applied to the cutterhead (kN·m)
u = Penetration rate (m/sec)

Table 6.20 Typical CLI values

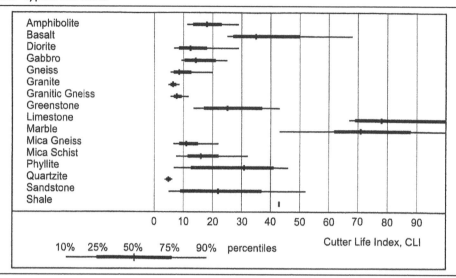

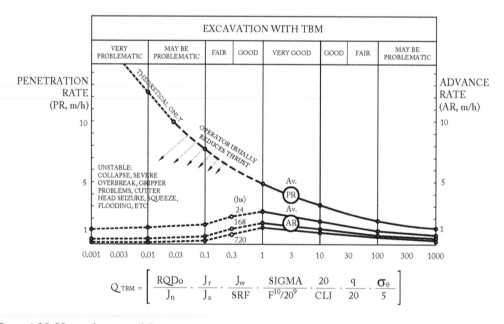

Figure 6.25 PR as a function of Q_{TBM}.

Source: Barton, 2000.

From Equation 6.15 it follows that SED has two components; one is the energy consumed by the thrust of the cutterhead on the ground and the other one is the energy dissipated by the rotation of the cutterhead.

Another interesting contribution of Teale's work in 1965 was that the energy spent on pushing the cutterhead represents only 1% of the SED.

During 2009 to 2011, Geocontrol (2011) developed, with partial funding from the Center for Industrial Technical Development (Centro para el Desarrollo Tecnologico Industrial,

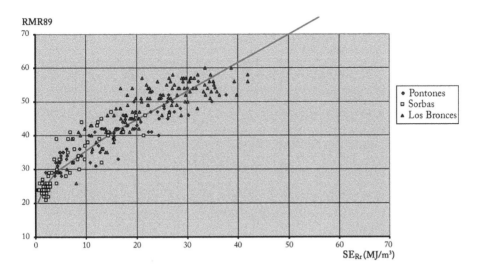

Figure 6.26 Correlation between SE_{Rr} and RMR_{89}.

CDTI in its Spanish acronym), a research project to develop a method to classify the ground excavated by the TBM using the TBM's operating parameters.

By similarity to Equation 6.15, it is considered that the SE of excavation is calculated by the expression:

$$SE = \frac{F}{A} + \frac{2\Pi \cdot T}{A \cdot P} \tag{6.16}$$

where P is the specific penetration achieved by the TBM in a tunnel per each revolution of the cutterhead, $P = u/\omega$.

In short, the SE can be expressed as:

$$SE = SE_T + SE_R \tag{6.17}$$

where SE_T is the specific energy consumed to advance the TBM (about 1% of the total) and SE_R is the specific energy of rotation consumed to rotate the cutterhead, which actually produces the excavation in the rock mass.

The research of Geocontrol (2011) was developed in two phases; an intense acquisition of data during the construction of the Pontones Tunnel (Asturias, Spain) and the development of the system carried out in the Sorbas (Almería, Spain) and South Vein Los Bronces Mine (Santiago de Chile) Tunnels.

Two important conclusions were obtained from the work carried out in the Pontones Tunnel (Asturias, Spain). The first one confirmed that the SE_T was only 1% of the SE, as found by Teale (1965). The second one was that the SE of rotation has three components:

$$SE_R = SE_{Rr} + SE_{Rf} + SE_{R\varepsilon} \tag{6.18}$$

where:

SE_{Rr} = Energy dissipated to press the TBM cutterhead to the tunnel face, which under normal conditions, is between 57 and 77% of the SE, the higher values correspond to the higher RMR ratings of the excavated rock mass

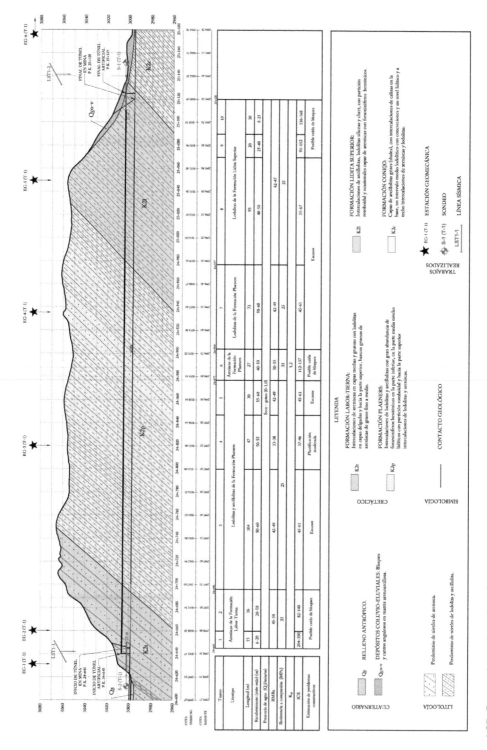

Figure 6.27 Geomechanical profile of a tunnel.

SE_{Rf}= Energy dissipated to rotate the TBM head during the cutting process which, under normal conditions, is between 41 and 28% of the SE

SE_{Re}= Energy dissipated to move the excavated ground through the cutterhead which does not exceed 1% of the SE

Since the SE_{Rr} is the SE component of greater importance, it was decided to correlate it with the RMR_{89} values of the excavated tunnel faces.

To do this, 270 mappings of the RMR_{89} at the tunnel faces of Pontones and Sorbas Tunnels as well as of the South Vein of Los Bronces Mine were used, calculating, in each case, the SE_r value.

Figure 6.26 shows the correlation obtained between the SE_{Rr} and the RMR_{89} which responds to the following equation:

$$RMR_{89} = \frac{5 \cdot Ln(SE_{Rr}/80 - 100)}{Ln(SE_{Rr}/80 - 1)} \qquad (6.19)$$

The correlation coefficient is $r^2 = 0.86$. The correlation found between the SE_{Rr} and the RMR_{89} has an average error of ± 5 points, quite reasonable to estimate automatically the ground quality at the tunnel face.

With this correlation, it is possible to detect, using the TBM's operating parameters, changes in the rock mass quality in real time, which is very useful to predict when the TBM will go through poor quality rock mass, thus preventing a costly situation when the TBM may be immobilized.

6.5 GEOMECHANICAL PROFILE

As a result of the classification of the grounds in which a tunnel is to be built, it is necessary to prepare the geomechanical profile containing the main data that characterize the ground behavior and also an estimation of the possible constructive problems that will be faced during the design phase.

An example of a geomechanical profile of a tunnel is shown in Figure 6.27.

BIBLIOGRAPHY

Alber, M., "Classifying TBM contracts", *Tunnels and Tunnelling*. December 1993, pp. 41–43.

Aydan, O., Dalgic, S., "Prediction of deformation behaviour of Bolu tunnels through squeezing rocks", *Proceedings of the Symposium on Sedimentary Rock Engineering*. Taipei, Taiwan. 1998.

Barton, N., "TBM performance estimation in rock using Q TBM", *Tunnels and Tunnelling*, September 1999, pp. 30–34.

Barton, N., "Rock mass classification for choosing between TBM and drill-and-blast or a hybrid solution". Keynote lecture, *Proceedings* of the *International Conference on Tunnels and Underground Structures*, ICTUS. Singapore, Balkema, Rotterdam, 2000.

Barton, N., Bieniawski, Z.T., "Setting the record straight about RMR and Q", *Tunnels and Tunnelling*. February 2008, pp. 26–29.

Barton, N., Grimstad, E., "The Q-System following twenty years of application in NMT support selection", *Proceedings of the 43rd Geomechanics Colloquy*. Salzburg, Austria. 1994.

Barton, N., Bandis, S., Bakhtar, K., "Strength, deformation and conductivity coupling of rock joints", *International Journal of Rock Mechanics and Mining Sciences & Geomechanics Abstracts*, Vol. 22. No. 3. 1985, pp. 121–140.

Barton, N.R., Lien, R., Lunde, J., "Engineering classification of rock masses for the design of tunnel support", *Rock Mechanics and Rock Engineering*. Vol. 6. No. 4. 1974, pp. 189–236.

Bieniawski, Z.T., "Engineering classification of jointed rock masses", *The Civil Engineer in South Africa*. Vol. 15. 1973, pp. 335–343.

Bieniawski, Z.T., *Engineering Rock Mass Classifications: A Complete Manual*. John Wiley and Sons, New York. 1989.

Bieniawski, Z.T., "Errores en la aplicación de las clasificaciones Geomecánicas y su corrección", *Jornada sobre la Caracterización Geotécnica del Terreno. Adif.* Madrid, Spain. 2011.

Bieniawski, Z.T., Celada, B., Galera, J.M., Alvárez, M., "Rock Mass Excavability (RME) index", *ITA World Tunnel Congress*. Seoul, Korea. 2006.

Bieniawski, Z.T., Celada, B., Rodríguez, A., Tardáguila, I., "Specific energy of excavation in detecting tunnelling conditions ahead of TBMs", *Tunnels & Tunnelling*. February. 2012, pp. 65–68.

Bieniawski, Z.T., Van Heerden, W.L., "The significance of in situ tests on large rock specimens", *International Journal of Rock Mechanics and Mining Sciences*. Vol. 12. No. 4. 1975, pp. 101–113.

Bruland, A., "The NTNU prediction model for TBM performance", *Norwegian Tunnelling Society*. Publication No. 23. Oslo, Norway. 2014.

Celada, B., Tardáguila, I., Rodríguez. A., Varona, P., Bieniawski, Z.T., "Actualización y mejora del RMR", *Ingeopres*, May–June. 2014, pp. 18–22.

Deere, D.U., "Technical description of rock cores for engineering purposes", *Felsmechanic und Ingenieur Geologie*. Vol. 1. 1963, pp. 16–22.

De Olivera, T., "Contribução à classificação geomecânica de maciços rochosos utilizando o Sistema RMR", Doctoral Thesis (TF-07/42). Universidad de São Paulo. Instituo de Geociencias, Brasil. 2007.

Geocontrol, S.A., *Actualización del Índice Rock Mass Rating (RMR) para mejorar sus prestaciones en la caracterización del terreno*. Centro para el Desarrollo Tecnológico Industrial (CDTI). Proyecto: IDI-20120658. Madrid, España. 2012.

Geocontrol, S.A., *Informe final sobre el proyecto de I+D para el desarrollo de un nuevo sistema para predecir los cambios del terreno por delante de las tuneladoras*. Centro para el Desarrollo Tecnológico Industrial (CDTI). Proyecto: IDI-20100374. Madrid. 2011.

Grimstad, E., Barton, N., "Updating of the Q-System", *Proceedings of the International Symposium on Sprayed Concrete*. Norwegian Concrete Association, Oslo, Norway. 1994.

Hoek, E., Brown, E.T., "Practical Estimates of Rock Mass Strength", *International Journal of Rock Mechanics and Mining Sciences*. Vol. 34. No. 8. 1997, pp. 1165–1186.

Hoek, E., Carranza Torres, C., Corkum, B., "Hoek–Brown failure criterion, 2002 edition", *Proceedings of the 5th North American Rock Mechanics Symposium*. Toronto, Canada. 2002.

Hoek, E., Carter, T.G., Diederichs, M.S., "Cuantificación del ábaco del Índice de Resistencia Geológica (GSI)", *Ingeotúneles*, No. 23. Madrid, Spain. 2015.

Hoek, E., Kaiser, P.K., Bawden, W.F., *Support of Underground Excavations in Hard Rock*. Taylor & Francis, London/New York. 1995.

Kalamaras, G., Bieniawski, Z.T., "A rock mass strength concept incorporating the effect of time", *Proc. 8th ISRM Congress*. Tokyo. September 1995, pp. 295–302.

Lauffer, H., "Classification for tunnel construction (in German)", *Beologie und Bauwesen*. Vol. 24. No. 1. 1985, pp. 46–51.

Lowson, A.R., Bieniawski, Z.T., "Critical Assessment of RMR based tunnel design practices: A practical engineer's approach", *RETC*, Washington, DC. 2013.

Lowson, A.R., Bieniawski, Z.T., "Validating the Yudhbir–Bieniawski rock mass strength criterion", *Proc. World Tunnel Congress*, ITA, Bangkok, Thailand. 2012.

Palmström, A., Singh, R., "The deformation modulus of rock masses. Comparisons between in situ tests and indirect estimates", *Tunnelling and Underground Space Technology*. Vol. 15, No. 3. 2013, pp. 115–131.

Palmström, A., Singh, R., "The deformation modulus of rock masses. Comparisons between in situ tests and indirect estimates". *Tunnelling and Underground Space Technology*, Vol.16. No. 3. 2001, pp. 115–131.

Priest, S.D., Brown, E.T., "Probabilistic stability analysis of variable rock slopes", *Transactions of the Institution of Mining and Metallurgy.* Vol. 92. 1983, pp. A1–A12.

Teale, R., "The concept of specific energy in rock drilling", *International Journal of Rock Mechanics and Mining Sciences.* Vol. 2. 1965, pp. 57–73.

Terzaghi, K., "Rock defects and loads on tunnel support", *Rock Tunnelling with Steel Supports.* Eds. R.V. Proctor and T. White. Commercial Shearing Co, Youngstown, OH. 1946.

Chapter 7

Methods for tunnel construction

Benjamín Celada Tamames

We will find the way, and if not, we will create it.

Hannibal, Carthage General (200 BC)

7.1 INTRODUCTION

The word method, derived from the Greek *"métodos"*, is defined as "the way of doing something in an organized way".

Adapting this definition, a construction method of a tunnel can be defined as the set of activities, performed under an integrating concept, which allows constructing a tunnel in a safe and economical way.

As discussed in Chapter 1, the ground is the predominant factor in the stabilization process during tunnel construction. According to this point of view the integrating concept of the activities necessary for tunnel construction must be the stress–strain behavior of the excavated rock mass. Table 7.1 shows the classification, based on this criterion, of the most commonly used methods in tunnel construction.

Table 7.1 Classification of the construction methods

Environment	Degree of mechanization	Behavior of the excavation	Construction methods
Underground	Partial	Elastic	Full face excavation
			Excavations of large sections
		Yielded Grounds	Top heading and bench excavation
			Forepoling placed prior to excavation
			Inverts
		Grounds with intense yielding	Sidewall galleries
			Foundation galleries
			Central temporary wall
			Reinforcement of the tunnel face
	Full	Elastic or with emerging yielding	Open TBMs
			Single Shields
			Double Shields
		Moderate to intense yielding	EPB Shields
			Hydroshields
			Hybrid Shields
Open cut	Partial	Elastic	Cut and cover
		Any	Excavation between diaphragm walls

Selection of the most appropriate tunnel construction method, out of many available, leads to the one which provides higher safety and lower construction costs.

Identifying that most appropriate method to construct a tunnel is a complex task, involving assessing the potential geological risks and other factors such as the environmental restrictions, the limitation of the impacts on the surface caused by the tunnel construction, the construction deadline, the accesses to the working place and to the equipment and the experience of the construction company.

This task is part of the structural design phase in the DEA, presented in Section 2.4.4, which begins after determining the tunnel geomechanical profile.

This chapter contains the concepts on which most construction methods are based, establishes their field of application and estimates the advance rates obtainable with them.

7.2 UNDERGROUND CONSTRUCTION METHODS WITH PARTIAL MECHANIZATION

The traditional method for underground tunnel construction is based on the repetition of elementary operations: excavation, load of the excavated grounds and execution of the tunnel support, which constitute the work cycle.

For this reason, these methods have been defined by the International Tunneling Association as sequential excavation methods, although given that the shotcrete is the dominant element of the support, they are also known as sprayed concrete lining (SCL).

7.2.1 Excavations with elastic behavior

According to what was stated in Section 1.3.3, an excavation is considered to have a fully elastic behavior when the index of elastic behavior ICE > 130.

Under these conditions, 66% of the displacement that the ground must undergo until stabilization will have already occurred at the tunnel face and the remaining 34% will occur only a few meters away from the tunnel face.

Ground displacements, referred to the origin, will be of a few mm and the convergence will be almost zero. In these cases, the excavation will be self-stable for a very long period of time and this state will only be affected by hypothetical external actions which reduce the ground quality.

When the excavation behaves elastically, the tunnel support will have a minimum impact on the work cycle that will mainly involve the excavation and extraction of the ground from the tunnel face.

In the following sections the excavation methods at full face are presented.

7.2.1.1 Full section excavation

Until a few decades ago the excavation at full section was only used in the construction of tunnels with widths smaller than 8 m. For larger widths, the large volume of excavated ground led, with the available equipment, to considerably longer work cycles.

Nowadays, with the use of heavier machinery, this method can be applied in the construction of tunnels with a width up to 12 m, which normally means an excavation height of about 9 m.

Figure 7.1 shows a standard tunnel section, 12 m wide, excavated at full section and Photograph 7.1 shows a snapshot of the loading process of the excavated ground.

In full face excavation, it is critical that there should not be rock falls at the tunnel face, which is consistent with the concept of elastic behavior of the excavation.

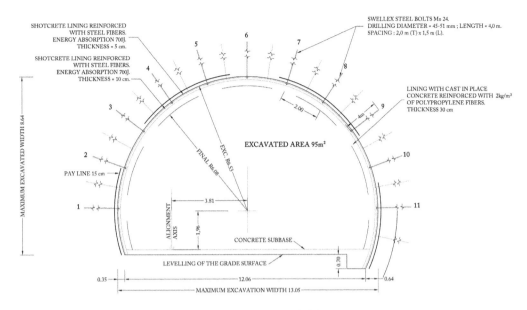

Figure 7.1 Standard tunnel section, 12 m wide excavated at full face.

Photograph 7.1 Ground loading with full face excavation.

Therefore, to design a tunnel to be excavated full face, it is essential to analyze the stability of the rock blocks that could appear at the tunnel front; for that purpose, the methodology presented in Chapter 10 should be followed.

The lower limit to apply the full face method corresponds to an excavation with an ICE = 70, which is the threshold of a moderate yielding.

Concerning the advance rates that can be reached when excavating at full face, Barton (2013) states that Norwegian constructors have achieved, during a 24-hour 7-day week, advances ranging from 150 to 176 m per week, the tunnel sections being between 35 and 45 m².

In the San Pedro Tunnel (Spain) with a width of 10 m, steady average rates of advance of 10 m/day were achieved, which is about 300 m/month; the cross section of the tunnel was 78 m².

The above figures confirm that, in the most favorable situation and applying the full face method, advance rates between 300 and 400 m/month can be reached, depending on the tunnel cross sections.

Under less favorable conditions, for tunnels of 10 to 12 m wide, average advance rates can be estimated at between 120 and 180 m/month.

7.2.1.2 Underground excavations with large sections

When a tunnel section to be excavated exceeds 120 m² it is difficult to excavate at full face not due to any kind of stress–strain constraints but due to operating reasons.

The solution in these cases consists in excavating in phases, fulfilling the operational constraints, because the elastic behavior of the excavation does not impose any constraint. The clearest example is the excavation of the Gjøvik cavern (Norway) which is presented in Section 1.2.3.3.

In this case, the section to be excavated was 1,525 m² and was divided into 14 phases, as shown in Figure 7.2, establishing the criterion that excavations progress downwards.

In excavations whose sections are close to 200 m² and have elastic behavior, the division of the section can be limited to only three phases, as was done in the construction of Manquehue I Tunnel (Chile) and in Roquetes Station (Barcelona, Spain).

Roquetes Station, which is part of Line 3 on the Barcelona Subway, has an excavated area of 194 m², with a width of 14.2 m and a height of 13.2 m.

Figure 7.3 shows the chosen excavation stages consisting of a top heading excavation, with a height of 8.0 m and an area of 103 m² and two phases to excavate the bench, with sections of 51 m² and 40 m² respectively.

Photograph 7.2 shows a view of the bench excavation at this station. The construction sequence adopted in the Manquehue I Tunnel, whose characteristics are presented in Section 1.2.3.1, is a clear example that there is no limitation posed by the ground when the excavation behaves elastically.

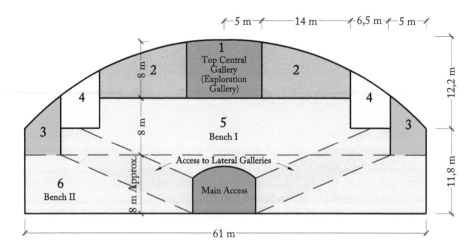

Figure 7.2 Division of the Gjøvik cavern section in Norway.

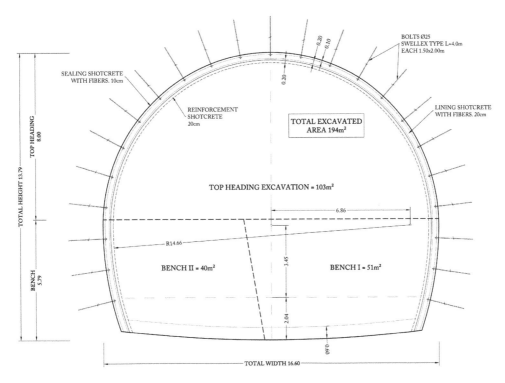

Figure 7.3 Section for the construction of Roquetes Station, Line 3 from the Barcelona Subway (Spain).

Photograph 7.2 Bench excavation of Roquetes Station.

As shown in Figure 7.4, the section of Manquehue I Tunnel was excavated in three stages, one for the top heading excavation and another two to excavate the bench.

The top heading excavation, which was 103 m² with a height of 23 m, was carried out without any difficulties, as shown in Photograph 7.3.

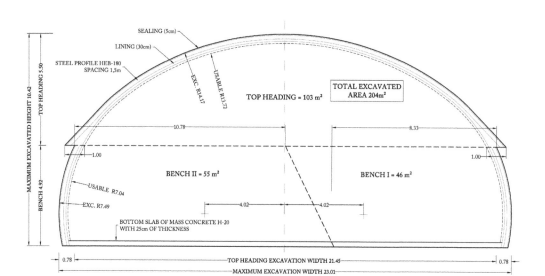

Figure 7.4 Construction stages in Manquehue I Tunnel.

Photograph 7.3 Top heading excavation of Manquehue I Tunnel.

7.2.2 Excavations in yielded grounds

Ground yielding in an excavation can start for ICE values below 130 points and becomes very intense for ICE values smaller than 39 points.

In these cases, the division of the section to be excavated cannot be only done according to some geometric criteria since, as the ICE values decrease, the yielding becomes more intense and the control of the tunnel face stability becomes more complicated.

This means that for lower ICE values it will be necessary to use heavier supports and decrease the volume excavated in each blasting which, initially, is achieved by decreasing the advance length.

Photograph 7.4 Top heading construction phase in a tunnel.

The effect of using heavier supports and reducing the advance length will lead to a decrease in the advance rates obtained, which can never be equal to those obtained when excavating in grounds with elastic behavior.

In yielded grounds, the most common construction method is based on the division of the section to be excavated into two stages: top heading and bench, although, as the yielding increases, improvements to the excavation stability are added such as the "elephant feet", a self-supported vault, or forepoling placed prior to the excavation and inverts.

7.2.2.1 Classical method of top heading and benching

The traditional method for tunnel constructions, with widths between 10 and 15 m, is the top heading and benching method, which divides the section to be excavated into two phases: the top heading, that usually is 6 m high to allow the full functionality of the regular machinery, and the bench, which completes the excavation section.

Photograph 7.4 shows the top heading construction phase in a tunnel of 14 m in width.

When the support is made with bolts and shotcrete, the bench phase can be excavated at once, but when steel arches are used in soils, it is necessary to construct the bench in two stages, with a lag of about 10 m between them.

The limit between the use of bolts or steel arches is close to $ICE = 40$.

When steel arches are used as support elements, it is advisable to support the steel arches placed at the top heading phase by widening bases, known as "elephant feet".

Figure 7.5 shows the arrangement of the elephant feet, in a tunnel 13.7 m wide, at the top heading excavation stage. The effect achieved with these elephant feet is to transfer the stress concentration, which appears in the steel arch supports, far away from the tunnel walls that must be excavated in the bench excavation phase; this improves the foundation conditions significantly.

7.2.2.2 Forepoling placed prior to the excavation

As ground yielding around the excavation increases, i.e., when ICE values are close to 15 points, it is necessary to introduce elements of support to improve the stability conditions in the top heading and bench method, such as forepoling placed prior to the excavation.

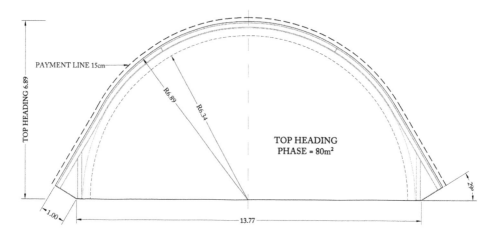

Figure 7.5 "Elephant feet" at the sides of top heading excavation of a 14 m wide tunnel.

This forepoling, also called "umbrellas" consists of micropiles drilled along the perimeter of the tunnel vault, with a spacing of between 9 and 15 m.

Micropiles are usually composed of steel tubes, with a diameter between 80 and 100 mm, which are placed in holes, drilled with an angle of 5° with respect to the tunnel axis, and are attached to the ground through cement injection. Photograph 7.5 shows an excavation under a micropile forepoling.

Successive forepoling installations usually have an overlap of 3 m. In grounds of poor quality, characterized by RMR values close to 20 points, it is recommended to place forepoling with self-drilling anchors, which are later grouted with cement; it is advisable to have an anchor head diameter of 70 mm.

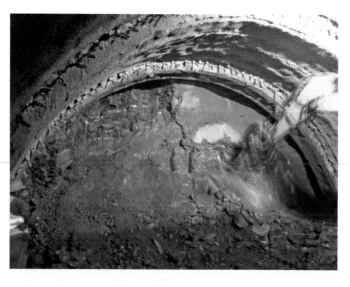

Photograph 7.5 Excavation under micropile forepoling.

7.2.2.3 Inverts

The construction of an invert in a tunnel results in an almost circular structural section, which has positive effects in countering significant ground yielding.

However, invert construction is an expensive and slow operation, and it is therefore advisable to limit its use to strictly necessary cases. To dimension the invert it is essential to model the stress–strain behavior of the excavation, following the methodology presented in Chapter 10.

Since inverts are used in sections with more unfavorable stress–strain behavior, this element of reinforcement should be combined with those already described, namely: elephant feet and forepoling placed prior to the excavation.

Figure 7.6 shows a tunnel section designed with an invert.

In case of unanticipated problems due to the presence of water in the ground and ICE values below 70 points, it is recommended to design the tunnel with an invert.

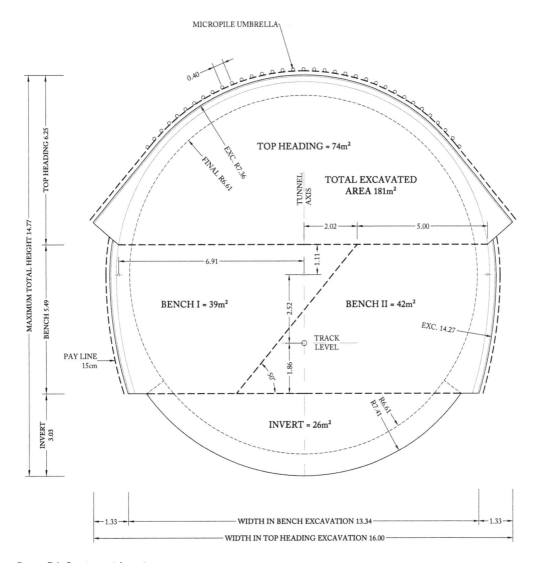

Figure 7.6 Section with an invert.

7.2.2.4 Self-supported vault

In shallow excavations, such as those with an overburden less than twice the excavation width, the arch effect may not be created.

Under these conditions, considering that shallow grounds are usually of poor quality, it is very likely that the tunnel support and lining will have to withstand the weight of the whole overburden over the tunnel vault.

To address this hazard, the concept of the self-supported vault method has been developed, which is a variant of the top heading and bench method, supplemented with elephant feet and forepoling placed prior to the excavation to support the vault structure.

The self-supported vault method is based on placing the lining as close as possible to the tunnel face, with the aim of integrating it with the support, so that they could withstand the weight of the whole overburden, as the arch effect is not possible.

Figure 7.7 shows a scheme of the application of the self-supported vault method to a 17 m wide tunnel, with an excavated section of 157 m².

With the self-supported vault method large section tunnels can be constructed; dividing it only into three phases where heavy machinery can be used in the excavation which provides high advance rates.

By using a variant of this method, 11 stations in Line 6 of the Santiago de Chile Subway have been designed and built, whose widths ranged from 15 to 20 m and the overburden from 12 to 20 m. This means that the ratio between the overburden thickness and the excavation width was between 1.3 and 0.75.

The support and lining thickness, considered as a whole, ranged between 50 and 70 cm, depending on the station width. The variant introduced in the self-supported vault method used in the stations of Line 6 in the Santiago de Chile Subway, consisted of replacing the forepoling by an intermediate temporary wall, which in the most complex situations was supported by an invert, also temporary, as shown in Figure 7.8.

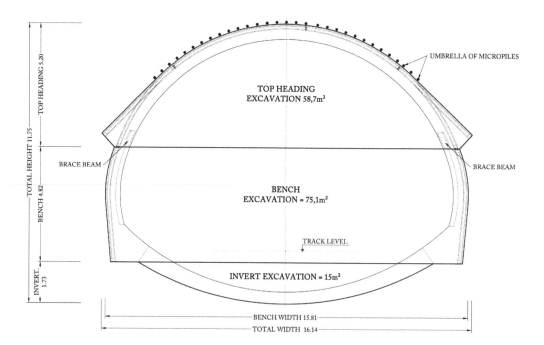

Figure 7.7 17 m wide tunnel designed with the self-supported vault method.

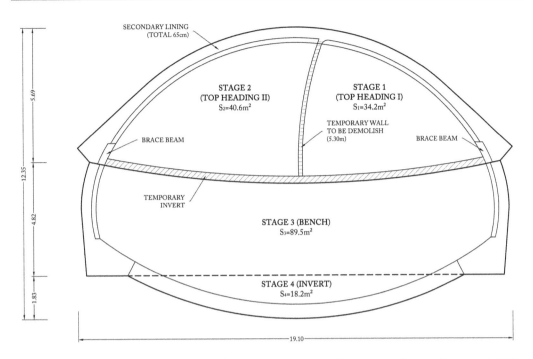

Figure 7.8 Self-supported vault method used in the construction of Los Leones Station, Santiago de Chile Subway Line 6.

Photograph 7.6 Vault excavation in two phases at Los Leones Station, Line 6, Santiago de Chile Subway.

Photograph 7.6 shows the excavation in two phases of the top heading in Los Leones Station.

It is important to note that the two longitudinal brace beams, built at the vault base, shown in detail in Figure 7.9, allow the safe excavation of the bench in a single phase, as illustrated in Photograph 7.7.

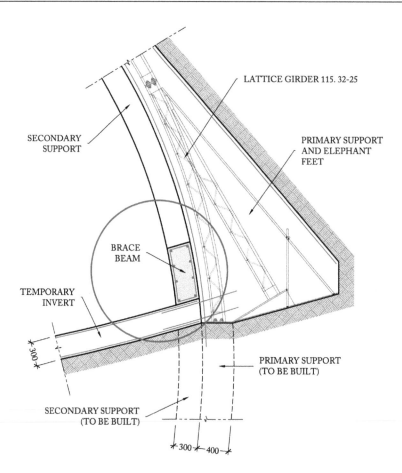

Figure 7.9 Detail of the brace beam.

If the ground where the elephant feet setting has not enough bearing capacity, it is necessary to underpin the vault with micropiles, which must be embedded in the brace beams, as shown in Figure 7.10.

7.2.3 Tunnel construction in grounds with intense yielding

In excavations with ICE values smaller than 40 points, the yielding of the ground around them will be intense, and, for excavation sections around 50 m², which are common in the construction methods presented in the previous section, tunnel stability will probably be affected.

To deal with extreme yielding, that is when reaching the ground bearing capacity, there are three possible strategies that can be followed:

I. In general, the number of excavation phases can be increased to reduce the excavation widths; as, according to Terzaghi and Protodyakonov's theories presented in Chapter 2, the loads on the support elements are proportional to the excavation width.

II. In shallow tunnels, the axial stresses acting on the support and lining will be small, due to the small overburden thickness. The bending moments can be minimized with a careful definition of the support and the lining geometry. In this context, it will be

Photograph 7.7 Bench excavation in one phase at Ñuble Station, Line 6 Santiago de Chile Subway.

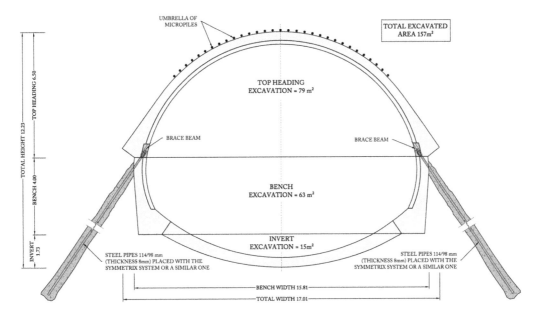

Figure 7.10 Self-supported vault method used in the construction of Los Leones Station.

economically feasible to design the support and the lining strong enough to be able to achieve the excavation balance with the smallest possible displacements. This involves constructing the tunnel including an invert as close as possible to the tunnel face.

III. In deep tunnels, the situation will become more complex, as axial stresses will be high and the excavation stabilization will require the design of very strong supports, which may be uneconomic.

To minimize the required strength of these supports, it is possible to use the highly ductile supports described in Chapter 9 and to design a section with over-excavation so that the over-excavation compensates the ground displacements before stabilization.

In fact, when a tunnel has to be designed in a highly yielded ground, a solution tailored to the characteristics of the ground, the tunnel geometry and the excavation depth must be found.

In all these cases it is essential to study the stress–strain interaction between all the excavation phases, following the DEA methodology, to ensure the stability of each phase and of the whole tunnel. This study should be done by solving three-dimensional geomechanical models, applying the methodology presented in Chapter 10.

7.2.3.1 Vertical and horizontal division of the section to be excavated

Among the construction methods that deal with intense ground yielding through a drastic reduction of the section to be excavated are the sidewall galleries, central temporary wall and foundation galleries methods.

7.2.3.1.1 Sidewall galleries method

The excavation method using sidewall galleries is based on the construction of two galleries about 4 m wide, which share a wall with the tunnel to be constructed. Once the two galleries have been excavated, the upper level of the tunnel is excavated. Then, the intermediate level, equivalent to the conventional bench, is excavated and finally an invert is constructed.

This is the most extreme method with regard to the excavation stages, because as shown in Figure 7.11 the excavation is divided into seven phases.

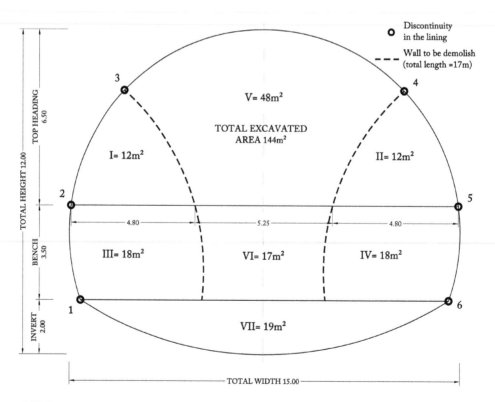

Figure 7.11 Stages of excavation using the lateral galleries method.

Photograph 7.8 Construction of a tunnel with a section of 150 m², excavated using sidewall galleries method.

This method can be considered as a variant of the top heading and bench excavation method, using an invert, in which both the top heading and the bench excavation have been divided into three phases each.

As shown in Figure 7.11 the excavated sections range between 12 and 48 m², although the average value is 20.5 m². Thanks to the small size of these sections the effects of intense yielding can be easily controlled, but there may be significant interference between the different working faces and the advance rates achieved are quite low.

Photograph 7.8 shows a tunnel with a section of 150 m², excavated using sidewall galleries.

In addition to the significant interference between the different working faces, the sidewall gallery method has three important drawbacks, derived from the joints created in the support, the temporary walls to be demolished and the invert excavation under the foundation galleries.

In Figure 7.11 it can be seen that, with this method, at least six joints are created in the support which are points of weakness, which must be carefully controlled during construction. It can also be observed that it is necessary to demolish the two temporary walls, whose total length is about 16 m.

Finally, it should be noted that with this method, the invert construction must be excavated under the level of support of the temporary walls, which may affect their stability.

7.2.3.1.2 Foundation galleries method

The foundation galleries method consists of the construction of two sidewall galleries placed at the lower part of the tunnel section, that serve as foundations for the tunnel vault and help to minimize its settlement. A scheme of this method is shown in Figure 7.12, applied to a tunnel of 19 m in width and an excavated section of 186 m².

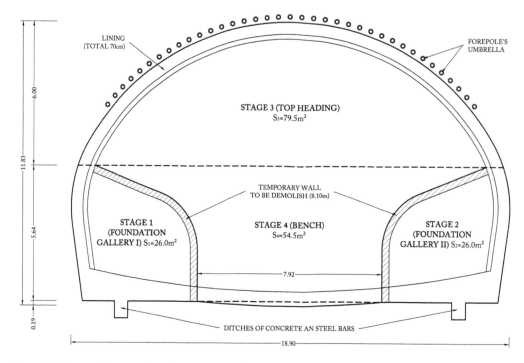

Figure 7.12 Stages of a tunnel section constructed using the foundation galleries method.

With the foundation galleries method, the division of the section is much less intense than with the sidewall galleries method, because, although the foundation galleries cross section is about 26 m², the remaining section is excavated in two phases with sections greater than 54 m².

In the foundation galleries, two walls must be built to support the tunnel vault, so the thickness of the reinforced concrete at the bottom slab ranges from 1.4 to 2.2 m.

Figure 7.13 shows the detail of the reinforced concrete at the bottom slab of the foundation gallery.

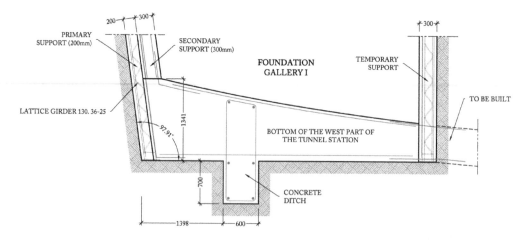

Figure 7.13 Detail of the bottom slab at the foundation gallery.

In this figure, a detail of a longitudinal bottom ditch can also be observed, filled with reinforced concrete, with the aim of minimizing the horizontal displacements of the foundation galleries, after the excavation of the vault.

As this method is used in grounds with high yielding, it is advisable to use forepoling umbrellas to control over-excavations at the tunnel vault.

If surface subsidence is to be minimized, the tunnel face can be reinforced using fiberglass bolts, anchored with cement grout and with lengths between 12 and 20 m. In these cases, the spacing between two sets of consecutive bolts is usually between 6 and 10 m.

This method was successfully implemented in the Teixonera Station construction, Line 5 of Barcelona Subway (Spain), which was 20 m wide and was excavated in graphite slates with a RMR of 14 points. As the overburden was about 80 m, the ICE value of the excavation was smaller than ten points.

Figure 7.14 shows the cross section of Teixonera Station and Photograph 7.9 shows a snapshot taken during its construction.

This method has been also used for the west part construction of Los Leones Station, Line 6 of the Santiago de Chile Subway, with the aim of reducing the surface subsidence (Celada et al., 2016).

Figure 7.15 shows the excavation sequence followed in the construction of Los Leones Station with the foundation galleries method.

The dimensions of the foundation galleries were carefully selected, to allow the movement of the conventional machinery, as shown in Photograph 7.10.

The demolition of the temporary walls in Los Leones Station was done without any problems, as shown in Photograph 7.11.

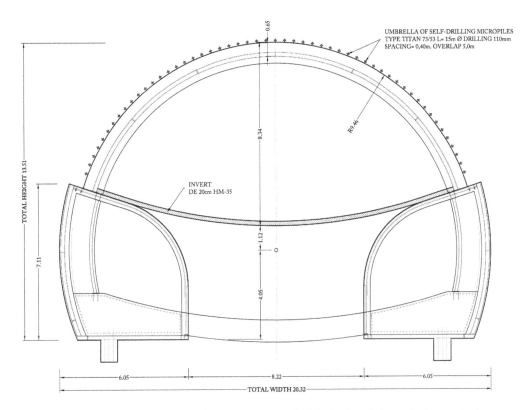

Figure 7.14 Cross section of Teixonera Station constructed with the foundation galleries method.

Photograph 7.9 Construction of Teixonera Station using the foundation galleries method.

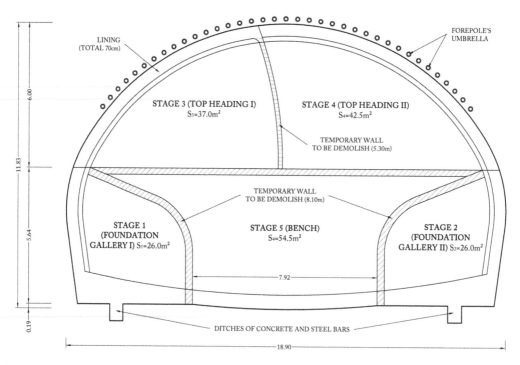

Figure 7.15 Excavation sequence at the west part of Los Leones Station constructed using the foundation galleries method.

The comparison between the results obtained with the self-supported vault method and the foundation galleries method, both used in the construction of Los Leones Station, shows that with the latter the subsidence was reduced by 32.3%, but the advance rates decreased by 28%, resulting in a cost increase of 38%.

Photograph 7.10 Construction of the foundation galleries at Los Leones Station. Line 6 of the Santiago de Chile Subway.

Photograph 7.11 Demolition of the temporary walls in Los Leones Station.

The foundation galleries method is especially recommended when it is desired to reduce the surface subsidence caused by the construction of underground works in areas with intense yielding.

In addition, this method brings a significant improvement in safety compared to the side-wall galleries method. This is because in the foundation galleries method, the final tunnel invert is embedded in the foundation galleries, which makes it unnecessary to excavate under them in contrast to the sidewall galleries method.

7.2.3.1.3 Central temporary wall method

When tunnel width is smaller than 14 m, the sidewall and the foundation galleries methods are difficult to perform due to the narrow width of the ground mass between both galleries.

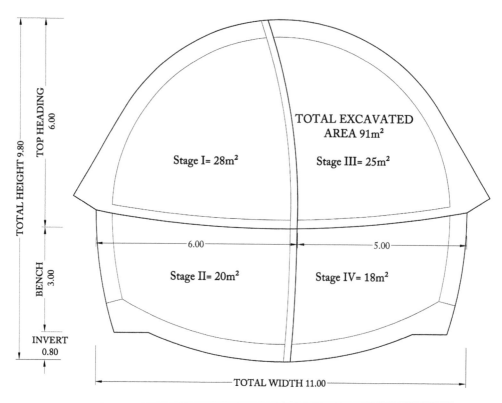

Figure 7.16 Stages of the excavation using the central temporary wall method.

This reduced width generates problems with stabilizing both galleries due to the intense ground yielding.

In these cases, one may use the central temporary wall method, which allows dividing the tunnel excavation into four phases, as illustrated in Figure 7.16.

Figure 7.16 shows that in a tunnel 11 m wide, the sections of the two phases into which the top heading excavation is divided are 28 and 25 m² and the widths are 6 and 5 m; this enables an easy excavation of both phases using conventional machinery.

Bench excavation is also divided into two phases, but smaller, with sections of 20 and 18 m².

The concept underlying the central temporal wall method is to construct two adjacent tunnels whose width is almost half of the final tunnel width.

7.2.3.2 Reinforcement of the tunnel face

In 2006, Pietro Lunardi published the book *Progetto e construzione di gallerie-analisi delle deformazioni controllate nelle rocce e nei suoli* which was published two years later in English under the title *Design and Construction of Tunnels. Analysis of Controlled Deformation in Rock and Soils* (ADECO-RS).

In the ADECO-RS method, Lunardi gives a predominant role to ground reinforcement at the tunnel face as it is the origin of a significant amount of the ground displacements generated during tunnel construction, as illustrated in Figure 7.17.

In this figure, it can be seen that at a distance ahead of the tunnel face, the ground is in a triaxial stress state (σ_1, σ_2 and σ_3).

However, at the tunnel face, no support elements are usually placed and therefore $\sigma_3 = 0$.

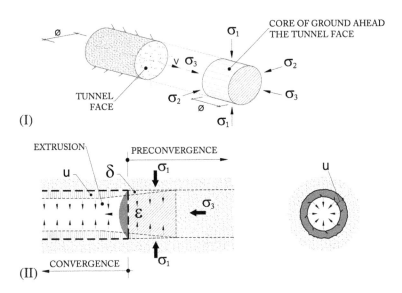

Figure 7.17 Stresses and displacements close to the tunnel face. I. Stress state ahead of the tunnel face. II. Ground displacement around the tunnel.

Source: Lunardi, 2006.

Since σ_3 at the tunnel face is reduced to zero, the new stress state created around the tunnel face leads to deformations on the tunnel face and on the lining, leading to three types of displacements.

1. **Pre-convergence**, takes place ahead of the tunnel face.
2. **Convergence** leads to a decrease of the excavated tunnel section.
3. **Extrusion** causes the displacement of the tunnel face toward the gap created by the tunnel construction.

Clearly, if the extrusion at the tunnel face exceeds the limit value that can be withstood by the ground at the tunnel face, it will collapse and seriously compromise the tunnel stability.

To control the displacements at the tunnel face in the ADECO-RS method, the ground at the tunnel face is reinforced with bolts of substantial length, normally around 20 m, with a longitudinal spacing of 10 m.

In order not to hamper the subsequent excavation of the ground at the tunnel face by the bolts, these are made of a polyester resin reinforced with glass fibers.

The rest of the activities in the ADECO-RS system are the same as those for the tunnel excavation at full face, including constructing an invert as close as possible to the tunnel face.

Photograph 7.12 shows the excavation of a tunnel face reinforced with fiberglass bolts. Figure 7.18 shows the phases that are part of the ADECO-RS construction method.

The ADECO-RS method uses conventional equipment, with the exception of the drilling machine which makes holes inside which bolts are placed, as shown in Photograph 7.13.

The ADECO-RS method is a suitable solution for the construction of tunnels with sections smaller than 120 m², in areas with high yielding.

Photograph 7.12 Excavation at full section in Vasto Tunnel (Italy) constructed using the ADECO-RS method in 1991.

Source: Lunardi, 2006.

7.3 UNDERGROUND CONSTRUCTION WITH FULL MECHANIZATION

TBMs allow mechanizing all the operations necessary in tunnel construction and for this reason they are considered a special option for tunnel design and construction.

The emergence of the first TBMs took place at the end of the 19th century, but they did not begin to be successfully used until the second half of the 20th century. Photograph 7.14 reproduces the TBM manufactured by John Fowler and Co. in Leeds, England, in 1881 designed to work in the tunnel under the English Channel, between France and England, Robbins (1976).

The first TBM capable of constructing a tunnel was manufactured by The Robbins Company, in Seattle, Washington, in 1997.

Since then, TBMs have had an impressive evolution that has led to the construction of tunnels in a wide variety of geological conditions, with large dimensions and, what is more remarkable, reaching advance rates impossible to attain with partially mechanized methods. One of the most illustrative example of the wide range of possibilities offered by the TBMs were the extensions in Madrid Underground between 1995 and 2003, Melis (2003).

Photograph 7.15 reproduces the TBM manufactured by Herrenknecht (Germany), which at the beginning of the 21st century was used to construct the North Tunnel of the Calle 30 South Bypass in Madrid (Spain). This TBM had a diameter of 15.15 m which, at that time, was the world record for its diameter (Melis et al., 2006).

The advantage of using TBMs lies in better work safety conditions by comparison with partially mechanized methods, as the operators who build the tunnel are usually protected from possible tunnel face instabilities.

In addition, impressive advance rates can be achieved with TBMs, which are unreachable with partially mechanized methods.

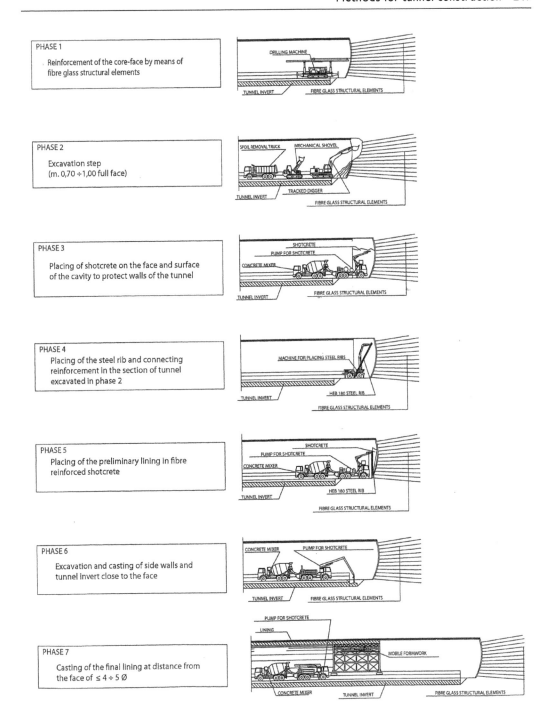

Figure 7.18 Construction phases in the ADECO-RS method.

Source: Lunardi, 2006.

Photograph 7.13 Drilling with hydraulic feed to drill holes of 20 m in length.

Photograph 7.14 TBM manufactured in 1881 by John Fowler & Co. (Leeds U.K.)
Source: Robbins, 1976.

According to Workman et al. (2016), during the construction of the Deep Rock Tunnel Connector, located in Indianapolis (USA), the following world records have been reached for tunnels with excavation diameters from 6 to 7 m:

- The best daily advance rate: 124.9 m
- The best weekly advance rate: 5151 m
- The best monthly advance rate: 1,754.0 m

In all highly mechanized processes, significant advance rates are achieved, provided that the work is carried out in conditions as assumed when the machine was designed; however, when the site conditions forecasts prove to be different, a TBM quickly becomes inefficient and, at the limit, it stops working.

Photograph 7.15 TBM of 15.15 m in diameter, manufactured by Herrenknecht, which constructed the North Tunnel of the Calle 30 South Bypass in Madrid (Spain).

Photograph 7.16 TBM trapped into ground.

This scenario applies to TBMs excavating in unexpected, different conditions than the one forecasted, particularly in the case of tunnels constructed in grounds with high yielding.

In these cases, the risk which can occur is that the ground surrounding the TBM could be excessively displaced and come into contact with the TBM shell and end up blocking the machine because it does not have the necessary force to overcome the friction between the ground and the shell, as illustrated in Photograph 7.16.

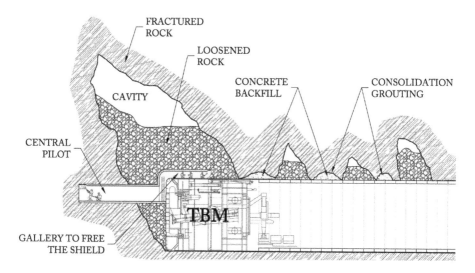

Figure 7.19 Detail of the pilot gallery excavated manually ahead one of the TBMs that worked on Pinglin Tunnel construction.

Source: Shen et al., 1999.

When the TBM gets stuck in the ground there is no other solution than to open a window in the TBM shell and, from it, to manually excavate a small gallery to reach the entrapment area and to remove, manually, the grounds in contact with the TBM. This process takes several weeks to release the TBM.

Probably the case that best illustrates this situation in the operation of the TBMs is the construction of the Pinglin Tunnels, in Taiwan, where two TBMs were immobilized several times in two years.

Figure 7.19, Shen et al. (1999), illustrates the process to release the TBMs from the Pinglin Tunnel, which consisted of manually digging a pilot gallery ahead of the TBM.

The following sections present the basic criteria to keep in mind to successfully use TBMs and describe the characteristics of the most important types of TBMs currently manufactured.

7.3.1 Basic issues with the use of TBMs

The main idea is that each TBM must be manufactured or adjusted according to the characteristics of the tunnel to be constructed, as Richard J. Robbins wrote in 1976:

> The idea of a universal mechanized tunneling system is not really convenient, as each tunnel has special requirements as a result of both the geological conditions and the structural design. Tunnel engineers must continue developing new concepts and combinations, which will turn the ideas that today seem definitive into obsolete ones.

Accordingly, here are the three basic concepts on the use of TBMs.

7.3.1.1 Ground investigation in tunnels constructed by TBMs

TBMs are machines in which it is very difficult to make changes once the tunnel excavation has begun, due to lack of space.

Table 7.2 Duration of the activities necessary to buy and assemble a TBM

Activity	Duration (Months)
Selection of the TBM and purchase order	3
Manufacturing of the TBM	12
Disassembly of the TBM and transport to the site	3
Assembly on site	3
Total	**21**

For the same reason, access to the tunnel face, to solve possible instability problems, is very constrained.

Accordingly, if, after the beginning of the construction, major modifications in a TBM are needed, these may require several weeks or months of work.

To avoid these problems, the best solution is to propose an effective ground investigation campaign that enables knowing, as accurately as possible, the characteristics of the tunnels sections with lower rock mass quality, so that the TBM could be prepared to face them.

7.3.1.2 Time required by a TBM to start working

Table 7.2 shows the average duration of the activities necessary for a TBM, with a diameter between 9 and 11 m, to begin constructing a tunnel. This requires 21 months.

In addition, it must be noted that a TBM requires a learning period of about two to four months for the crew to get used to TBM operation and so that reasonable advance rates can be achieved.

If the use of a reconditioned TBM is decided upon, the manufacturing time can be reduced to about four months.

7.3.1.3 Cost of the tunnels constructed with a TBM

TBMs have a cost of several tens of millions of dollars, which must be amortized in at least 75% of the tunnel construction, making a significant impact on the construction cost.

The tunnel section made with a TBM is always circular which requires the excavation and filling of the lower part of the section, except in tunnels with a large section in which this space can be used. Figure 7.20 shows the use of the excavated space under the pavement of Calle 30 South Bypass, built with the TBM that was shown in Photograph 7.15.

TBMs are able to reduce the costs in tunnel construction, when compared to partially mechanized methods, due to the shorter execution times in urban tunnels because it is not necessary to build intermediate accesses in long tunnels outside the cities.

Considering the above circumstances, there is a minimum length for each tunnel in which the construction costs using TBM are equal to those with traditional methods. Table 7.3 shows the tunnel length ranges at which the construction costs using TBMs or conventional methods break even, depending on the tunnel diameter.

As a general rule, it can be stated that it is necessary to have a tunnel longer than 3 km in order to achieve lower construction costs with TBM than with a partially mechanized method.

In tunnels of large diameter, which allow for accommodating overlapped roads, TBMs start being profitable in tunnels longer than 1.5 km.

For tunnels with diameters between 6 and 8 m, construction costs with TBMs begin to be more attractive in tunnels longer than 2.5 km.

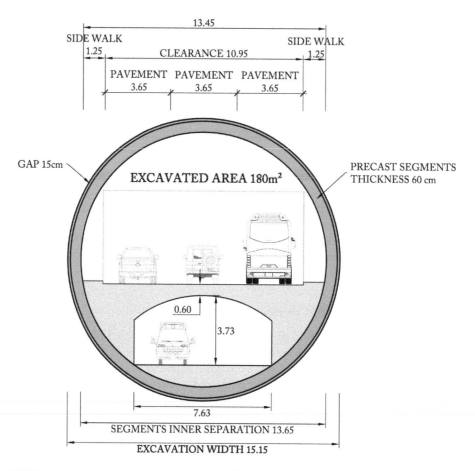

Figure 7.20 Evacuation gallery under the road of the Calle 30 South Bypass in Madrid.

7.3.2 TBMs for stable tunnel faces

In a tunnel, the tunnel face will not have stability problems, from a stress–strain point of view, when the excavated ground has an ICE > 130.

In these cases, the only task to be faced by the TBMs is to excavate the ground, because the stability of the excavation will be assured by having sufficient stand-up time.

The types of TBMs suitable to construct tunnels with stable tunnel faces are open TBMs, single shields and double shields, whose most relevant characteristics are presented in the following sections.

7.3.2.1 Open TBMs

Open TBMs are the oldest known such machines and feature a rotating cutterhead, equipped with disk cutters, which is pushed against the tunnel face by hydraulic cylinders, which also serve to move the TBM. The thrust force in the cutterhead is obtained by supporting the TBM against the ground by means of grippers driven by hydraulic cylinders.

These TBMs are called open TBMs because the excavated ground can be seen from behind the cutterhead, although they are also called TBM with gripper typology or, simply,

Photograph 7.17 Cutterhead of an open TBM.

Table 7.3 Length at which the construction costs using TBMs and conventional methods break even

Tunnel diameter (m)	Length at which the costs are balanced (km)
4–6	2.5–3.5
6–8	2.0–3.0
8–11	3.0–4.0
11–14	2.5–3.5
> 14	1.5–2.5

TBM for hard rocks. Photograph 7.17 shows the cutterhead of an old open TBM of small diameter and Photograph 7.18 shows one of the grippers from this machine.

The excavated ground falls to the lower part of the tunnel face and there it is loaded by perimeter buckets of the cutterhead that, after being raised, they are discharged on a belt conveyor located close to the TBM axis. As the excavated ground does not go through the cutterhead, but is taken by the perimeter buckets, the cutterhead is a continuous surface where the cutters are located and its open surface is less than 10% of its surface, as shown in Photograph 7.19.

In these TBMs the rock mass support is made with bolts, metallic mesh and shotcrete, which are placed behind the cutterhead. When a liner has to be placed, it is usually made of cast-in-place concrete. All systems required to operate the TBM are placed in back-up trailers, whose total length usually exceeds 100 m, which are hooked to the cutterhead and are pulled by it.

Figure 7.21 shows the structure of an open TBM manufactured by Herrenknecht AG (Germany).

Photograph 7.20 shows a view of the two TBMs which constructed the Lotschberg Tunnel (Switzerland), of 34 km in length and which, once reconditioned, were used in the San Pedro Tunnel (Spain), of 8.9 km in length.

Photograph 7.18 Gripper of an open TBM.

Photograph 7.19 Cutterhead of the open TBM which excavated the Lotschberg Tunnel (Switzerland).

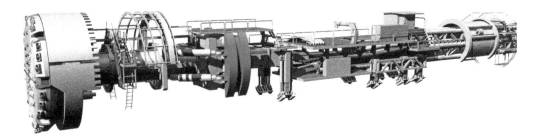

Figure 7.21 Structure of an Open TBM manufactured by Herrenknecht AG.

Source: Herrenknecht AG.

Photograph 7.20 Open TBM used in Lotschberg and San Pedro Tunnels.

Photograph 7.21 shows a detail of the front part of one of the TBMs used in the San Pedro Tunnel, in which the location of the bolt drilling system and one of the grippers can be seen.

This photograph clearly illustrates that bolting can be done from a distance of only 3.5 m from the tunnel face.

Open TBMs have serious difficulties in going through unstable ground, particularly in the presence of water, so, in these cases, the unstable ground must be treated before excavating.

To do this, as shown in Figure 7.22, modern open TBMs are equipped to execute umbrellas and grouting in the ground ahead of the TBM.

In open TBMs, the excavation process is only stopped to push the grippers forward, since the operations to make the support are simultaneous with the excavation.

The stress induced by the grippers to the ground is one of the causes that limits open TBMs' use, since this support produces pressures of several MPa, that sometimes exceeds the ground bearing capacity.

Photograph 7.21 Front part of one of the open TBMs used in San Pedro Tunnel.

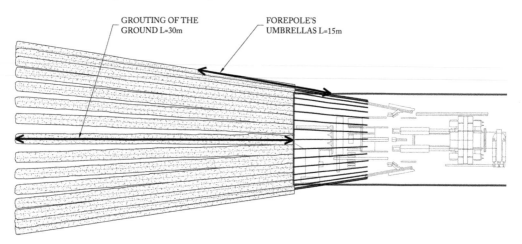

Figure 7.22 Example of drilling ahead of an open TBM, manufactured by Herrenknecht (Germany), to improve the ground to be excavated.

When open TBMs excavate in suitable grounds, very high advance rates are achieved, which enables obtaining records such as the ones already mentioned in Section 7.3. In the construction of the Gotthard Base Tunnel, which with its 57 km was the longest tunnel in the world for quite some time, open TBMs were used to build 50 of its 57 km.

Photograph 7.22 presents a view of a section of the Gotthard Base Tunnel, constructed with an open TBM.

7.3.2.2 Shields

One of the problems in tunnels constructed with open TBMs is that, when a liner needs to be placed, it must be done with cast-inplace concrete.

Photograph 7.22 Gotthard Base Tunnel section, constructed with an open TBM. Note the perfect execution of the concrete despite the presence of water in the ground.

To solve this problem, idea of making a TBM that would allow placing rings consisting of precast segments made of reinforced concrete that serve as both the support and the lining arose.

The solution consisted of providing open TBMs with a steel perimeter shield and installing a rotating erector to assemble the segment rings inside of it, which is the reason why these TBMs are known as shields. Photograph 7.23 shows the view of a shield.

Photograph 7.23 Shield at Pontones Tunnel (Asturias, Spain).

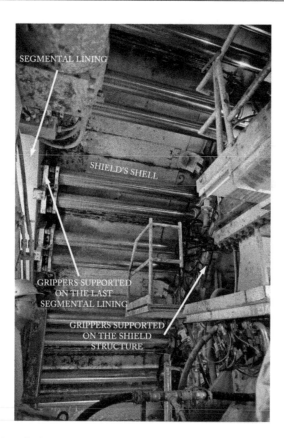

Photograph 7.24 Hydraulic cylinders to push the shield forward.

The perimeter shield of these TBMs is incompatible with the grippers in open TBMs; the reason is that the thrust required by the shields to excavate is provided by hydraulic cylinders, which rest on the last segment ring placed, as illustrated in Photograph 7.24.

As the shield moves forward, the fully constructed tunnel emerges, as illustrated in Photograph 7.25.

Due to the perimeter shell, shields have some protection against small instability problems that can occur in the excavated ground, which means that with this type of TBMs, grounds of lesser strength than those considered optimal for open TBMs can be excavated.

For this reason, shields often excavate both hard and soft rocks, which makes it necessary to change the cutterheads in order to avoid the entrapment of the cutters in the presence of fine particles.

To excavate in soft grounds, the cutterheads have fixed tools, like chisels, scrapers and openings to let the finest ground particles pass through them, in order to avoid the entrapment of the disk cutters. This means that, depending on the type of ground to be excavated, the sum of the openings in the cutterhead can be up to 40% of its surface area.

Photograph 7.26 shows a cutterhead for hard and soft rocks.

Shields allow an increase of the field of application over open TBMs, widening their application to softer grounds, but they also present some important disadvantages.

The main disadvantage of shields is that the excavation process must be stopped while the segment ring is being assembled, which takes between 15 and 30 minutes for each ring.

Photograph 7.25 Tunnel constructed with a shield.

Photograph 7.26 Cutterhead to excavate mainly in hard rocks but also in soft rocks.

The second drawback is that in shields there is no effective support until the segmental rings are in contact with the ground, which occurs, in the most favorable cases, at a distance from the tunnel face of between 10 and 12 m.

This requires grounds with significant stand-up times, which are not always achieved in lesser quality grounds.

The third drawback is because the ring of segments is assembled inside the shield, which forces the outer radius of the ring to be about 15–20 cm smaller than that of the excavation.

This results in a radial distance between the outer surface of the rings and the ground which is called a "gap". The gap complicates the loading of the segmental rings and increases the subsidence in shallow tunnels, the reason why it has to be filled.

Until recently, the gap was filled with gravel if the tunnel was excavated in rock or with cement mortar if it was excavated in soft ground.

In both cases, due to the angle of repose of the gravel or mortar, it was never possible to completely fill the gap which considerably increased the distance to the tunnel face without support.

This problem is currently solved with filling materials composed of two components: a mortar of sand, fly ashes, cement and bentonite, and a catalyst additive, which ensures the quick hardening of the filler and enables the completely filling of the gap (Mendaña et al., 2014).

7.3.2.3 Double shields

Double shields were created to solve the problems found with shields, due to the impossibility of excavating during the segmental ring assembly.

The solution that was found consisted of adding another shield with grippers to the single shield TBM, so that during the ring assembly, the double shield could work as an open TBM, supported against the ground.

Photograph 7.27 shows a view of one of the two double shields, manufactured by Herrenknecht AG, which were used in the construction of the Guadarrama Tunnel, consisting of two tubes of 28 km length and a diameter of 9.5 m.

In this photograph, the front and rear shields, the telescoping zone between them and the gripper to support the double shield when excavating and assembling the segmental ring are clearly seen.

Theoretically, double shield TBMs would be the perfect TBMs, but these machines have two significant drawbacks.

One of these is caused by the fall of small stones into the telescopic area when the TBM is supported by the grippers and a ring of segments is being assembled; as this makes the operation to move the rear shield forward, known as regripping, significantly harder.

Photograph 7.27 Double shield, manufactured by Herrenknecht AG (Germany), used in Guadarrama Tunnel.

Photograph 7.28 Double shield used in Sorbas Tunnel.

In fact, out of the four double shields used in the Guadarrama Tunnel construction, the one that achieved the best performance as a double shield worked for 84.5% of the excavated length and the worst worked for 64.7% (Cobreros, 2005).

The second major problem that double shields have is derived from their greater length when compared to single shields, as in fact, double shields are the result of combining an open TBM and a shield.

Photograph 7.28 shows a view of the double shield used in the Sorbas Tunnel construction (Almería, Spain), 16.6 m in length, which represents 1.7 times its diameter.

The fact that double shields are longer than open TBMs or single shields, for the same diameter, is a major problem because the excavated ground has to be self-stable over a length longer than when using open TBMs or single shields.

In order to mitigate this problem, compact double shields have been manufactured, as the one shown in Photograph 7.29, which was used in the construction of the Abdalajís Tunnel (Málaga, Spain). The double shields used in the Abdalajís Tunnel had a length of only 10.0 m for an excavation diameter of 9.8 m which enabled significant reduction of the distance between the tunnel face and the point at which the segmental ring was out of the machine.

In contrast, the greater compactness of this double shield caused functionality problems during construction, due to the smaller space available inside the double shield.

7.3.2.4 Fields of application of TBMs in stable tunnel faces

In order to determine the field of application of TBMs when a tunnel face is stable, the RME index, presented in Section 3.4.1, is useful.

Based on the RME, the concept of Average Rate of Advance (ARA) was defined as the advance rate, expressed in m/day, achieved in sections with lengths over 30 m and similar RME values, and without exceptional stops of the TBM during the construction of each section.

Photograph 7.29 Compact double shield used in Abdalajís Tunnel construction (Málaga, Spain).

The ARA values obtained were standardized for a tunnel with an excavation diameter of 10 m, which was called ARA$_{Typified}$ (ARA$_T$).

To correlate the ARA$_T$ values with the type of TBM used in each case, it was concluded that the analyzed sections had to be grouped into two typologies; one defined by excavated grounds with a uniaxial compressive strength lower than 45 MPa and the other by grounds with a uniaxial compressive strength greater than 45 MPa (Bieniawski et al., 2009).

For grounds whose intact rock has a uniaxial compressive strength (σ_{ci}) lower than 45 MPa, the correlations shown in Figure 7.23 were proposed.

From the graphs shown in Figure 7.23 the following conclusions are obtained:

1. Open TBMs are the least suitable TBMs to construct tunnels in rocks with $\sigma_{ci} < 45$ MPa.
2. The most suitable TBMs to construct tunnels in rocks with $\sigma_{ci} < 45$ MPa, are Double Shields. These TBMs will be able to work in double shield mode on grounds with a RME > 77 and in single shield mode when the RME < 77.
3. If RME < 77 in all tunnel sections, it will be more profitable to use a single shield.

In the case of tunnels excavated in rocks with a uniaxial compressive strength (σ_{ci}) exceeding 45 MPa, the correlations shown in Figure 7.24 were proposed.

From the graphs shown in Figure 7.24 the following conclusions are obtained:

1. With the existing TBMs, tunnel sections in rocks with $\sigma_{ci} > 45$ MPa and RME < 45 cannot be economically excavated. These grounds correspond to the combination of very hard and abrasive rocks in a rock mass with little fracturing.
2. In the range 70 < RME < 80, the three types of TBMs studied (open TBMs, single shields and double shields) give similar performances.

3. In grounds with RME > 80, open TBMs provide the best advance rates.
4. For grounds with 45 < RME < 75, single shields provide the best results.
5. Double shields working as double shields in grounds with RME > 75 and as single shields for RME < 75, would give reasonably good performances in both grounds.

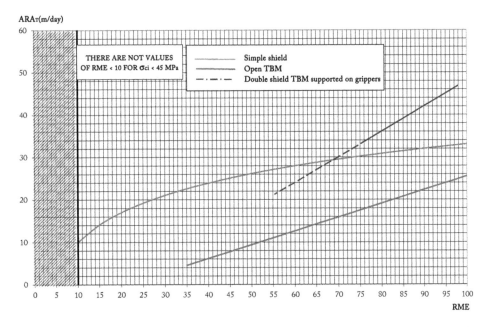

Figure 7.23 Correlation between the RME and the ARA_T for tunnel sections excavated in rocks with $\sigma_{ci} < 45$ MPa.

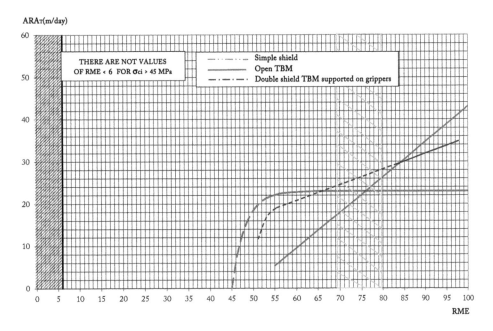

Figure 7.24 Correlation between the RME and the ARA_T for tunnel sections excavated in rocks with $\sigma_{ci} > 45$ MPa.

7.3.3 Forecasting the advance rates of TBMs in stable tunnel faces

The forecast of the advance rates that can be obtained with a TBM when constructing a particular tunnel is a difficult task, as the solution to the problem is influenced by the characteristics of the rock mass to be excavated, the TBM chosen and the professional training of the team that must handle the TBM.

Therefore, it is not surprising that, despite the numerous publications on the prediction of the advance rates of TBMs, there is no methodology to solve this issue in a completely satisfactory and definitive way.

Nevertheless, four methodologies have made important contributions to this task.

7.3.3.1 Penetration Rate Index by the Colorado School of Mines

In 1993 Rostami and Ozdemir proposed "A new model for performance prediction of hard rock TBMs", based on the calculation of the penetration that is achieved in the ground in one rotation of the cutterhead.

This index is known as the Penetration Rate and should be expressed in millimeters penetrated in the ground by one revolution of the cutterhead, although it is often expressed in mm/hour. The PR index is closely related to the work done by the cutters and the spacing between them.

Figure 7.25 illustrates the effect of an isolated cutter when it indents on a rock surface with a force F.

The force that pushes the cutter creates high pressures on the ground surface so that the cutter penetrates the rock surface (indentation) creating a distribution of pressures in the ground that has its highest values in the contact area between the cutter and the ground.

These pressures produce two effects on the ground; on the one hand, they create an area of crushed ground and on the other hand, they induce cracks by tensile forces.

The effects produced when two adjacent cutters indent the ground, as happens when the cutterhead of the TBM rotates are illustrated in Figure 7.26 and depend on the spacing between the cutters.

If the spacing between both cutters is large enough, they act individually, according to the model presented in Figure 7.26, creating a ridge between them.

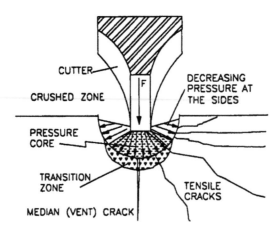

Figure 7.25 Effect produced by the indent of a single cutter.

Source: Rostami and Ozdemir, 1993.

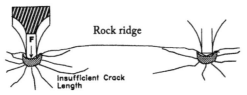

a. Ridge formation due to lack of pressure and length of cracks

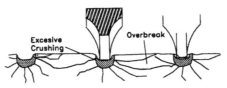

b. Over break due to excessive loading and longer cracks.

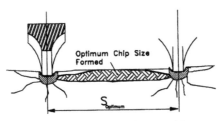

c. Normal cutting with optimum crack length and direction.

Figure 7.26 Indent of two adjacent cutters.

Source: Rostami and Ozdemir, 1993.

However, if the spacing between both cutters is very small, the tensile cracks generated by both cutters are linked to each other, resulting in an excessive ground fracturing.

There is an optimum spacing between cutters when the traction cracks induced by each have a minimum connection that induces the formation, by tensile rupture, of a flat piece of ground usually called "chip".

Photograph 7.30 shows a granite chip produced with a TBM which is about 14 cm long, 5 cm wide and 1 cm thick.

The average thickness of the chips produced with the optimal spacing between cutters, is known as the PR and is measured in mm/rev. or in m/h.

To calculate the maximum indentation pressure value (P′) Rostami and Ozdemir (1993) proposed the following correlation:

$$P' = 100500 + 12170S + 7.888\sigma_c - 28830\sigma_t^{0.1} - 192S^3 - 0.000147\sigma_c^2 - 29450T - 13000R$$

where:

S = spacing between cutters (inch)
σ_c = uniaxial compressive strength of the rock (psi)
σ_t = indirect tensile strength of the rock (Brazilian Test) (psi)
T = minimum thickness of the disk cutter (inch)
R = disk cutter radius (inch)

From P′ value, using the methodology proposed by Rastami and Ozdemir, the PR and other characteristics of the TBM, such as the number of cutters, the total thrust, the torque and the necessary power can be calculated.

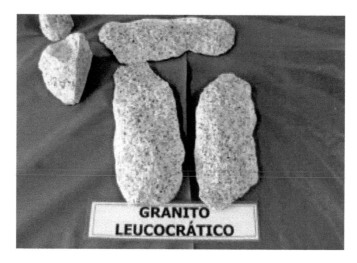

Photograph 7.30 Granite chips from Guadarrama Tunnel.

Once the value of PR has been calculated, the Advance Rate (AR) is calculated by the expression:

$$RA = PR \cdot U$$

where U is the percentage of time in which the TBM is excavating.

The method proposed by the Colorado School of Mines to predict the PR of a TBM has a physical meaning, since it handles both data from the ground and from the TBM. However, the AR calculation, established for a period of time of several hours, is quite subjective, since the percentage of time that the TBM is available for excavation is very variable and difficult to know accurately.

7.3.3.2 Q_{TBM} method

The Q_{TBM} index, already presented in Section 3.4.2, allows estimation of the PR in ground by the expression proposed by Barton (2003):

$$PR\,(m\,/\,h) = 5\,(Q_{TBM})^{-0.2}$$

Barton makes the assumption that as the period in which the AR of a TBM is measured increases, the AR decreases. Therefore, to calculate the AR the following equation is proposed:

$$AR\left(\frac{m}{h}\right) = PR\left(\frac{m}{h}\right) \cdot T^{m}$$

where:
    T = period of time in which AR is calculated.
    m = coefficient which controls the decrease in the AR over time.

AR values are classified into five groups, as shown in Figure 7.27.

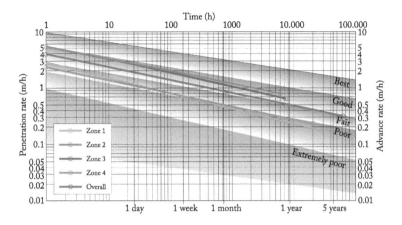

Figure 7.27 AR values classification.

The methodology of predicting the TBM performance based on the Q_{TBM} index has much less physical meaning than the methodology proposed by the Colorado School of Mines, and the hypothesis that as the tunnel length increases the performance decreases has had a significant disagreement, because of insufficient evidence in practice.

7.3.3.3 Rate of Advance by the Norwegian University of Science and Technology

The method proposed by the Norwegian University of Science and Technology (NTNU) enables estimation of the AR of a TBM when excavating in hard rocks, the cutters' lifetime and the excavation costs.

This method was developed in 1976 but was described in detail much later by Bruland (2014). The software program FullProf has been developed to enable its application.

Just like the methods described previously, the NTNU methodology first determines the PR, based on the parameters of the rock mass and the TBM.

To take into account the influence of the rock mass factors, the equivalent fracturing factor is calculated as a function of the cracks and joints present in the ground, its orientation with respect to the tunnel axis and the rock drillability according to the drilling rate index (DRI), which is presented in Section 3.4.1.

The factors related to the TBM are taken into account through the equivalent thrust, calculated using the data from the thrust per cutter, diameter and the spacing between them.

The PR (i_0), expressed in mm/revolution, can be determined using the chart in Figure 7.28. To calculate the PR, expressed in m/h, the following expression can be used:

$$PR = i_0 \cdot RPH \cdot \left(\frac{60}{1.000} \right)$$

where RPH is the rotation speed of the cutterhead.

From the PR, the NTNU method enables calculation of the AR taking into account the following factors:

- Operating time of the TBM
- Time to push the TBM
- Inspection and replacement of the cutters

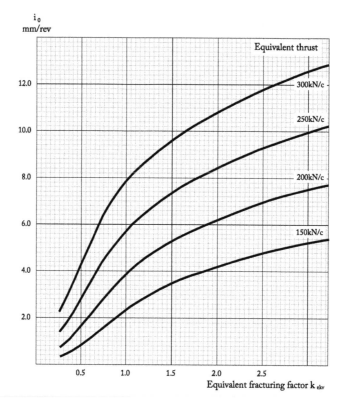

Figure 7.28 Chart to determine the PR (i_0).

Sources: Norwegian Tunneling Technology; Bruland, 2014.

- Maintenance time of the TBM
- Maintenance time of the auxiliary installations
- Time spent on other unproductive activities

The NTNU methodology enables estimating the cutter consumption, as a function of the TBM diameter, the quartz content in the ground and the CLI, presented in Section 3.4.2.

The calculation of the PR with the NTNU methodology is close to the methodology proposed by the Colorado School of Mines, but the calculation process of the AR through the NTNU methodology is less subjective than the one followed by the Colorado School of Mines.

7.3.3.4 RME methodology

As discussed in Section 7.3.2.5.4, the ARA_T can be estimated with the RME, for open TBMs, single and double shields.

ARA_T values are determined for the specific characteristics of each tunnel, taking into account the experience of the tunneling crew, the adaptation to the excavated ground and the tunnel diameter, for which the following expression is used:

$$ARA_R = ARA_T \cdot F_E \cdot F_A \cdot F_D$$

where:

ARA_R = Average advance rate in tunnel sections with lengths larger than 30 m when no extraordinary stops or incidents occur

F_E = Factor that depends on the experience of the crew in driving the TBM

F_A = Factor that depends on the adaptation to the ground to be excavated

F_D = Factor that depends on the tunnel diameter

The F_E factor, which can vary between 0.7 and 1.2, is calculated by the expression:

$$F_E = F_{E1} + F_{E2} + F_{E3} + 0.7$$

F_{E1}, F_{E2} and F_{E3} are calculated with the criteria shown in Table 7.4. F_A ranges between 0 and 1.2, the value which is reached when 15 km of tunnel have been excavated and it is rated according to the graph in Figure 7.29. F_D is rated using the graph in Figure 7.30 and its value ranges between 0.6 and 1.6 depending on the diameter of the excavation.

This methodology cannot be applied to predict the ARs with TBM in stable tunnel faces when excavating fault zones; but in such cases the criteria shown in Table 7.5 can be applied.

7.3.3.5 Comparison of the methods to predict TBM advance rates for stable tunnel faces

The results obtained with a TBM for stable tunnel faces are the result of the combination of three basic factors:

I. Reliable knowledge of the ground to be excavated
II. Appropriate selection of the TBM
III. Experience of the crew that drives the TBM

Accordingly, the methodology to predict ARs with open TBMs should take into account these three basic factors. However, the four methodologies, presented in the previous sections, make some simplifications that do not take these three factors fully into account.

The methodologies of the Colorado School of Mines and the NTNU rigorously consider the ground excavability, but they do not distinguish the type of TBM used, and do not consider the experience of the crew.

Table 7.4 Rating of F_{E1}, F_{E2} and F_{E3}

Contractor experience	Without experience	1 to 5 tunnels constructed with TBM	6 to 10 tunnels constructed with TBM	11 to 20 tunnels constructed with TBM	> 21 tunnels constructed with TBM
Value of F_{E1}	0	0.05	0.10	0.15	0.2
Skills of the TBM staff	Little trained and without experience in TBMS	Trained, but without experience in TBMS		Trained and with experience in TBMS	
Value of F_{E2}	0	0.1		0.15	
Spare parts supply	Existence of a local representative of the TBM manufacturer			Time necessary to receive the spare parts	
	Yes	No		≤ 1 month	> 1 month
Value of F_{E3}	0.075	0		0.075	0

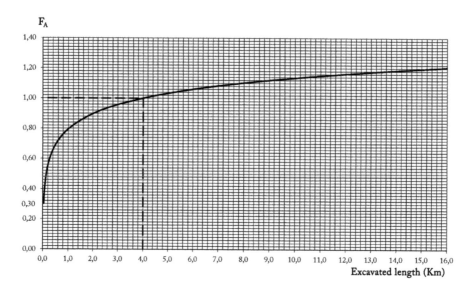

Figure 7.29 Rating of F_A.

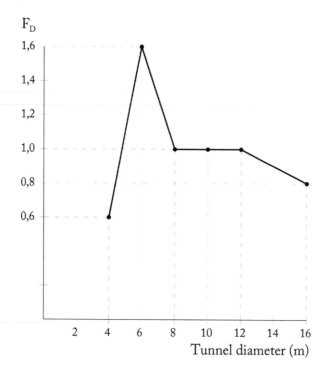

Figure 7.30 Rating of F_D.

The methodology based on the RME considers statistically the ARs obtained by three types of TBMs for stable tunnel faces but only gives some subjective recommendations about the ARs in fault zones.

Salehi and Shahandeh (2013) have compared the four predictive methods presented in previous sections, applying them to a 16 km long and 4.56 m diameter hydraulic tunnel, which is part of the Karaj–Tehran water supply.

Table 7.5 Evaluation of the possible advance rates with TBM for stable tunnel faces in fault zones

Overburden (m)	< 50				50–200				> 200 (No extreme squeezing)			
Water at the tunnel face $\left(\dfrac{1}{min.m^2}\right)$	Dry		< 0.1	> 0.1	Dry		<0.1	> 0.1	Dry		<0.1	> 0.1
Type of ground	Blocks	Clay	Any		Blocks	Clay	Any		Blocks	Clay	Any	
Advance rate with Shield TBM $\left(\dfrac{m}{day}\right)$	10–20	> 40	5–20	1–10	10–20	> 40	5–15	1–5	5–15	20–40	1–10	< 1
Advance rate with TBM $\left(\dfrac{m}{day}\right)$	10–15		5–15	1–5	5–15		1–10	1–5	5–10		1–5	< 1

In conclusion, Salehi and Shahendeh point out that the application of the RME-based methodology gave a construction period estimation of 21 months, which was the closest value to the actual period of time for the tunnel construction.

7.3.4 Pressurized shields

When an excavation face shows instability problems it has to be immediately reinforced; if not, the ground at the tunnel crown will quite probably collapse and the incident can cause unpredictable consequences.

Shields pressurized at the tunnel face have been developed to prevent these instabilities, which enable the creation of a direct pressure at the tunnel face to stabilize it.

Probably the first pressurized shield was used in the construction of the first tunnel under the Thames River (London), built by M.I. Brunel, in the first half of the 19th century that has a length of about 430 meters.

The construction works of the first tunnel under the Thames began in 1804, but they were abandoned due to the difficulty in controlling the water inflow into the shaft which was built to provide an access to the tunnel level.

In 1823 M.I. Brunel presented a rectangular shield which was 12.5 m wide and 7.2 m high, that is, with a section around 90 m², that should enable the construction of two twin tunnels, with 4.6 m of clear width each, as the ones illustrated in Figure 7.31.

As shown in Figure 7.32 Brunel's shield was divided into three floors and each floor was split again into 12 cells with one worker per cell.

The protection against tunnel face instabilities was achieved by covering it with wooden boards supported on an iron casting structure, as illustrated in Figure 7.33.

Figure 7.34 reproduces a section of the tunnel longitudinal profile, in which the construction process can be appreciated. It is important to realize that the tunnel lining, made with bricks, was completed at only 3 m from the excavation face.

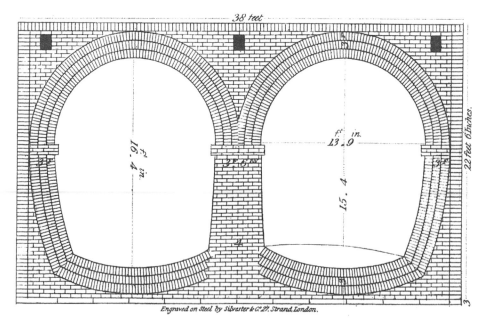

Figure 7.31 Section of the first tunnel under the Thames River.

Source: Thames Tunnel Co., 1838.

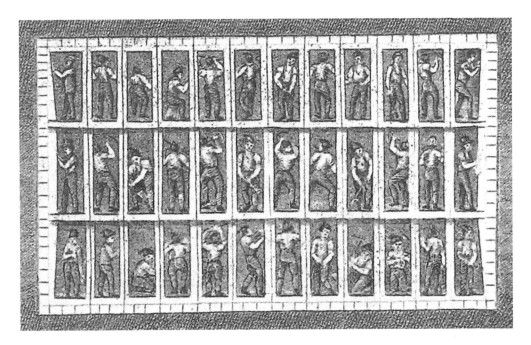

Figure 7.32 Front view of Marc Brunel's shield.

Source: Thames Tunnel Co., 1838.

The tunnel construction began in November 1825 and finished in November 1841, although just after the start of the works M.I. Brunel was replaced by his son Isambard Kingdom Brunel who was the engineer responsible for the construction management.

During the 16 years that the construction period lasted, it was necessary to overcome many difficulties, including five floods at the tunnel face, three of them concentrated on a stretch of only 12 m.

Brunel's idea of working with a protective shield at the tunnel face is still valid today, almost 200 years later.

As described in the following sections, currently, there are two procedures to apply pressure at the tunnel face: by kneading the excavated soil to create a mass with plastic consistency or carrying the excavated soil by mixing it with a bentonite suspension and water, creating a pumpable fluid. Both the ground plastic mass and the fluid with the suspended soils are compressed and that pressure is transferred to stabilize the tunnel face.

In both cases, there is a closed chamber behind the TBM cutterhead where the muck or fluidized soils are compressed. To isolate the excavation chamber, which is under pressure, a double door lock is installed. This technology is always applied to the shield TBMs and is called earth pressure balanced (EPB) shields or hydroshields, depending on how the pressure at the tunnel face is created.

7.3.4.1 EPB shields

EPB means that, at the tunnel face, balance between the pressures that try to destabilize it (water and earth pressures) and the stabilizing pressures of the muck, is achieved.

To keep the muck pressure in the excavation chamber, the ground mass is carried out of the chamber with a screw conveyor, located inside a steel casing to maintain the pressure.

Figure 7.33 Detail of Marc Brunel's shield.

Source: Thames Tunnel Co., 1838.

Figure 7.35 shows the longitudinal section of an EPB shield manufactured by Herrenknecht AG As shown in Figure 7.36, the pressure of the muck can be varied by modifying the force applied to the cutterhead or by varying the flow of the extracted ground, changing the rotation rate of the screw conveyor.

EPB shields are especially suitable for excavating in grounds without gravels, but their field of application can be extended to grounds with gravels by using polymer-based chemical additives.

Based on the studies carried out by Thewes (2009), Figure 7.37 shows the characteristics of the ground conditioning required, depending on their grain size distribution, to excavate with EPB-shields.

Zone 1, striped in green, corresponds to grounds which are ideal to be excavated with EPB shields. As indicated, those grounds must have more than 30% of fines and must not contain gravels.

Zone 5, shaded in red, corresponds to grounds that should not be excavated with EPB shields whose granulometry is characterized by the lack of fines and contains medium and coarse gravels in a proportion greater than 40%.

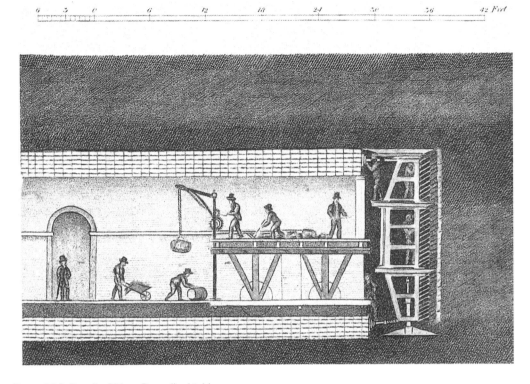

Figure 7.34 Detail of Marc Brunel's shield.

Source: Thames Tunnel Co., 1838.

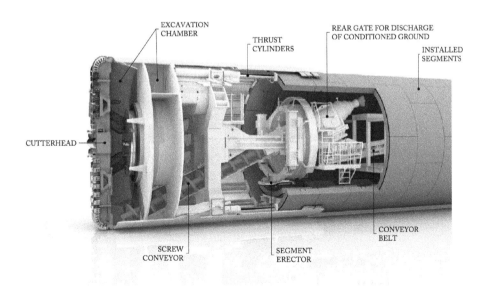

Figure 7.35 Longitudinal section of an EPB shield manufactured by Herrenknecht AG.

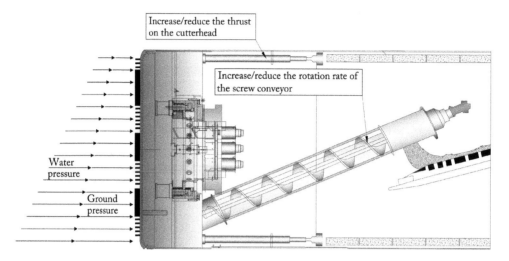

Figure 7.36 Actions to modify the pressures generated by the muck in an EPB shield.

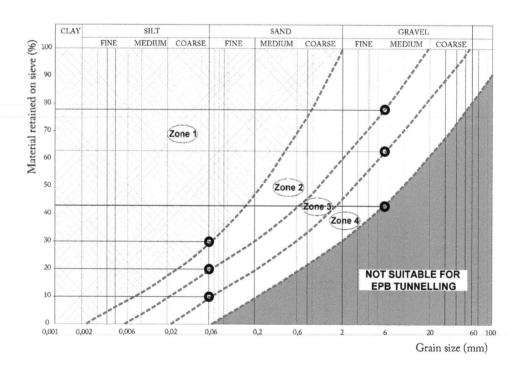

Figure 7.37 Ground conditioning when a tunnel is excavated with EPB-Shields.

In the three remaining zones (2, 3 and 4) the ground conditioning intensity in the excavation chamber, by adding chemical products, varies as it is shown in Table 7.6.

Ground conditioning generates an increase in the costs and a considerable decrease in the excavation rate if not performed in the appropriate way and for this reason it is very important to accurately know the proportion of the grounds to be excavated and their grain size distribution.

Table 7.6 Types of ground conditioning

Granulometric zone	Fines (%)	Medium gravels (%)	Products added through the cutterhead
2	20-30	<20	Foams
3	20-10	<40	Foams and Polymers
4	<10	<60	Foams, Polymers and fines

Figure 7.37 represents these grounds according to their granulometry.

The cutterheads of EPB shields have two main differences when compared to the TBM cutterhead for stable tunnel faces; one is the cutting elements and the other is the open surface of the cutterhead.

In general, EPB shields will have to excavate in low resistance zones, mainly in soils as the tunnel face is more unstable in this kind of ground.

To excavate in soils, the TBM does not use disk cutters, which make the ground work in compression and traction, but it uses chisels or scrapers that excavate the ground by cutting.

Photograph 7.31 shows the scrapers and buckets of an EPB shield.

Unlike TBMs in stable tunnel faces, where the holes in the cutterhead surface to extract the excavated grounds are less than 10% of its surface, EPB shields have a larger amount of holes, usually between 30 and 60% of the total cutterhead surface.

The reason is that EPB shields look for the best conditions to knead the excavated ground so letting the ground pass through the cutterhead makes it easier to achieve this goal.

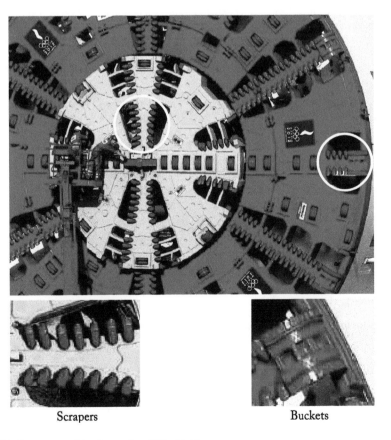

Scrapers   Buckets

Photograph 7.31 Scrapers and buckets of an EPB shield.

Photograph 7.32 EPB Shield cutterhead that constructed the Gorg–Bon Pastor section in Line 9, Barcelona Metro, equipped with disks and scrapers.

When an EPB shield may have to excavate in rocks and soils, either on mixed-faces conditions or in separate sections, the cutterhead must be equipped with disks and scrapers, as illustrated in Photograph 7.32.

7.3.4.2 Hydroshields

Hydroshields differ from EPB shields in the way the excavated ground is carried out and how the pressure is transferred to the ground in the excavation chamber.

The excavated ground transportation is done hydraulically by mixing it with bentonite slurry that almost fills the whole excavation chamber. In this way, as happens in boreholes, the bentonite slurry makes the ground surface less permeable. The pressure of the bentonite slurry in the excavation chamber is controlled by regulating the pressure of the compressed air reservoir created in its upper part.

Accordingly, along a tunnel constructed with hydroshields there must be three pipes to ensure the following three functions:

1. Supply of compressed air to the excavation chamber
2. Bentonite slurry feed line
3. Pumping of the excavated grounds suspended in bentonite slurry

Figure 7.38 shows the longitudinal section of a hydroshield manufactured by Herrenknecht AG in which the screw conveyor, typical of an EPB shield, has been replaced by the pipe that hydraulically transports the excavated ground. Photograph 7.33 shows the feed lines and pumping lines in the A-86 Motorway Tunnel, the Paris ring motorway.

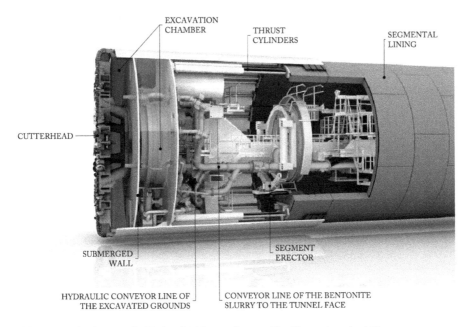

Figure 7.38 Longitudinal view of a Hydroshield manufactured by Herrenknecht AG.

Source: Herrenknecht AG.

Photograph 7.33 Lines of bentonite slurry and hydraulic transport in the A-86 Motorway Tunnel, Paris.

Once outside, the slurry line and excavated ground line are sent to a separation plant where the granular fractions are separated from the slurry, and the bentonite is recovered.

Figure 7.39 shows a simplified scheme of a hydroshield pipeline network.

The separation plant must have a minimum surface area of about 4,000 m², which is a major problem for the use of hydroshields in urban areas.

Photograph 7.34 shows the separation plant used in the construction of the section in sands of the A-86 Tunnel, Paris.

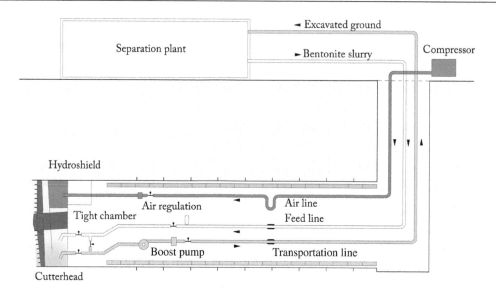

Figure 7.39 Scheme of a hydroshield pipeline network.

Photograph 7.34 Separation plant in the A-86 Motorway Tunnel, Paris.

Due to the waterproof wall effect created with the bentonite slurry, hydroshields are suitable to work on sandy and gravelly grounds, as illustrated in Figure 7.40, Thewes (2007).

In Figure 7.40, three areas for the application of hydroshields with the following characteristics are distinguished:

- **Area A.** The soil has no clay and the percentage of sand is greater than 10%. It is the ideal ground in which to use a hydroshield.
- **Area B.** The soil has clay and up to 40% of fine particles. In these grounds, bentonite recovery is more difficult as the clay content increases.

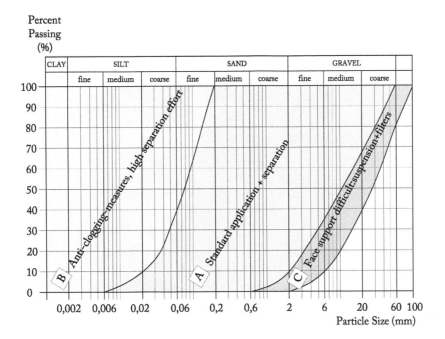

Figure 7.40 Areas for the application of hydroshields.

Source: Thewes, 2009.

- **Area C.** The ground has less than a 10% of sand and the rest are gravels; this makes it very permeable and requires the addition of fine filler materials so that the bentonite slurry could be effective.

A hydroshield cutterhead can be seen as a large mixer of the fluid composed by water, bentonite and ground, hence why the percentage of open surface in the cutterhead can be up to 60%. Photograph 7.35 shows a cutterhead, of 14 m in diameter, belonging to the hydroshield which built the IV Tunnel under the Elba River (Germany) and, once reconditioned, the Lefortovo Tunnel in Russia. This cutterhead has an open surface of 46%.

The hydraulic transport of the excavated grounds considerably influences the diameter of the fragments to be transported and, therefore, it is common to place a jaw crusher at the hydroshield cutterhead, as it is shown in Photograph 7.36.

7.3.4.3 Remarks on the pressures at the excavation chamber

When a shield works with the chamber at pressure, more energy is needed at higher pressures and, in addition, as the pressure in the chamber increases, so does the wear on the shield elements that are subjected to friction caused by the weight of the excavated grounds. This leads to lower ARs and higher costs.

To calculate the pressure necessary at the excavation chamber, when a specific tunnel section is excavated, several methods exist, among which the one proposed by Anagnostou and Kovari (1996) is probably the most commonly used.

The Anagnostou–Kovari method considers that the equilibrium of the excavation face of a shield depends on the prism illustrated in Figure 7.41.

Photograph 7.35 Hydroshield with 14.2 m of diameter.

Photograph 7.36 Jaw crusher at the hydroshield cutterhead.

The pressure to be applied at the excavation face is the one necessary to ensure the stability of the ground prism in Figure 7.41, which is affected by the water table.

The Anagnostou–Kovari model is a deterministic model so, as indicated in Chapter 2, this method is very conservative.

It is of great importance to correctly define the pressure to be applied in the excavation chamber of a shield, for that reason, it is recommended to follow DEA methodology to define the pressure, which is presented in Section 2.4.4.

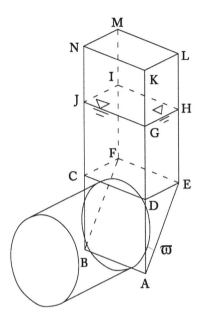

Figure 7.41 Model proposed by Anagnostou–Kovari.

In particular, the pressure in the excavation chamber should be calculated by solving a three-dimensional geomechanical model and the result should be checked during the tunnel construction, verifying that the movements predicted for the calculations correspond reasonably with those measured by monitoring, both on the surface and in areas close to the excavation.

7.3.5 Hybrid shields

As mentioned before, TBMs are very rigid systems, so it is quite difficult and expensive to introduce changes during the excavation to adapt to the ground characteristics.

Moreover, as the main field of application of TBMs is long tunnels, it is highly probable to find grounds with characteristics which differ a lot with respect to those found in the rest of the tunnel, which leads to construction difficulties.

Accordingly, the sentence written by Richard J. Robbins in 1976, quoted earlier, has much relevance again:

> The idea of a universal mechanized tunneling system is not really convenient, as each tunnel has special requirements as a result of both the geological conditions and the structural design. Tunnel engineers must continue developing new concepts and combinations, which will have a broad effect to turn the ideas that today seem definitive into obsolete ones.

The idea of developing a TBM with several operational modes has been adopted and efficiently developed by Herrenknecht in the last decades, which has led to hybrid or multimodal TBMs.

The type of TBM for which hybrid TBMs have been developed are open shields, which must be able to work as hydroshields or EPB shields, as illustrated in Figure 7.42.

To have a functional hybrid TBM, the change in the operational mode has to be carried out in the shortest time possible, which requires limiting the interventions in the excavation

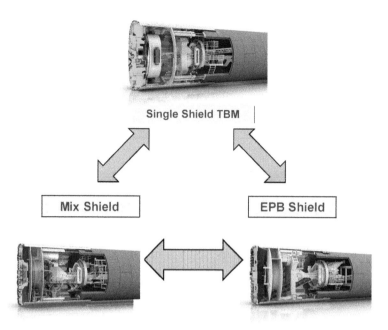

Figure 7.42 Concept of hybrid TBMs.

Source: Herrenknecht AG.

chamber as much as possible, since the mechanical wear produced by the excavated ground, as it passes through the holes in the cutterhead, make these interventions extremely difficult.

The option of using hybrid TBMs is very attractive, since the field of application of TBMs is widened. As shown in Figure 7.43 (Hourtovenko, 2015), a shield that can work as an EPB shield and hydroshield would practically cover all common types of soils.

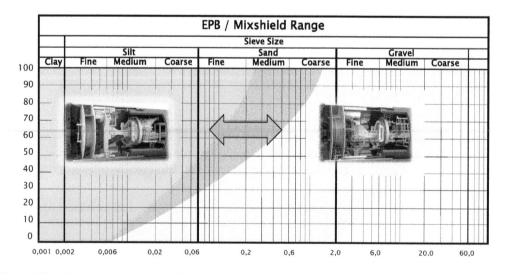

Figure 7.43 Field of application of an EPB–hydro shield.

Source: Hourtovenko, 2015.

The following sections present some examples of bimodal shields, which can work in two of the three possible modes: open, EPB and hydroshield, and multimodal shields, which can work in the three modes.

7.3.5.1 Bimodal shields

Bimodal, EPB–hydro or open shields have already been used for years and have two of the three primary transport systems for the excavated ground:

- Conveyor belt, for the open shield
- Screw conveyor, for the EPB shield
- Steel pipe lines, for the hydroshield

The coexistence of two transport systems requires a minimum diameter of the TBM, which is around 10 m.

The A-86 Motorway Tunnel, built in the 1990s, is a good example of the use of an EPB–hydro Shield.

As shown in Figure 7.44, the tunnel had a length of about 10 km and an excavation diameter of 11.6 m. It had to be excavated into two very different types of ground; the well-known Fontainebleau sands were present in the central part and at the Colbert portal, which were excavated in hydroshield mode; and the rest of the tunnel was excavated in EPB mode to excavate loams and clay soils.

Two operational modes of this shield are schematically shown in Figure 7.45.

In the EPB-Hydro Shield used in the A-86 Tunnel, the screw conveyor was always assembled to the TBM and to switch from EPB mode to Hydro mode, the screw conveyor was partially removed, as illustrated in Figure 7.46.

The Katzenberg Tunnel, which consists of two tubes with an outer diameter of 11.12 m and has a length of 8.9 km, is part of the new high-speed railway line between Mannheim (Germany) and Basel (Switzerland).

This tunnel was completed in 2011 by excavating limestones, sandstones and marls with two shields, which could work in open or EPB mode, as shown in Photograph 7.37.

The difficulties faced in the Katzenberg Tunnel were very different from those of the A-86 Tunnel, as most grounds had no instability problems and it was necessary to work in EPB mode only when going through some faults, in the presence of water.

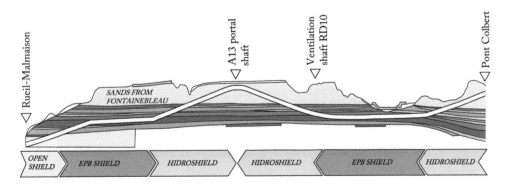

Figure 7.44 Operational modes used for the construction of the A-86 Motorway Tunnel, Paris.

Source: Herrenknecht AG.

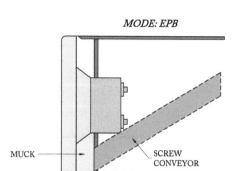
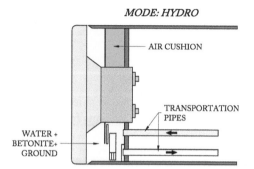

Figure 7.45 Scheme of the EPB or hydro operational mode.

Source: Herrenknecht AG.

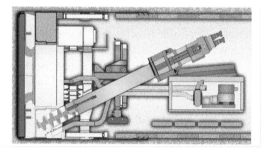
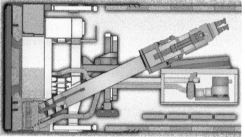

Figure 7.46 Partial removal of the screw conveyor to switch from EPB to hydro mode.

Source: Herrenknecht AG.

Photograph 7.37 TBMs in the Katzenberg Tunnel. With permission of Wayss & Freytag Ingenieurbau AG.

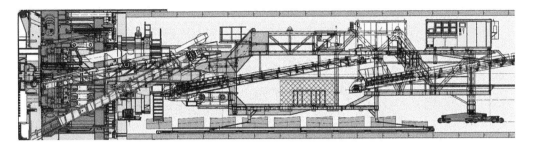

Figure 7.47 One of the shields used in Katzenberg Tunnel.

Source: Herrenknecht AG.

In this case, as shown in Figure 7.47, a conveyor belt to work as open shield and a screw conveyor to work as EPB shield coexisted as the main mode of transportation for the excavated ground.

7.3.5.2 Multimodal shields

Multimodal shields are TBMs that can work in the three possible modes: open, EPB and hydro.

Herrenknecht AG has developed this technology in recent years, which is based on the idea of carrying out, in the three modes, the primary transportation of the excavated grounds by means of a screw conveyor.

As illustrated in Figure 7.48, the screw conveyor transports the excavated ground when operating in open shield mode; the ground is transformed into muck when in EPB mode and is mixed with water and bentonite if working as a hydroshield.

When the shield works in open or EPB mode, the screw conveyor discharges the excavated grounds on a conveyor belt through a first gate, which closes when it works as a hydroshield to lead the mixture of ground, bentonite and water, through a second gate, to the pipe that transports the mix outside.

Herrenknecht AG calls this type of TBM variable density shields as the shield can adapt to the characteristics of the ground when the ground conditions change.

Photograph 7.38 shows the first tunnel constructed with a variable density shield, manufactured by Herrenknecht AG, which was used in the construction of the Klang Valley Mass Rapid Transit (KVMRT) line between Sungai Buloh and Kajan, in Kuala Lumpur (Malaysia).

The excavated grounds in the KVMRT were highly karstified limestones. During the construction of the Stormwater Management and Road Tunnel (SMART), 41 sinkholes developed in the same section in which the KVMRT was going to be constructed.

However, during the KVMRT construction only two sinkholes were produced in that section, which is a reduction of 95% in comparison to the frequency of sinkholes recorded during the SMART construction.

In the middle of this photograph, the shell to accommodate the screw conveyor can be seen which is one of the essential elements of a variable density TBM.

7.3.6 Remarks on the selection of the TBM

The technology of TBMs has been developed significantly in recent years to adapt them to very different types of ground and to construction conditions of great difficulty.

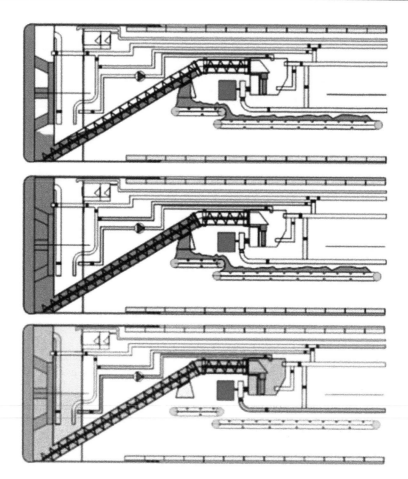

Figure 7.48 Operation mode of a multimodal shield. (Herrenknecht AG). 1. Excavation face without pressure (open shield). 2. Excavation face with the pressure generated by the grounds (EPB shield). 3. Excavation face with hydraulic pressure (hydroshield).

Nevertheless, the most important premise in successfully using a TBM is to accurately know the ground to be excavated, for which it is recommended to use the ground investigation methods presented in Chapter 4.

In tunnels excavated in rock, it is necessary to evaluate the excavability difficulty using the RME index, presented in Section 3.4.1.

Also, it is necessary to evaluate if there are sections along the tunnel, with highly yielded grounds, which can be identified by the ICE index, described in Section 1.3.3.

Once a detailed longitudinal tunnel profile is available, the most suitable type of TBM can be selected. There are three options: open TBM, single shield and double shield; the type of treatment in areas with high yielding can be decided from these two alternatives: treatment of the ground ny the TBM or working with pressure options at the excavation face.

It should not be forgotten that the Gotthard Base Tunnel, the longest and deepest tunnel ever built, was mostly excavated with open TBMs.

In tunnels with unstable tunnel faces, the most important thing is to accurately know the granulometry of the ground to be excavated; if it is a clearly cohesive ground, EPB shields will be the most appropriate, and if there are coarse granular grounds, hydroshields will be the best alternative.

Photograph 7.38 Breakthrough of the first tunnel constructed with a variable density TBM, June 2014. (Herrenknecht AG).

The selection of a dual shield, EPB and hydro, is very attractive; but to know if is the best option, it is necessary to evaluate the consequences and costs derived from the necessity of having a bentonite recovery plant.

Finally, the most complex problem are tunnels that must be excavated both in rock and soils; although again, the best starting point is to have an accurate longitudinal geotechnical profile of the grounds to be excavated.

In these cases, the most important choice is whether to use a TBM for rocks adapted to work on soils or to use a TBM for soils adapted to excavate rocks. In both cases it is necessary to consider the possibility of changing the head of the TBM, taking advantage of a station or other similar large excavation, although it is also a reasonable option to design a mixed cutterhead.

7.4 OPEN CUT CONSTRUCTION METHODS

Open cut tunnels are often referred to as "cut and cover tunnels" because the ground above the tunnel is completely excavated and it is later replaced by ground compacted over the tunnel or by a concrete structure.

In any case, cut and cover tunnels are not tunnels in the strict sense, because no arch effect can be developed on the ground, and the filler placed on the tunnel must be considered as a dead load.

Among the construction methods for open cut tunnels there are two which are most widely used: cut and cover and excavation between diaphragm walls. The following sections present their main characteristics.

7.4.1 Cut and cover method

The cut and cover method consists of excavating a trench with self-stable slopes, constructing the tunnel structure with cast-in-place concrete and restoring the original profile of ground by backfilling the void created above the structure to the original surface.

This is a very simple and safe method, but is hardly used in urban areas, because much of the surface is affected during its construction.

A 14 m wide tunnel located at an average depth of 25 m, constructed using the cut and cover method, is shown in Figure 7.49.

In this figure it can be seen that in order to build a tunnel 14 m wide it is necessary to affect an area of 93 m wide.

Photograph 7.39 shows some tunnels used to store water in Carthage (Tunisia) constructed with the cut and cover method.

Most tunnels portals are cut and cover tunnels, constructed with this method, as shown in Photograph 7.40.

Photograph 7.41 shows the tunnel, constructed with the cut and cover method, in the access to Madrid through O'Donnell Street.

7.4.2 Excavation between diaphragm walls

The construction of tunnels using the excavation between diaphragm walls method solves the problem of the great surface impact produced with the cut and cover method, because instead of excavating between two self-stable slopes it is excavated between two walls of reinforced concrete.

With this solution, the impact on the surface is limited to a strip with the tunnel width as illustrated in Figure 7.50.

The walls usually consist of reinforced concrete modules with a length of 2.5 m and a thickness of between 0.8 and 1.2 m, although when there is no water flow in the ground, the walls can be constructed using piles, with diameters between 1.0 and 1.5 m.

Photograph 7.42 shows a tunnel constructed by excavating between pilewalls.

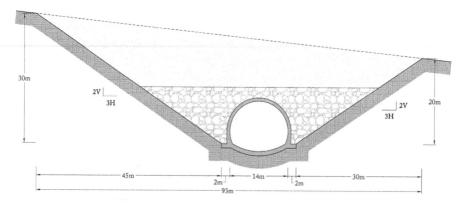

Figure 7.49 Tunnel constructed using the cut and cover method.

Photograph 7.39 Tunnels to store water in Carthage (Tunisia) constructed using the cut and cover method.

Photograph 7.40 Construction of a tunnel using the cut and cover method.

Photograph 7.43 illustrates the construction of the tunnel vault, which will be later excavated between two walls braced by the vault tunnel.

This construction method is normally used in urban areas, although it has three major drawbacks. One is the impact on traffic and commercial activity, the second is derived from the need to divert all services which go through the excavation and the third is related to the possible dam effect generated by the tunnel when the water table is close to the surface.

Photograph 7.41 Tunnel constructed using the cut and cover method in the access to Madrid through O'Donnell Street.

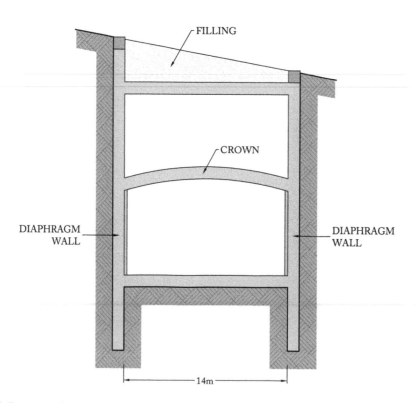

Figure 7.50 Excavation between diaphragm walls.

Photograph 7.42 Tunnel excavated between pilewalls.

Photograph 7.43 Vault construction of a tunnel that will be excavated between diaphragm walls.

To minimize the impact on the traffic, the surface slab can be constructed in two stages as shown in Photograph 7.44, which shows the construction between diaphragm walls of the underground parking of Serrano Street in Madrid.

When the water table is close to the surface, the construction of a tunnel between diaphragm walls will create a dam effect that will alter the water table distribution.

To avoid this drawback, transverse siphons may be designed, such as those shown in Figure 7.51.

Photograph 7.44 Construction in two stages of the slab of the underground parking in Serrano street, Madrid.

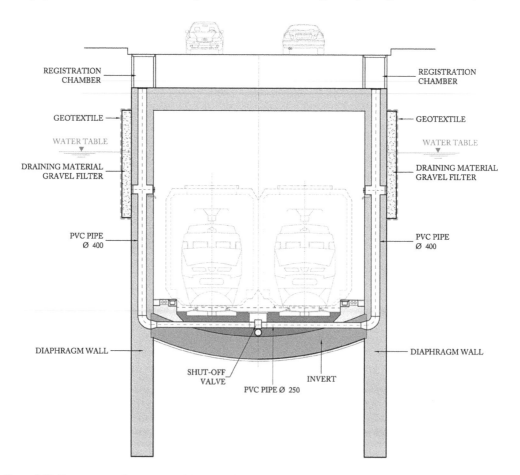

Figure 7.51 Transverse siphons to avoid the dam effect after the tunnel construction between diaphragm walls.

BIBLIOGRAPHY

Anagnostou, G., Kovari, K., "Face stability conditions with earth-pressure-balanced shields", *Tunnelling & Underground Space Technology*. Vol. 11, No. 2. 1996.

Bäppler, K., *Different TBM's – Different Applications*. Herrenknecht, Bergen, Norway. 2016.

Barton, N., "TBM prognoses for open-gripper and double-shield machines: Challenges and solutions for weakness zones and water", *Bergmekanink-Geoteknikk, Oslo*. No. 21. 2013, pp. 1–17.

Barton, N., Abrahão, R.A., "Employing the QTBM prognosis model", *Tunnels & Tunnelling International*. December. 2003, pp. 20–23.

Bieniawski, R., Celada, B., Tardáguila, I., "Selección de tuneladoras para macizos rocosos con frentes estables", *Ingeopres*. No. 181. Madrid, Spain. 2009.

Bruland, A., "The NTNU prediction model for TBM performance", *Norwegian Tunnelling Society*. Publication No 23. 2014.

Celada, B., Adasme, J., González, H., "Design and engineering during the construction of Los Leones station (Line 6, Santiago Subway, Chile)", *Proceedings of ITA WTC*. San Francisco, CA. 2016.

Cobreros Aranguren, J.A., "El proyecto del túnel ferroviario de alta velocidad a través de la Sierra de Guadarrama", *Túnel de Guadarrama*. ADIF, Madrid, Spain. 2005.

Expositions NLFA, Fondation Hänggiturm à Ennende, Zurich, Switzerland. 2006.

Hourtovenko, S., "Multi Mode TBMs", *Challenges and Innovations in Tunnelling*. Tunnelling Association of Canada. October 2015.

Lunardi, P., *Design and construction of tunnels. Analysis of controlled deformation in rock and soils (ADECO-SR)*. Springer, 2006.

Melis, M., *Madrid Metro and Railway Infraestructure. (1995–2003)*. Comunidad de Madrid, Madrid. 2003.

Melis, M., Arnaiz, M., Fernández, E., "Mega EPBMs lead the way for Madrid's renewal", *Tunnels & Tunnelling International*. June. 2006, pp. 23–25.

Mendaña, F., Villanueva, J.P., "Impermeabilidad y calidad general de revestimiento en los túneles construidos con TBM", *Revista de Obras Públicas*. No. 2. 2014, p. 3557.

Robbins, R.J., "Mechanised tunneling-progress and expectations", Sir Julius Wernher Memorial Lecture. *Tunnelling '76: Proceeding of an International Symposium*. London. 1976.

Rostami, J., Ozdemir, L., "A new model for performance prediction of hard rock TBMs", *Rapid Excavation and Tunneling Conference (RETC) Proceedings*. Boston, MA. 1993.

Salehi, B., Shahandeh, M., "Forward March", *Tunnels & Tunnelling International*. December. 2013.

Shen, C.P., Tsai, H.C., Hsieh, Y.S., Chu, B., "The methodology through adverse geology ahead of Pinglin Large TBM", *Rapid Excavation and Tunneling Conference (RETC) Proceedings*. Orlando, FL. 1999.

Thames Tunnel Co., *An Explanation of the Works of the Tunnel under The Thames*. W. Warrington, London. 1838.

Thewes, M. "TBM Tunelling Challenges-Redefining the State-of-the-Art". Keynote lecture at the 2007 ITA World Tunnel Congress, Prague, Magazine Tunel, Vol. 16, extra issue, pp. 13–21. 2007.

Thewes, M., "Bentonite slurry shield machines: State of the art and important aspects for application", *Operación y mantenimiento de escudos: presente y futuro*. Aula Payma Cotas, Barcelona, Spain. 2009.

Workman, D., Martz, D., Lipofsky, S., "A novel continuous conveyor system and its role in reccord-setting rales at the indianapolis deep rock tunnel connector", *ITA World Tunnel Congress Proceedings*. San Francisco, CA. 2016.

Chapter 8

Constitutive models to characterize the ground behavior

Benjamín Celada Tamames and Pedro Varona Eraso

I would give everything I know for half of what I ignore.

René Descartes

8.1 INTRODUCTION

Previous chapters have focused on the description of underground works typology, reviewing basic issues, proposing design methodologies for underground works, describing methods for tunnel construction and, above all, proposing the appropriate ground characterization to accurately define its stress–strain behavior.

This chapter deals with the selection of the constitutive models that must represent, as accurately as possible, the stress–strain behavior of an excavation. This activity is the main phase of the Active Structural Design methodology, presented in Section 2.4.4.

8.2 THE OBJECTIVE OF A CONSTITUTIVE MODEL

The main objective of a constitutive model, also referred to as failure criteria, is to define all possible limit stress–strain states of the ground.

For this purpose, constitutive models are expressed as a mathematical function, formulated in stresses and/or strains, adjusted for each type of ground according to its stress–strain characteristics.

8.3 SELECTION OF A CONSTITUTIVE MODEL

The selection of the constitutive model for a given ground should be made by choosing the one that best represents its stress–strain behavior, both in the elastic domain, in the yielding phase before and after failure; and using the results available from laboratory tests.

The following sections present the stability mechanisms of the ground during tunnel construction and the available constitutive models; whose clear understanding is essential for the selection of the appropriate constitutive model.

8.3.1 Stability mechanisms in underground excavations

Hoek and Brown (1980) proposed new stability mechanisms for the ground around tunnels, shown in Figure 8.1, defining four categories:

- Shallow tunnels
- Fractured rock mass at medium or low depth
- Massive rock mass at medium or low depth
- Massive rock mass at great depth

Subsequently, Kaiser et al. (2000) investigated Hoek and Brown's ideas further by quantifying the stability mechanisms of the ground in nine cases, as illustrated in Figure 8.2.

In this case, the fracture condition of the rock mass is evaluated depending on its rock mass rating (RMR) value and to quantify the effect of stresses on the ground behavior, the major principal stress or the stress induced by mining activities close to the tunnel are considered.

Recently, Loring (2009) qualitatively defined 20 cases of instability in underground excavations, which correspond to the combination of five fracturing states and four stress levels, as shown in Figure 8.3.

Representation	Features
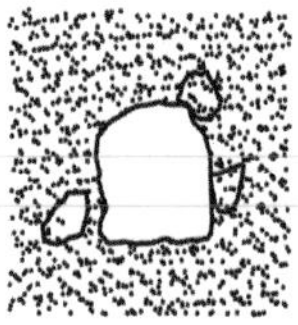	• Shallow tunnels in soils or weathered rock mass • Ground yielding at the tunnel face and footings • Risk of instability and short stand-up time • Cut and cover construction techniques, tunnels in soils and urban areas
	• Blocky jointed rock partially weathered • Medium-low stress level • Instability associated to the falls of blocks from roof and sidewalls
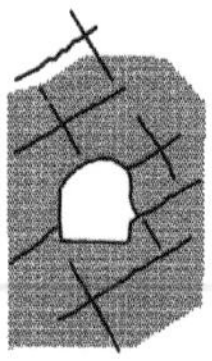	• Massive rock with few unweathered joints • No serious stability problems
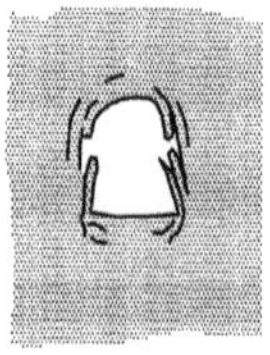	• Massive rock mass at great depth • Brittle stress induced failures, spalling and popping with possible rockbursts

Figure 8.1 Stability mechanisms for underground excavations.

Source: Hoek and Brown, 1980.

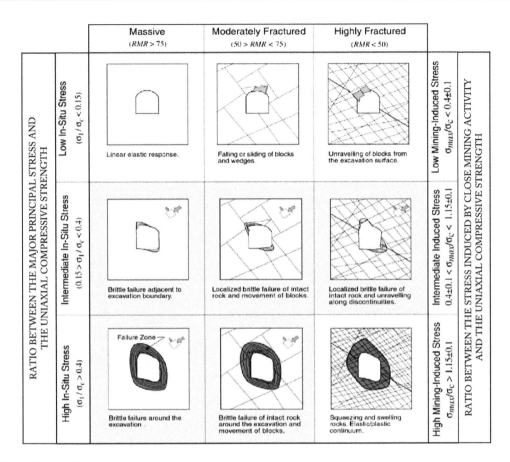

Figure 8.2 Stability mechanisms of the ground.

Source: Kaiser et al., 2000.

In 2015, the GT30 Working Group from the AFTES presented ten stability mechanisms in underground works by combining four basic parameters: fracturing, diameter of the excavation, rock strength and in-situ stresses.

To sum up, it can be concluded that, basically, there are four stability mechanisms in underground works:

- **Very intense yielding,** produced in excavated sections with ICE < 15, whose solution at the design stage will require the use of specific constitutive models.
- **Intense to zero yielding,** which will appear in sections with ICE values in the range 15 < ICE < 130. In these cases, conventional constitutive models can be used.
- **Fall of rock blocks,** which will occur in excavations located at low to medium depth in areas with little yielding; defined by ICE values > 70. These mechanisms require specific models, based on the displacement mechanisms of rock blocks.
- **Rockbursts,** are produced in fully elastic grounds, defined by ICE values > 130, in very deep excavations. The rockburst phenomena are relatively local and require very specific constitutive models.

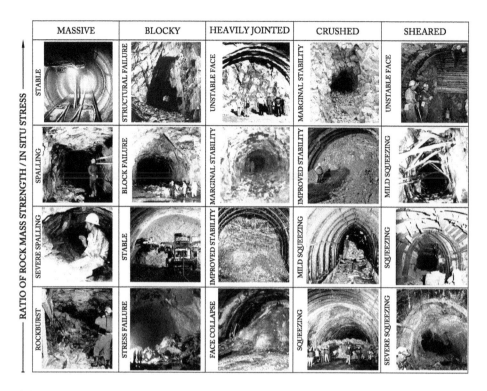

Figure 8.3 Illustration of the stability mechanisms in underground excavations.

Source: Lorig, 2009.

8.3.2 Main constitutive models

The loading process of the ground starts with an elastic response which, according to Section 5.2, continues until the applied load reaches 40 to 50% of its maximum resistance.

At that moment the yielding phase begins, which continues until the peak strength is reached, next, the post-failure phase begins and develops until the sample disintegrates. This process varies depending on the type of ground. Four different behaviors, are distinguished which are shown in Figure 8.4.

Brittle behavior, typical in hard and very hard rocks ($\sigma_{ci} > 70$ MPa), is characterized by the complete loss of strength after reaching the peak value, with an axial strain smaller than 1%.

The perfectly ductile behavior, obtained when the peak stress is maintained until axial strains above 5% are reached, is unusual and only occurs in some evaporite rocks, such as rock salt.

Ductile behavior with softening, which is typical of rocks with low to medium strength, $10 < \sigma_{ci} < 70$ MPa, is an intermediate behavior between brittle and perfectly ductile behavior.

For soft rocks or soils, $\sigma_{ci} < 10$ MPa, it is common to have a decrease in the modulus of deformation when approaching the peak stress at the elastic phase, which is usually reached at axial strains larger than 3%.

Brittle failure, perfectly ductile failure and that achieved after bilinear elasticity, can be explained by constitutive models formulated on stresses, because in these cases it can be assumed that there is a unique relationship between stresses and strains.

However, as shown in Figure 8.5, for ductile fracture with softening, a given value of the stress corresponds to two different strains, which requires expressing the constitutive models on strains for this type of ground behavior.

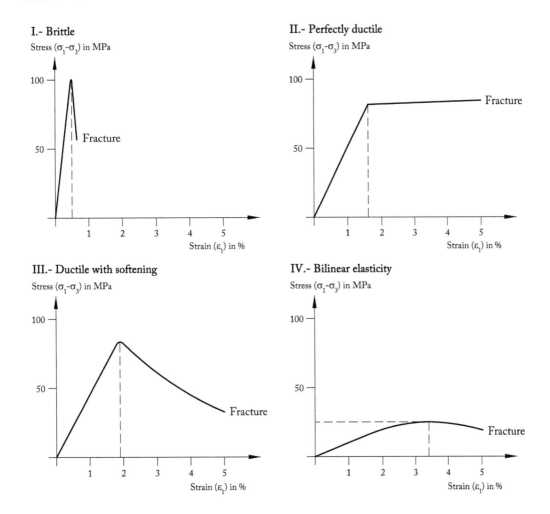

Figure 8.4 Types of failure in rocks.

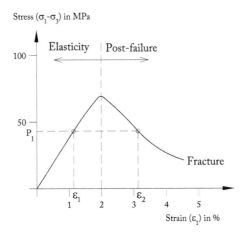

Figure 8.5 Relationship between stresses and strains in ductile fracture with softening.

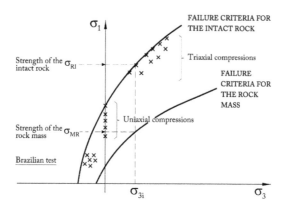

Figure 8.6 Failure criteria for the intact rock and the rock mass.

8.4 CONSTITUTIVE MODELS FOR THE INTACT ROCK AND THE ROCK MASS

As noted previously, the equation that defines the constitutive model of a ground must be adjusted to its specific characteristics based on the results from the laboratory tests; but as these tests are performed in samples without discontinuities, the results obtained will only represent the constitutive model for the intact rock.

According to the scale effect, described in Section 1.3.2., the failure criterion for the rock mass should enable less intense stress states than for the intact rock, as illustrated in Figure 8.6.

In this figure, the curves of the intact rock and of the rock mass are represented and, as it can be appreciated, for a certain value of the confinement stress (σ_{3i}) there is a value of the rock mass strength (σ_{MR}) which is smaller than that of the intact rock (σ_{RI}).

Nowadays, the only constitutive model that allows the failure criterion to change from that of the intact rock to that of the rock mass is Hoek and Brown's model (1980), which is presented in Section 8.5.1.3.

8.5 MOST COMMONLY USED CONSTITUTIVE MODELS

Constitutive models for rocks have been extensively developed throughout the 20th century, so, currently, there are many available constitutive models, although most of them are not commonly used.

As an example, the characteristics of 15 constitutive models which are part of FLAC3D V5.0 program library, Itasca Consulting Group (2012), are presented in Table 8.1.

In addition, it should be taken into account that FLAC3D enables creating constitutive models not included in its library, although they have to be programmed in FISH language which is part of the FLAC program.

In light of the current excess of available constitutive models, it must be understood that as the complexity of a constitutive model increases, so does the difficulty in obtaining real and representative data that could be adjusted to a given ground.

Accordingly, in the calculations for feasibility and detail projects, it makes no sense to use constitutive models which are not based on laboratory tests.

Table 8.1 Constitutive models available in Version 5.0 of FLAC3D

No.	Name	Utility
1	Null	Elements of a geomechanical model that are excavated
2	Elastic, isotropic	Elastic and homogeneous grounds
3	Elastic, orthotropic	Elastic grounds with three orthogonal planes of symmetry
4	Elastic, transversely isotropic	Layered grounds with different properties in normal and parallel directions to the layers
5	Drucker–Prager	Soft clays with low friction angles
6	Mohr–Coulomb	General use
7	Ubiquitous Joint	Layered grounds
8	Yielding with softening	Modeling of the ductile post-failure with softening
9	Ubiquitous Joint with bilinear yielding	Grounds with softening, matrix yielding and ubiquitous joints
10	Double Yield	Grounds with shear yielding and compaction
11	Modified Cam-Clay	Grounds with significant volume change during the loading process
12	Hoek–Brown	General use in rock masses
13	Modified Hoek–Brown	Characterization of the plastic strains through dilation
14	Cysoil	Soils with nonlinear behavior
15	Simplified Cysoil	Hyperbolic friction-hardening law and Mohr–Coulomb with two dilation models

The following sections present the constitutive models which are most commonly used today.

8.5.1 Constitutive models formulated on stresses

Constitutive models formulated on stresses are the most numerous, and among them, Mohr–Coulomb, Hoek–Brown and Cam-Clay are mostly used for the design of underground excavations, whose most important characteristics are explained in the following sections.

8.5.1.1 Mohr–Coulomb model

The Mohr–Coulomb failure criterion was presented by Christian Otto Mohr in 1882, based on the works developed by Coulomb (1776) affirming that the ground fails under a shear stress, which can be calculated by the expression:

$$\tau = c + \sigma_n \cdot tg\phi \tag{8.1}$$

where

τ = shear stress (MPa)
c = cohesion of the ground (MPa)
σ_n = stress normal to the failure surface (MPa)
ϕ = angle of internal friction (°)

Figure 8.7 shows the failure surface at shear stress which, according to the Mohr–Coulomb criterion, would be created inside a test specimen subjected to triaxial compression. This plane is defined by the angle $\phi = 45 + \phi/2$.

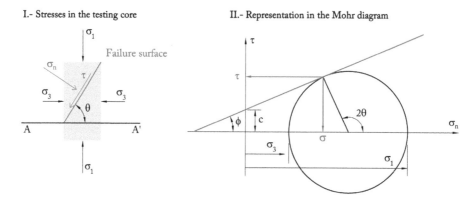

I.- Stresses in the testing core   II.- Representation in the Mohr diagram

Figure 8.7 Stresses developed in a test specimen tested at triaxial compression, according to Mohr–Coulomb.

Equation 8.1 can be written in principal stresses:

$$\sigma_1 = \sigma_c + \frac{1 + \text{sen}\phi}{1 - \text{sen}\phi} \cdot \sigma_3 \tag{8.2}$$

where

σ_1 = major principal stress
σ_c = uniaxial compressive strength
σ_3 = minor principal stress

Equation 8.2 should be used to check the suitability of the cohesion and friction angle adopted in each case, as the compressive strength calculated with this expression should be similar to the results from the uniaxial compression tests.

In order to adjust the Mohr–Coulomb failure criterion to a specific ground, all available tests should be used, representing them in a Mohr stress diagram, defined by τ and σ_n.

For this, it should be taken into account that, in most common tests, the principal stresses at failure are the ones given in Table 8.2.

Using the criteria from this table, Figure 8.8 shows the adjustment of Mohr–Coulomb failure envelope by five tests: two triaxial tests, one uniaxial compression test, one Brazilian test (indirect tension) and one direct tensile test.

The angle of internal friction of the ground (ϕ) is the one that the Mohr–Coulomb failure envelope forms with the σ_n axis and the cohesion is the ordinate at which the failure envelope cuts the τ axis.

Table 8.2 Principal stresses in various tests

	Principal stresses	
Test	σ_1	σ_3
Uniaxial compression	σ_c	0
Indirect tension (Brazilian test)	$3\sigma_T$	$-\sigma_T$
Triaxial compression	σ_1	σ_3 (constant)

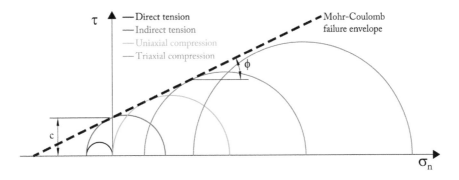

Figure 8.8 Determination of Mohr–Coulomb failure envelope.

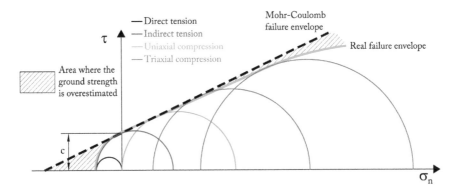

Figure 8.9 Comparison between Mohr–Coulomb failure envelope and the real failure envelope.

8.5.1.2 Improved Mohr–Coulomb model

The main problem of the Mohr–Coulomb model is that it assumes linear ground behavior, both in tensions and compressions, which differs from the real ground behavior, as can be appreciated in Figure 8.9.

In this figure it can be seen that the Mohr–Coulomb criterion overestimates the real ground strength, especially in the area where it works in tension and to a lesser extent in compression. To solve these shortcomings, the Mohr–Coulomb criterion has been improved by adopting two variants:

1. It is considered that in the area at tension, the maximum admissible value of σ_n corresponds to the ground real tensile strength.
2. It is considered that the cohesion (c) and the friction angle (ϕ) of the ground are not constant but depend on the normal tension acting on the failure surface.

From a practical point of view, this means that several pairs of values (c, ϕ) can be defined from the tangent to the real failure envelope for a given value of σ_n.

Figure 8.10 illustrates these two variants, which substantially improve the Mohr–Coulomb model provided that the value of σ_n is properly chosen.

The assumption that both the cohesion and the angle of friction of the ground depend on the normal stress is still questionable after more than 130 years using the Mohr–Coulomb criterion.

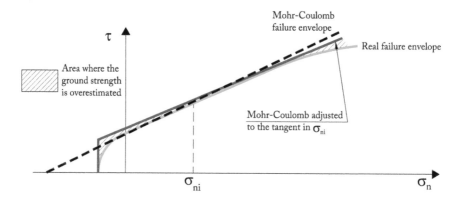

Figure 8.10 Improvement of Mohr–Coulomb model, by limiting the stresses at tension and estimating c and ø depending on σ_n.

8.5.1.3 Hoek–Brown model

The Hoek–Brown model was developed during their research for the book *Underground Excavations in Rock*, which was published in 1980.

This model solves the two major errors of the Mohr–Coulomb model; taking into account the value of the ground tensile strength and considering that the dependence between σ_1 and σ_3 is not linear.

According to Hoek and Brown (1980), the initial formulation of the Hoek–Brown criterion, which was aimed at defining the behavior of intact rock samples, is:

$$\sigma_1 = \sigma_3 + \sqrt{m \cdot \sigma_c \cdot \sigma_3 + s \sigma_c^2} \tag{8.3}$$

where:

  σ_1 = major principal stress at failure
  σ_3 = minor principal stress at failure
  m = material constant of the ground
  σ_c = uniaxial compressive strength of the intact rock
  s = material constant of the ground

Figure 8.11 shows the Hoek–Brown failure criterion, showing the position of representative points from triaxial tests, uniaxial compression tests and direct tension.

When using this criterion for a rock mass, Equation 8.3 changes to:

$$\sigma_1 = \sigma_3 + \sqrt{m \cdot \sigma_{CMR} \cdot \sigma_3 + s \sigma_c^2} \tag{8.4}$$

where

  σ_{CMR} = uniaxial compressive strength of the rock mass.
  Making $\sigma_3 = 0$ in Equation 8.4, which corresponds to uniaxial compression, results:

$$s = \left(\frac{\sigma_{CMR}}{\sigma_c} \right)^2 \tag{8.5}$$

For the intact rock $\sigma_{CMR} = \sigma_c$, i.e., $S = 1$ and the Hoek–Brown criterion is written as Equation 8.4 with $m = m_i$.

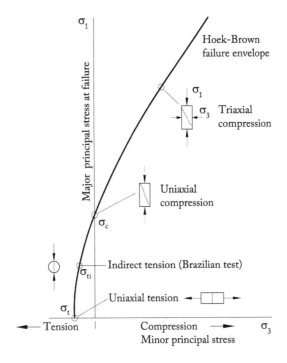

Figure 8.11 Illustration of the Hoek–Brown failure criterion, 1980.

Accordingly, to define the Hoek–Brown failure envelope for the intact rock, it is only necessary to know its compressive strength (σ_c) and the value of m_i by using the results of the laboratory tests.

To adjust the value of m_i contained in Expression 8.4, results:

$$m_i = \frac{(\sigma_1 - \sigma_3)^2 - \sigma_c^2}{\sigma_c \cdot \sigma_3} \tag{8.6}$$

And making a change of variables

$$\left. \begin{aligned} y &= (\sigma_1 - \sigma_3)^2 \\ x &= \sigma_3 \end{aligned} \right\} \tag{8.7}$$

results

$$m_i = \frac{y - \sigma_c^2}{\sigma_c \cdot x}$$

and also

$$y = \sigma_c^2 + m_i \cdot \sigma_c \cdot x \tag{8.8}$$

Equation 8.8 is a linear equation in (x, y) which, knowing σ_c, allows determining m_i by linear regression.

8.5.1.4 Modified Hoek–Brown model

The Hoek–Brown criterion has had two important modifications: Hoek–Brown (1988) and Hoek–Brown (2002).

The most relevant modification, presented in 1988, was the use of Priest–Brown expressions (1983) to calculate the parameters m and s of a rock mass using the RMR and the value of m_i for the intact rock.

Priest–Brown equations, for underground works, are:

$$\left.\begin{array}{l} m = m_i \cdot e^{\frac{RMR-100}{28}} \\[2mm] s = 1 \cdot e^{\frac{RMR-100}{9}} \end{array}\right\} \tag{8.9}$$

Another interesting contribution of the 1988 version is the possibility of estimating the values of m and s in the preliminary phases of the projects, using the data included in Table 8.3.

Afterward, *Hoek* et al. (1995) published the values of m_i for the most important types of rocks, which are shown in Table 8.4.

The values included in Table 8.4 have been obtained from laboratory tests, applying the load perpendicularly to the cleavage and the values indicated in brackets have been estimated.

With the modifications introduced in 2002, the expression for the Hoek–Brown constitutive model changed to:

$$\sigma'_1 = \sigma'_3 + \sigma_{ci}\left(m_b \frac{\sigma'_3}{\sigma_{ci}} + s \right)^a \tag{8.10}$$

Where

σ'_1 = major principal effective stress at failure
σ'_3 = minor principal effective stress at failure
σ_{ci} = uniaxial compressive strength of the intact rock
m_b; s and a = material constants of the ground

The values of m_b and s are calculated by the expressions:

$$\left.\begin{array}{l} m_b = m_i \cdot e^{\frac{GSI-100}{28-14D}} \\[2mm] s = 1 \cdot e^{\frac{GSI-100}{9-3D}} \end{array}\right\} \tag{8.11}$$

where:

GSI = Geological Strength Index
D = Factor associated with the degree of disturbance to which the ground has been subjected by blasting

The parameter a is calculated by the expression:

$$a = \frac{1}{2} + \frac{1}{6}\left(e^{\frac{-GSI}{15}} - e^{\frac{-20}{3}} \right) \tag{8.12}$$

Table 8.3 Estimation of the parameters m and s in preliminary studies

Table 1: Approximate relationship between rock mass quality and material constants

Disturbed rock mass m and s values

undisturbed rock mass m and s values

EMPIRICAL FAILURE CRITERION		CARBONATE ROCKS WITH WELL DEVELOPED CRYSTAL CLEAVAGE dolomite, limestone and marble	LITHFIELD ARGILLACEOUS ROCKS mudstone, siltstone, shale and slate (normal to cleavage)	ARENACEOUS ROCKS WITH STRONG CRYSTALS AND POORLY DEVELOPED CRYSTAL CLEAVAGE sandstone and quartzite	FINE GRAINED POLYMINERALLIC IGNEOUS CRYSTALLINE ROCKS andesite, dolerite, diabase and rhyolite	COARSE GRAINED POLYMINERALLIC & GNEOUS L METAMORPHIC CRYSTALLINE ROCKS amphibolite, gabbro gneiss, granite, norite, quartz-diorite
$\sigma_1' = \sigma_3' + \sqrt{m\sigma_c\sigma_3' + s\sigma_c^2}$ σ_1' = major principal effective stress σ_3' = minor principal effective stress σ_c = uniaxial compressive strength of intact rock, and m and s are empirical constants.						
INTACT ROCK SAMPLES Laboratory size specimens free	m	7.00	10.00	15.00	17.00	25.00
from discontinuities	s	1.00	1.00	1.00	1.00	1.00
CSIR rating: RMR=100	m	7.00	10.00	15.00	17.00	25.00
NGI rating: Q=500	s	1.00	1.00	1.00	1.00	1.00
VERY GOOD QUALITY ROCK MASS Tightly interlocking undisturbed rock	m	2.40	3.43	5.14	5.82	8.56
with unweathered joints it 1 to 3 m.	s	0.082	0.082	0.082	0.082	0.082
CSIR rating: RMR=85	m	4.10	5.85	8.78	9.95	14.63
NGI rating: Q=100	s	0.189	0.189	0.189	0.189	0.189
GOOD QUALITY ROCK MASS Fresh to slightly weathered rock, slightly	m	0.575	0.821	1.231	1.395	2.052
disturbed with joints at 1 to 3 m.	s	0.00293	0.00293	0.00293	0.00293	0.00293
CSIR rating: RMR=65	m	2.006	2.865	4.298	4.871	7.163
NGI rating: Q=10	s	0.0205	0.0205	0.0205	0.0205	0.0205

(Continued)

Table 8.3 (Continued) Estimation of the parameters m and s in preliminary studies

FAIR QUALITY ROCK MASS						
Several sets of moderately weathered	m	0.128	0.183	0.275	0.311	0.458
joints spaced at 0.3 to 1 m.	s	0.00009	0.00009	0.00009	0.00009	0.00009
CSIR rating: RMR = 44	m	0.947	1.353	2.030	2.301	3.383
NGI rating: Q = 1	s	0.00198	0.00198	0.00198	0.00198	0.00198
POOR QUALITY ROCK MASS						
Numerous weathered joints at 30–500 mm,	m	0.029	0.041	0.061	0.069	0.102
some gouge. Clean compacted waste rock	s	0.000003	0.000003	0.000003	0.000003	0.000003
CSIR rating: RMR = 23	m	0.447	0.639	0.959	1.087	1.598
NGI rating: Q = 0.1	s	0.00019	0.00019	0.00019	0.00019	0.00019
VERY POOR QUALITY ROCK MASS						
Numerous heavily weathered joints spaced	m	0.007	0.010	0.015	0.017	0.025
<50 mm with gouge. Waste rock with fines.	s	0.0000001	0.0000001	0.0000001	0.0000001	0.0000001
CSIR rating: RMR = 3	m	0.219	0.313	0.469	0.532	0.782
NGI rating: Q = 0.01	s	0.00002	0.00002	0.00002	0.00002	0.00002

Source: Hoek–Brown (1988).

Table 8.4 Typical values of m_i for various rocks

Rock type	Class	Group	Texture			
			Coarse	Medium	Fine	Very fine
SEDIMENTARY	Clastic		Conglomerate (22)	Sandstone 19	Siltstone 9	Claystone 4
				—— Greywacke —— (18)		
	Non-Clastic	Organic		—— Chalk —— 7		
				—— Coal —— (8-21)		
		Carbonate	Breccia (20)	Sparitic Limestone (10)	Micritic Limestone 8	
		Chemical		Gypstone 16	Anhydrite 13	
METAMORPHIC	Non Foliated		Marble 9	Hornfels (19)	Quartzite 24	
	Slightly foliated		Migmatite (30)	Amphibolite 25 - 31	Mylonites (6)	
	Foliated*		Gneiss 33	Schists 4 - 8	Phyllites (10)	Slate 9
IGNEOUS	Light		Granite 33		Rhyolite (16)	Obsidian (19)
			Granodiorite (30)		Dacite (17)	
			Diorite (28)		Andesite 19	
	Dark		Gabbro 27	Dolerite (19)	Basalt (17)	
			Norite 22			
	Extrusive pyroclastic type		Agglomerate (20)	Breccia (18)	Tuff (15)	

The GSI was presented by Hoek and Brown (1998) with the sole purpose of enabling the parameters m and s from the Hoek–Brown criterion to be obtained.

To calculate the GSI, it is only necessary to evaluate the surface condition of the joints present in the rock mass and their structure, as indicated in the chart in Figure 8.12.

To decrease the subjectivity in determining the GSI, Cai et al. (2004) and Hoek et al. (2015) have presented their publications in which the GSI evaluation is quantified.

Figure 8.13 shows the approach of Hoek et al. (2015) to calculating the GSI using the RQD and the parameter $JCond_{89}$, which corresponds to the parameter of the RMR_{89} that evaluates the condition of the joints.

The parameter D takes into account the degree of disturbance produced in the ground due to blasting operations, Hoek et al. (2002), and it is evaluated according to the criteria presented in Figure 8.14.

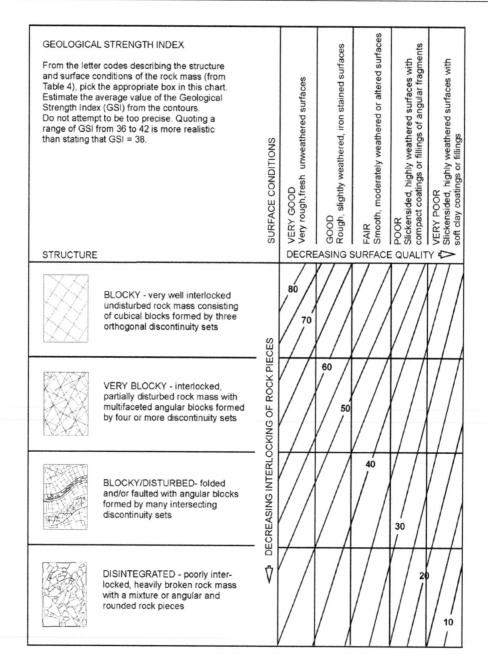

Figure 8.12 GSI Estimation using geological descriptions.

In the work of Hoek et al. (2002), special attention is given to the determination of the D parameter, which has a great influence on the estimation of the ground properties and whose application to underground works must be done with caution.

This is because many underground sections are excavated without blasting and, when used, the disturbance of the remaining ground is very low.

Another aspect to consider is that when there is no ground disturbance by blasting, D = 0, Equations 8.11 and 8.9 are identical and in this case RMR = GSI, a dubious result.

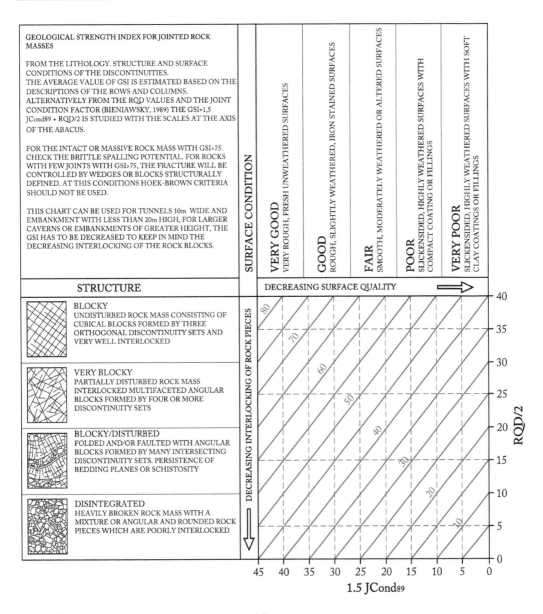

Figure 8.13 Quantitative chart to determine the GSI.

Source: Hoek et al., 2015.

Probably the most interesting contribution of the Hoek–Brown criterion update in 2002 is its relationship with the Mohr–Coulomb criterion, presented in the following section.

8.5.1.5 Linearization of the Hoek–Brown criterion

The Mohr–Coulomb criterion, which in its original version has significant shortcomings when applied to rocks, has been used since 1882 and still is the most used method, despite the advantages derived from using the Hoek–Brown criterion.

Appearance of rock mass	Description of rock mass	Suggested value of D
	Excellent quality controlled blasting or excavation by Tunnel Boring Machine results in minimal disturbance to the confined rock mass surrounding a tunnel.	$D = 0$
	Mechanical or hand excavation in poor quality rock masses (no blasting) results in minimal disturbance to the surrounding rock mass. Where squeezing problems result in significant floor heave, disturbance can be severe unless a temporary invert, as shown in the photograph, is placed.	$D = 0$ $D = 0.5$ No invert
	Very poor quality blasting in a hard rock tunnel results in severe local damage, extending 2 or 3 m, in the surrounding rock mass.	$D = 0.8$
	Small scale blasting in civil engineering slopes results in modest rock mass damage, particularly if controlled blasting is used as shown on the left hand side of the photograph. However, stress relief results in some disturbance.	$D = 0.7$ Good blasting $D = 1.0$ Poor blasting
	Very large open pit mine slopes suffer significant disturbance due to heavy production blasting and also due to stress relief from overburden removal. In some softer rocks excavation can be carried out by ripping and dozing and the degree of damage to the slopes is less.	$D = 1.0$ Production blasting $D = 0.7$ Mechanical excavation

Figure 8.14 Evaluation of the D Factor.

Source: Hoek et al., 2002.

Therefore, it is appropriate to establish a correlation between the results provided by Hoek–Brown and Mohr–Coulomb criteria, for a given range of the minor principal stress (σ_3).

Figure 8.15 shows three ways to linearize the Hoek–Brown criterion.

The linearization through the tangent to the Hoek–Brown curve at the point σ_{3i} gives an excellent adjustment close to this point, but the error is unacceptable in the zone at tension and also, the value of the adjusted uniaxial compressive strength is much larger than the real value.

The linearization through the secant to the value of the uniaxial compressive strength and to the Hoek–Brown curve at point σ_{3i} has a physical meaning, as the adjusted uniaxial

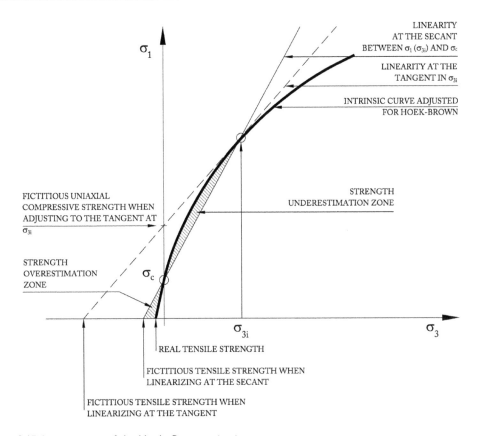

Figure 8.15 Linearization of the Hoek–Brown criterion.

compressive strength matches with the real one, but there are two zones with errors, one underestimating the ground strength and another overestimating it.

Hoek et al. (2002) have proposed an adjustment procedure that reduces both areas of error, whose approach is shown in Figure 8.16 for the range of σ_3 between the ground tensile strength, $\sigma_3 = \sigma_t$ and the maximum value of σ_3.

With this adjustment, the values of the friction angle ($\varnothing'$) and the cohesion (c') are calculated through the complex expressions:

$$\varnothing' = \sin^{-1}\left[\frac{6am_b(s + m_b\sigma'_{3n})^{a-1}}{2(1+a)(2+a) + 6am_b(s + m_b\sigma'_{3n})^{a-1}}\right] \tag{8.13}$$

$$c' = \left[\frac{\sigma_{ci}\left[(1+2a)s + (1-a)m_b\sigma'_{3n}\right](s + m_b\sigma'_{3n})^{a-1}}{(1+a)(2+a)\sqrt{1 + (6am_b(s + m_b\sigma'_{3n})^{a-1}/((1+a)(2+a))}}\right] \tag{8.14}$$

where

$$\sigma_{3n} = \sigma'_{3max}/\sigma_{ci} \tag{8.15}$$

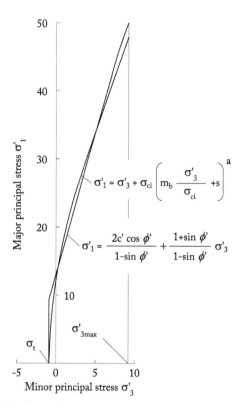

Figure 8.16 Adjustment proposed by Hoek et al. (2002) reducing the areas of error.

This adjustment is valid for the range of σ'_3 between σ_t and σ'_{max}; where

$$\sigma'_{3max} = 0.47 \cdot \sigma'_{CM} \left(\frac{\sigma'_{CM}}{\gamma \cdot H} \right)^{-0.94} \tag{8.16}$$

and σ'_{CM} are calculated by the expression:

$$\sigma'_{CM} = \sigma_{ci} \cdot \frac{(m_b + 4s - a(m_b - 8s))(m_b/4 + s)^{a-1}}{2(1 + a)(2 + a)} \tag{8.17}$$

This methodology is implemented in the RocData program, produced by Rocscience (2014).

8.5.1.6 Cam-Clay model

The Cam-Clay model was developed in the 1960s, by Roscoe–Schofield (1963), Schofield–Wroth (1968) and Roscoe–Burland (1968), to reproduce the stress–strain behavior of grounds with high void ratios, and was further improved by Britto and Gunn (1987).

The Cam-Clay model was initially developed for normally consolidated or slightly over-consolidated clays. This model reproduces a nonlinear elastic behavior with hardening as the average stress acting on the ground increases, in other words, its compaction.

The main behaviors explained with this model are the following ones:

- During the elastic phase the ground modulus is not linear.
- During the consolidation process, the ground has a different modulus of deformation depending on whether the stress is higher or smaller than the overconsolidation stress.
- The existence of critical states is considered, which are those states produced by an increase in the strain without an increase of the effective stress.

Although the Cam-Clay model was originally developed to explain the behavior of clay, subsequent experiments have shown that this model explains, reasonably well, the stress–strain behavior of those materials that, due to their high void ratio and low strength, are likely to have a packing of their matrix during the loading process.

To apply the Cam-Clay criterion to a ground with a void ratio, e, subjected to principal stresses, σ_1 and σ_r, and to a pore pressure u, the following three state variables are used: p', q' and n, defined by:

$$p' = \frac{\sigma_1' + \sigma_r' + \sigma_r'}{3} \tag{8.18}$$

as the following is satisfied

$$\left.\begin{aligned}\sigma_1' &= \sigma_1 - u \\ \sigma_r' &= \sigma_r - u\end{aligned}\right\} \tag{8.19}$$

where u is the pore pressure, results:

$$p' = \frac{\sigma_1 - 2\sigma_r}{3} - u \tag{8.20}$$

$$q' = \sigma_1' - \sigma_r' \tag{8.21}$$

and, when Equation 8.20 is applied, it changes to:

$$q' = \sigma_1' - \sigma_r' \tag{8.22}$$

$$n = 1 + e \tag{8.23}$$

where e is the void ratio of the ground.

According to the critical state theory, if a soil sample is subjected to an isotropic compression test (q = 0) and the trajectory obtained is represented in a diagram (Ln p'; V), a straight line with slope $-\lambda$ or $-\kappa$ will be obtained, depending on whether it is a virgin compression or in an unloading-reloading state.

Figure 8.17 shows the change in volume of a sample tested at isotropic compression, using an odometer test to illustrate the meaning of λ and κ.

The equation for the virgin compression line is:

$$v = N - \lambda \cdot Ln\,p' \tag{8.24}$$

where N is a parameter of the ground.

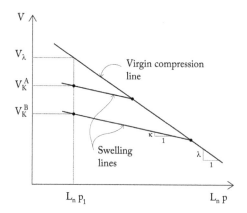

Figure 8.17 Virgin compression and unloading-reloading lines of an isotropic compression test.

The equation for the reloading or unloading line is:

$$v = v_\kappa - \kappa \operatorname{Ln} p' \tag{8.25}$$

The Cam-Clay constitutive model of a ground is defined by the following four parameters:
M: is related to the effective friction angle of the material, ø', by the expressions:

$$\text{In compression } M = \frac{6 \operatorname{sen} ø'}{3 - \operatorname{sen} ø'} \tag{8.26}$$

$$\text{In tension } M = \frac{6 \operatorname{sen} ø'}{3 + \operatorname{sen} ø'} \tag{8.27}$$

N: corresponds to the specific volume of the ground for $p' = 1$. This parameter depends on the unit system chosen.
κ: is the slope of the overconsolidation line.
λ: is the slope of the virgin compression line.
   In addition, to apply the Clam-Clay criterion, it is necessary to know the preconsolidation stress of the ground, in order to choose the consolidation line followed by the stress state of the ground.
   To obtain representative parameters of the real situation, it is recommended to perform the oedometer tests up to a pressure equal to or greater than four times the preconsolidation pressure, which may require the use of nonconventional devices.
   In spite of the great complexity of using a Cam-Clay constitutive model in a stress–strain calculation program, this constitutive model is probably one of the best to study the behavior of grounds that can be rearranged under the effect of stresses.

8.5.2 Constitutive models formulated on strains

Both the Mohr–Coulomb and the Hoek–Brown models prevent the pressure inside an element from exceeding the values given by the failure envelopes for each value of σ_3. However, both models, enable maintaining the maximum stress constant when reached.
   This, in practice, means that when using Mohr–Coulomb or Hoek–Brown criteria, the ground has an elasto-plastic behavior which, in general, does not correspond to the real ground stress–strain behavior, as discussed in Section 8.3.

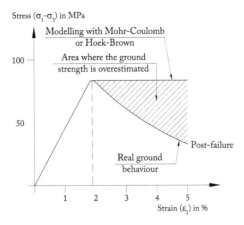

Figure 8.18 Overestimation of the ground strength.

Figure 8.18 illustrates the error made when modeling a ground behavior, which has a ductile post-failure with softening, using Mohr–Coulomb or Hoek–Brown models.

To avoid these problems, the strain softening constitutive model has been developed, which in FLAC is associated with the Mohr–Coulomb model. In this way, the values of the tensile strength, cohesion, friction and dilatancy can be defined, at post-failure, depending on the ground strain, as shown in Figure 8.19.

The strain softening constitutive model allows reproduction of the post-failure behavior, but before doing so it is necessary to perform uniaxial compression tests until failure.

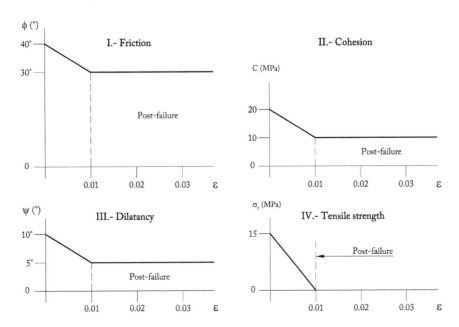

Figure 8.19 Parameter definition for a strain softening model.

8.6 TIME DEPENDENT MODELS

Sometimes it is necessary to solve time dependant problems in tunnel design, for example the phenomena of creeping or anhydrite swelling. However, the models described previously do not take into account the effect of time to solve these problems, so it is necessary to use specific constitutive models.

Accordingly, the following sections contain two constitutive models to solve the problems of creep and anhydrite swelling.

8.6.1 Constitutive model for creep phenomenon

The phenomenon of creep, described in Section 5.7.3, is characterized by ground displacements over time when a constant pressure is applied to the ground; with a constant strain phase, which can last for decades before ground failure. Figure 8.20 shows the evolution of the vertical and horizontal convergence, measured in an investigation gallery during 6,910 days, that is, almost 19 years.

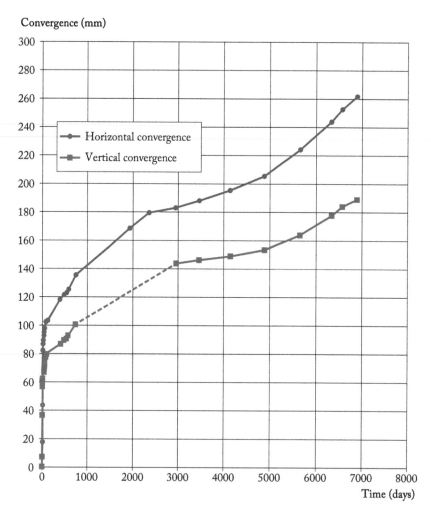

Figure 8.20 Convergences due to creep in an investigation gallery measured during 6,910 days.

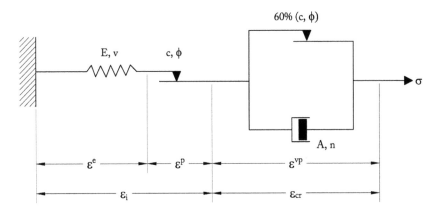

Figure 8.21 Rousset model (1998) for creeping.

One of the most widely used models to simulate ground creep is the one-dimensional visco-plastic model, developed by Rousset (1988); its sketch is shown in Figure 8.21.

In this model, the total strain (ε) is the sum of an instantaneous component, consisting of an elastic component (ε^e) and a plastic component (ε^p), plus the creep component (ε_{cr}), this means,

$$\varepsilon = \varepsilon_i + \varepsilon_{cr} = (\varepsilon^e + \varepsilon^p) + \varepsilon_{cr} \tag{8.28}$$

The elastic strain depends on the modulus of deformation and the Poisson coefficient of the ground and is simulated by the spring in Figure 8.21.

The plastic strain is governed by the cohesion and angle of friction of the ground and is simulated by the slider in Figure 8.21.

The creep strain is simulated by a slider and a damper placed in parallel.

This slider is calibrated so that it does not move until the stress applied on it exceeds 60% of the ground strength, defined by the cohesion and the angle of friction. When this value is reached, the slider allows the movement of the damper and creep strains appear, according to the following law:

$$\dot{\varepsilon}_{cr} = A \cdot \bar{\sigma}^n \tag{8.29}$$

where A and n are ground properties and $\bar{\sigma}^n$ depends on the second stress invariant, which quantifies when the stress state is no longer isotropic and allows creeping; and $\bar{\sigma}^n$ is calculated by the expression:

$$\bar{\sigma} = \sqrt{3J_2} \tag{8.30}$$

where

$$J_2 = \frac{1}{6}\left[(\sigma_1 - \sigma_2)^2 + (\sigma_1 - \sigma_3)^2 + (\sigma_2 - \sigma_3)^2\right] \tag{8.31}$$

A and n can be determined by the tests described in Section 5.7.3, or by adjusting them to real convergence measures if the tunnel is already built.

The Rousset creep model (1988) is implemented in the FLAC program as the CPOW constitutive model.

8.6.2 Constitutive model for anhydrite swelling

Anhydrite swelling in underground excavations, which was explained in Section 5.7.2.2, is a really serious problem because it leads to additional ground stresses on the tunnel support-lining reaching values up to 7 MPa. These values are quite high but, in addition, they usually appear in specific areas at the excavation perimeter, which is especially damaging for the tunnel support-lining.

These pressures are produced by a chemical reaction whose speed rate varies depending on the ground characteristics and the way in which the water, for the start and development of the reaction, is added.

Wittke (1999) has studied, in detail, the anhydrite swelling process and has formulated the following laws of behavior:

$$
\varepsilon_{i\infty}^q = \begin{bmatrix} 0 \\ K_q \log(\sigma_i/\sigma_0) \\ K_q \log(\sigma_c/\sigma_0) \end{bmatrix} \quad para \quad \begin{cases} \sigma_i \geq \sigma_0 \\ \sigma_c \leq \sigma_i < \sigma_0 \\ \sigma_i < \sigma_c \end{cases}
\tag{8.32}
$$

where:

$\varepsilon_{i\infty}^q$ = final axial strain, due to swelling, in "i" direction
σ_0 = axial swelling stress
σ_c = lower limit of the axial stress

This constitutive model considers that the axes of the swelling principal directions $(\varepsilon_{1\infty}^q, \varepsilon_{2\infty}^q, \varepsilon_{3\infty}^q)$ match with the axes of the principal directions $(\sigma_1, \sigma_2, \sigma_3)$ and, therefore, the value of the swelling principal stresses depend only on their corresponding principal stresses.

Figure 8.22 shows the evolution of the vertical swelling stress depending on the constraint imposed on the vertical displacement.

In this case, it can be seen that if the swelling is fully confined, the swelling stress is $\sigma_0 = 750$ Kp and if the anhydrite can freely swell, the swelling stress is zero and reaches 5%.

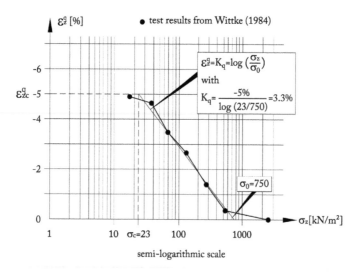

Figure 8.22 Evolution of the swelling stress depending on the constraint imposed on the vertical displacement.

Source: Wittke, 1999.

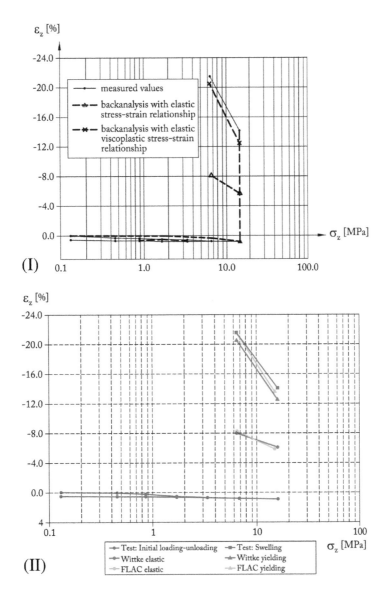

Figure 8.23 Comparison between the data from Huder–Amberg test in yielding with the data adjusted using FLAC program.

Source: Rodríguez–Ortiz et al., 2003. I. Input data (Wittke, 1999). II. Comparison between adjusted and input data.

To define the kinetics of the swelling process, Wittke (1999) uses the following equation:

$$\frac{\partial \varepsilon_i^q(t)}{\partial t} = \frac{1}{\eta_q} \cdot \left[\varepsilon_{i\infty}^q - \varepsilon_i^q(t) \right] \tag{8.33}$$

where:

η_q = parameter that controls the swelling time

$\varepsilon_{i\infty}^q$ = swelling principal stresses for $t = \infty$

$\varepsilon_i^q(t)$ = principal swelling strain at time t

Equation 8.33 is valid as long as the ground strength is not exceeded because, in that case, the increase in volume at post-failure will enable the entry of water and the anhydrite moisture will increase.

To take this effect into consideration, Wittke (1999) proposed to calculate the parameter η_q by the expression:

$$\frac{1}{\eta_q} = a_0 + a_{el} \cdot \varepsilon_v^{el} + a_{vp} \cdot \min\left\{\varepsilon_v^{pl}, \text{maxEVP}\right\} \tag{8.34}$$

where a_0, a_{el} and a_{vp} are constants.

The parameter a_0 represents the change in the swelling rate as a function of the anhydrite content, ε_v^{el} is the volumetric strain before swelling, ε_v^{pl} is the plastic volumetric strain and maxEVP is the upper limit of the plastic volumetric strain.

Rodríguez–Ortiz et al. (2003) have developed an algorithm, programmed in FISH language, to simulate anhydrite swelling using the FLAC program.

Figure 8.23 compares the results obtained with FLAC and the data provided by Wittke (1999), which shows the excellent adjustment obtained with FLAC.

BIBLIOGRAPHY

Association Française des Tunnels et de l'Espace Souterrain (AFTES), *Rapport général du thème A3.* Tunnels et Espace Souterrain, No. 247. 2015.

Britto, A.M., Gunn, M.J., *Critical State Soil Mechanics via Finite Elements.* Ellis Horwood Ltd., London. 1987.

Cai, M.M., Uno, P., Tasaka, Y., Minami, M., "Estimation of rock mass strength and deformation modulus of jointed hard rock masses using the GIS System", *International Journal of Rock Mechanics and Science*, Vol. 41. No. 1. 2004, pp. 3–19.

Coulomb, C.A., "Essai sur une application des regles des maximin et minimis a quelques problemes de statique relatifs à la architecture". *Memoires de l'Academie. Royale pres Divers Savants.* Vol. 7. 1776, pp. 343–387.

Hoek, E., Brown, E.T., "The Hoek–Brown failure criterion – a 1988 update", *Proceedings of 15th Canadian Rock Mechanics Symposium.* University of Toronto, Canada. 1988.

Hoek, E., Brown, E.T., *Underground Excavations in Rock.* Institution of Mining and Metallurgy, London. 1980.

Hoek, E., Carranza-Torres, C., Corkum, B., "Hoek–Brow failure criterion–2002 edition", *Proceedings North America Rock Mechanics Society.* TAC Conference. Toronto, Canada. 2002.

Hoek, E., Carter, T.G., Diederichs, "MS. Cuantificación del ábaco del Índice de Resistencia Geológica GSI", *Ingeotúneles.* No. 23. Madrid, Spain. 2015.

Hoek, E., Kaiser, P.K., Bawden, W.F., *Support of Underground Excavations in Hard Rock.* Taylor & Francis, New York. 1995.

Itasca Consulting Group-Ine, *FLAC 3D Constitutive Models.* Mineapolis, MN. 2012.

Kaiser, P.K., Diederichs, M.S., Martín, C.D., Sharp, J., Steiner, W., "Underground works in hard rock tunnelling and mining". *Proc. Geo Eng.* Melbourne, Australia. 2000.

Loring, L.J., "Aplicación de modelos numéricos continuos y discontinuos al análisis de excavaciones subterráneas en macizos rocosos", *Jornada Técnica de Cálculo de Túneles, Sociedad Española de Mecánica de Rocas (SEMR).* Madrid. 2009.

Priest, S.D., Brown, E.T., "Probabilistic stability analysis of variable rock slopes". *Transactions of the Institution of Mining and Metallurgy.* Vol. 92. 1983.

RocScience Inc, RocData 5.0. "*Analysis of rock, soil and discontinuity strength data*". Toronto, Canada. 2014.

Rodríguez-Ortiz, J.M., Varona, P., Velasco, P., "Modeling of anhydrite swelling with FLAC", *FLAC & Numerical Modeling in Geomechanics*. A. A. Balkema, Avereest, the Netherlands. 2003.

Roscoe, K.H., Burland, J.B., "On the generalised stress–strain behaviour of 'wet clay'", *Engineering Plasticity*. Eds. J. Heyman, F.A. Leckie. Cambridge University Press, Cambridge, UK. 1968.

Roscoe, K.H., Schofield, A.N., "Mechanical behaviour of an idealised wet-clay", *Proceedings of 2nd. European Conference on Soil Mechanics and Foundation Engineering*. Wiesbaden, Germany. 1963.

Rousset, G., "Comportement Mécanique des Argiles Profondes. Application au Stockage de Déchets Radioactifs", Thèse de Doctorat de l'Ecole Nationale des Ponts et Chaussées. Paris. 1988.

Schofield, A.N., Wroth, C.P., *Critical State Soil Mechanics*. Mcgraw-Hill, London. 1968.

Wittke, W., *Stability Analysis for Tunnels – Fundamentals*. Geotechnical Engineering in Research and Practice, WBIPrint 4, Essen, Germany. 1999.

Chapter 9

Types of tunnel supports

*Mario Fernández Pérez, Benjamín Celada Tamames,
and Isidoro Tardáguila Vicente*

Success is only found by one who acts!

Miguel de Cervantes

9.1 INTRODUCTION

As stated in Section 1.3.1. tunnel support is placed to reinforce the excavated ground to withstand the increase in shear stresses created during the redistribution of the stresses during tunnel excavation.

This concept of tunnel support means that its elements should quickly interact with the excavated ground and perform together until the tunnel section is stabilized.

Currently, the most common types of tunnel support are steel arches, rock bolts and shotcrete, whose characteristics and fields of application are presented in the following sections.

The document *Specifications for Tunnelling*, published by the British Tunnelling Society and the Institution of Civil Engineers (2010), contains complete specifications for the support types used for tunnel construction in the United Kingdom.

9.2 STEEL ARCHES

In the past, a temporary support of complex tunnel sections was provided by using frames made of wooden logs.

Then followed a permanent lining made of bricks or masonry stones, which was placed as shown in Photograph 9.1 depicting the Blue Ridge Tunnel, 1,302 m long, open to traffic in 1858 in central Virginia (USA) and which stopped its service in 1944.

At the beginning of the 20th century there was a double development; when steel arches started to be used as the main support and the lining began to be made of cast-in-place concrete.

The interaction between the arches, either of wood or steel, and the ground is very low due to the difficulty in ensuring the contact between the arches and the ground, caused by the overbreak.

Moreover, the radial stress transferred to the ground by the arches is also very low, even when they are spaced at a distance of only 0.5 m.

Steel arches are not used in tunnel sections with little yielding (ICE > 70) because these sections are almost self-stable. For this reason, they are mainly used as a support for the short term in tunnel sections with worse behavior, especially in fault zones.

Photograph 9.1 Blue Ridge Tunnel, Central Virginia (USA) lined with bricks.

Source: Tunnels and Tunnelling International, August 2016.

9.2.1 Technology of the steel arches

Steel arches are made up of pieces, as shown in Figure 9.1, which are assembled at the tunnel face and placed manually or with the aid of equipment used at the tunnel face, as shown in Photograph 9.2.

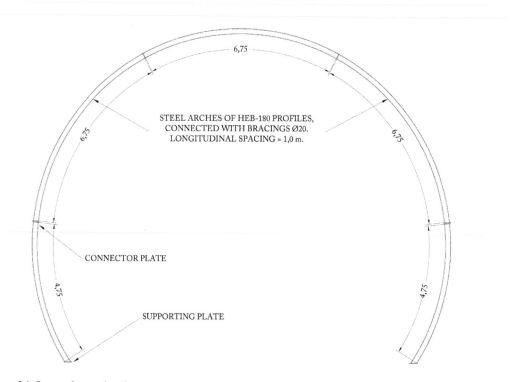

Figure 9.1 Parts of a steel arch.

Photograph 9.2 Positioning of a steel arch at the tunnel face.

9.2.1.1 Yieldable steel arches

In 1932, the Toussant–Heinzman profiles were commercially produced, known since then as TH or Ω profiles, which were characterized by their ability to slide and to prevent profile bending.

TH profiles are weight between 29 and 36 kg/m, although the most commonly used for tunnel supports are the ones between 25 and 36 kg/m. In Figure 9.2 the connection of two TH profiles is shown.

9.2.1.2 Rigid arches

Rigid arches are composed of HEB steel profiles and are joined together by steel connector plates, which are screwed, as shown in Figure 9.3.

In regular tunnel constructions, arches made with HEB-180 profiles or heavier are used. Photograph 9.3 shows the storage of these arches.

Arches made of HEB profile are not suitable in tunnel sections with large displacements, because, as shown in Photograph 9.4, these arches bend when the elements that compose them cannot slide.

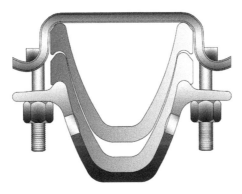

Figure 9.2 Commercialized TH profiles.

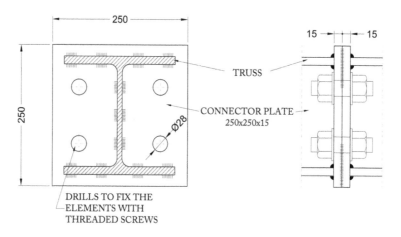

Figure 9.3 Detail of the connection plate between HEB profiles.

Photograph 9.3 Arches of HEB profiles.

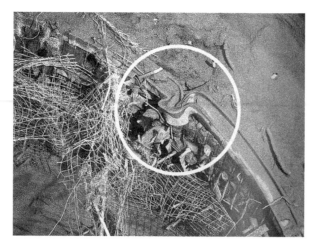

Photograph 9.4 Arch with HEB profiles bent by large ground displacements in Gotthard Base Tunnel.

Source: Ehrbar et al., 2016.

9.2.1.3 Arch lattice girders

Both, the steel arches made of TH or HEB profiles produce significant obstructions when concrete is projected on them, so, to avoid these problems, arches made of lattice girders were developed, which are made of steel bars, as shown in Figure 9.4.

The lattice girders are joined together by screwed plates; in a similar way as HEB profile arches. Photograph 9.5 shows the arch lattice girders.

9.2.1.4 Auxiliary elements

To keep arches parallel and to contribute to distributing the ground pressures, bracings made of steel bars are used, as shown in Figure 9.5.

When arches are used in grounds of a very poor quality it is normal to support them against the ground on plates, around one meter wide, which are called "elephant feet".

If the excavation is done in two phases, usually referred to as top-heading and bench, the elephant feet are placed at the top-heading excavation, as shown in Figure 9.6.

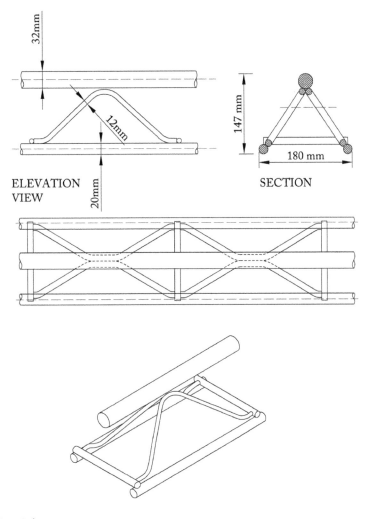

Figure 9.4 Lattice girders.

Photograph 9.5 Lattice arches.

9.2.2 Design calculations of the steel arches

Like other support elements, steel arches are calculated using geomechanical models with specialized software.

These software programs have a library with different structural elements that allow modeling the behavior of the arches and obtaining, after solving the model, the axial forces and bending moments on them.

In FLAC software, the most commonly used program for tunnel design, arches are modeled with *beam* elements, which require the following calculation properties:

- Area of the cross section (m²)
- Moment of inertia and polar moment of the cross section (m⁴)
- Modulus of deformation (N/m²)
- Poisson ratio
- Density (kg/m³)

Figure 9.7 shows the distribution of the axial forces on a steel arch, calculated using FLAC software.

9.2.3 Fields of application

The common field of application of steel arches are tunnel sections with intense yielding, with ICE < 40, in which overbreaks can appear due to the lack of self-stability in the short term.

In these cases, the arches are used as a passive support to prevent local ground falls.

As the arches bearing capacity is very small, they must be used in combination with shotcrete to achieve the tunnel stabilization.

In addition, steel arches can be used as resistant elements to construct protection canopies in the tunnel portals or to construct cut-and-cover sections, acting as a lost formwork, as illustrated (see Photograph 9.6).

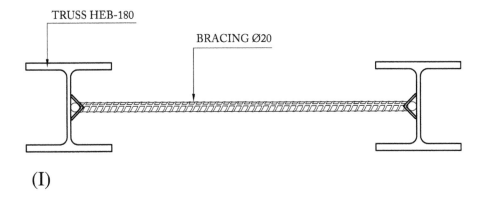

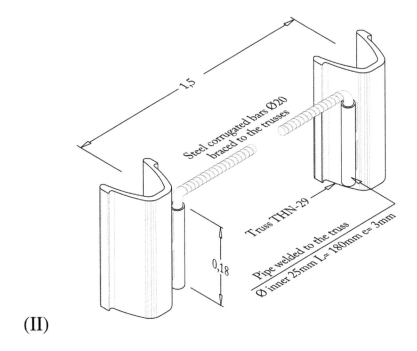

Figure 9.5 Bracings between arches. I. In rigid steel arches. II. In yieldable arches.

9.3 ROCK BOLTS

Rock bolts are bars made of steel or polyester resin with glass fibers, with a diameter of around 25 mm that are placed inside bores drilled in the ground and fixed in it by chemical action or by friction.

In general, the bolt length is usually around one-third of the excavation width in which they are placed and are longitudinally and transversely spaced at distances between 1 and 2.5 m depending on the spacing between the discontinuities.

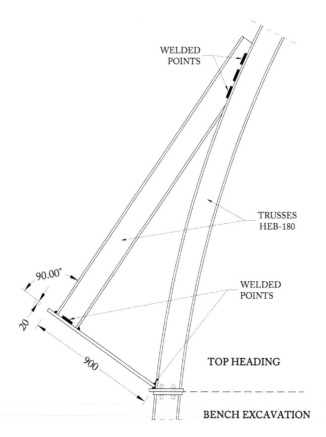

Figure 9.6 Elephant foot details.

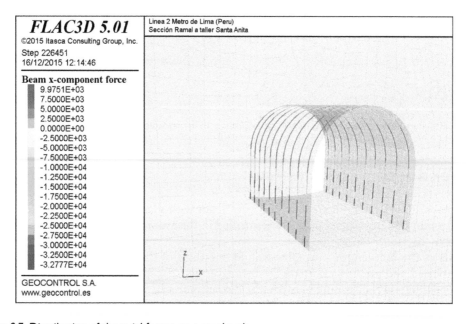

Figure 9.7 Distribution of the axial forces on a steel arch.

Photograph 9.6 Protection canopy, made of steel arches, at the tunnel portal.

The support achieved with bolts provides the best ground interaction, when compared to other supports, and can be used in all types of ground, although it is more efficient and economical in grounds where it is possible to drill bores, with diameters between 30 to 40 mm, without their collapse.

The radial stress reached with bolts is much higher than with steel arches, but is still low. It is known that a steel bolt with an elastic limit of 520 MPa and 25 mm in diameter, breaks at an axial force around 25 tons; so, if they are placed with a spacing of 1 m × 1 m, which is a very high bolt density in tunnels, the stress achieved is 25 t/m² (0.25 MPa).

Under these conditions, the ability of the bolts to modify the stress distribution around the tunnel is small, as shown in Figure 9.8.

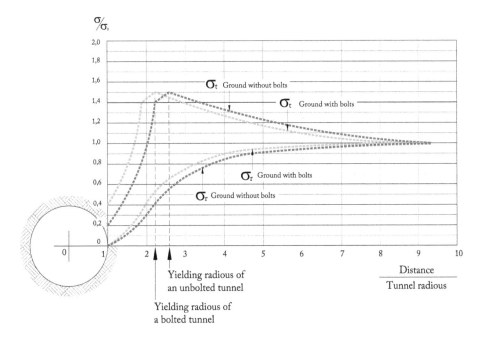

Figure 9.8 Effect of the bolts on the distribution of the stresses around the tunnel.

Source: Celada, 2011.

However, the excellent interaction achieved between bolts and the ground contributes to the ground reinforcement, which increases its strength in contrast to the fully passive effect achieved with steel arches.

9.3.1 Technology of the bolts

A bolt assembly placed in the ground consists of a bar, an anchoring system and a faceplate; the current technology is explained below.

9.3.1.1 Bars

The bars from the bolts are made of corrugated steel, with an elastic limit of 520 MPa, although for self-drilled bolts, steels that reach 750 MPa are used.

When bolts are placed in grounds that must be later excavated, steel bars create significant problems; so they are usually replaced by bars made of polyester resin, reinforced with fiberglass. This material has a slightly higher tensile strength than steel, but its shear strength is much lower, which avoids problems during excavation after their installation.

Photograph 9.7 shows several bolt bars, made of polyester resin and Photograph 9.8 shows a detail of their connection.

Photograph 9.9 shows the excavation of a tunnel face reinforced with polyester bolts.

9.3.1.2 Anchoring system

Currently, there are three anchoring systems for bolts: by cement, by resin or by friction. Their characteristics are explained in the following sections.

9.3.1.2.1 Anchoring with cement

Rock bolts anchored with cement were already used in the 1960s but the installation procedure was purely manual, the performance rates were low and the anchor quality was difficult to control.

Recently, the marketing of additives which transform the cement into a gel, enable easy installation of the bolts anchored with cement.

This anchoring system has not been adopted in other applications, except for long bolts, used to stabilize the slopes. In these cases, bolts of more than 9 m long are used and, at these lengths, the anchorage with resin or by friction is not practical.

To ensure the correct distribution of the cement grout along long bolts, it is necessary to attach a plastic tube along the whole bar, through which the grout is injected using a pump. Photograph 9.10 shows some bolts ready to be anchored with cement.

9.3.1.2.2 Anchoring with resin

In the late 1960s, resin cartridges to anchor bolts were marketed to eliminate the drawbacks of cement anchoring.

The resin cartridges were 25 to 30 mm in diameter and 500 mm in length, and had two compartments. One contained the polyester resin embedded in an inert granular material, and in the other one was a catalyst which after mixing with the resin caused hardening in a few minutes.

Photograph 9.7 Bolts made of polyester bars.

Source: Weidmann Electrical Technology AG, courtesy of Gabriella Williams.

Photograph 9.8 Connection of polyester bars.

Source: Celada, 2011.

Photograph 9.9 Tunnel face reinforced with polyester bolts.

Source: Celada, 2011.

Photograph 9.10 Bolts ready to be anchored with cement.

Source: Celada, 2011.

The installation process of a bolt anchored with resin is as follows:

1. Drilling of a borehole to accommodate the bar of the bolt.
2. Introduce the resin cartridges into the bore.
3. Introduce the bar into the bore using a rotating movement to destroy the cartridges, so that the catalyst and the resin mix.
4. Install the face plate.

Figure 9.9 illustrates the installation process of a bolt anchored with resin.

Bolts anchored with resin quickly prevailed over bolts anchored with cement, both in tunnel construction and for the support of mining excavations; and fully automatic rock bolting rigs were commercialized.

However, resin cartridges could not be stocked for more than six months due to catalyst deterioration.

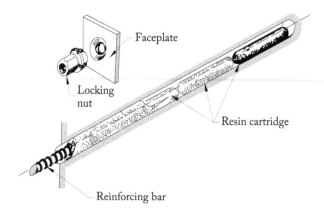

Figure 9.9 Installation of a bolt anchored with resin.

Source: Hoek and Brown, 1980.

Photograph 9.11 Tests about the resin distribution along the bolts.

Source: Celada, 1971.

Moreover, and more importantly, the mixing process between the resin and the catalyst was not always performed correctly, which led to defective anchors.

Bolts longer than 4 m, which were common in tunnels, required a significant effort to mix the resin and the catalyst. This resulted in incorrect anchoring.

To find the optimal rotation of the bars, in order to mix and correctly distribute the resin, specific tests were carried out by anchoring bolts of 1.5 m in length into concrete prisms of 180 cm×20 cm×20 cm, which were later longitudinally broken to observe the resin distribution, Celada (1971). Some of these tests are shown in Photograph 9.11.

Despite the efforts made to establish an efficient methodology to install the bolts anchored with resin, the installation difficulties were not eliminated and, in some cases, collapses due to an inefficient bolt installation took place, such as the one illustrated in Photograph 9.12.

Due to these problems, from the 1990s, the rock bolts anchored with resin were gradually replaced by bolts anchored by friction and, currently, the resin is only used in bolts shorter than 4 m.

9.3.1.2.3 Anchoring by friction

To solve the problems created by bolts anchored with resin, friction anchors were developed. The bar of these bolts was a tube made of a steel plate that exerts a radial pressure on the sidewalls of the borehole where the bolt was placed. Invented in 1973 by James Scott of

Photograph 9.12 Collapse of the tunnel lining due to a poor resin distribution along the bolts.

Colorado (Scott, 1977), the most popular rock bolt with this type of anchoring system is the Split Set®, later followed by Swellex®.

In Split Set bolts, the tube diameter is larger than the borehole in which it will be placed and for this reason the anchor tube is longitudinally slotted, as shown in Figure 9.10.

Split Set bolts are installed into the borehole as shown in Figure 9.11. This avoids the problems associated with the handling of the resin cartridges and creates, due to the decrease in diameter of the Split Set tube when placed into the borehole, a radial pressure in the bore which, although small, creates an initial loading.

The major disadvantage of Split Set bolts is that they are made of a 2.3 mm thick steel sheet, which reduces the maximum axial force that can be withstood to less than 10 t.

This constraint has limited the use of Split Set bolts in the support of mine drifts and its use in tunnel construction is limited.

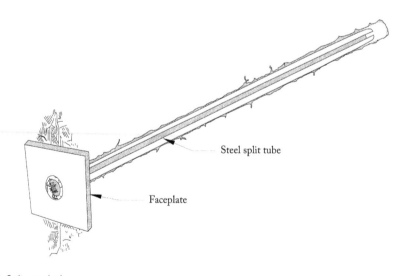

Steel split tube

Faceplate

Figure 9.10 Split-set bolt.

Source: Scott, 1977.

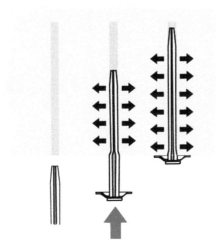

Figure 9.11 Installation process of a Split-set bolt.

Figure 9.12 Manufacturing process of a Swellex bolt.

Source: Epiroc.

Swellex bolts started being sold in the early 1980s and, since then, they have been widely accepted for tunnel construction because they have solved the problems derived from the installation of bolts anchored by chemical procedures and provide a similar strength.

Both Swellex and Split Set bolts consist of a tube made of a steel sheet, but in Swellex bolts, the sheet is folded twice forming a chamber closed at its ends by two steel tube segments.

The manufacturing process of a Swellex bolt is illustrated in Figure 9.12.

As the steel sheet is folded, the strength of the Swellex cross section is much greater than the one of the Split Set bolts, which combined with the use of high-strength steel, gives very large axial strengths.

The installation of Swellex bolts is very simple, as once placed into the drill, it is enough to expand them by injecting water, at 30 MPa, with a small electric pump or driven by compressed air, as shown in Figure 9.13.

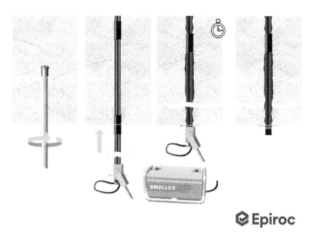

Figure 9.13 Installation of a Swellex bolt.

Source: Epiroc.

Since its commercial release in 1982, Swellex technology has undergone significant development, so that currently there are bolts of this type which can withstand an axial force of 24 tons with strains exceeding 20% of the anchored length, an amount close to what is considered ideal for rock bolts.

In Figure 9.14 i a Swellex bolt is shown.

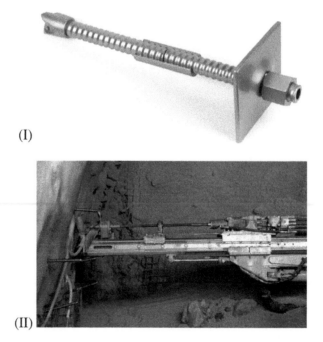

Figure 9.14 Self-drilling bolt and its installation. I. Self-drilling bolt. II. Installation of a self-drilling bolt.

Source: Dolsak, 2008.

Figure 9.15 Swellex bolt.

Source: Epiroc.

9.3.1.3 Self-drilling bolts

The installation of the bolts described above requires that the drillhole inside where they are to be placed does not collapse during the bolt installation. This cannot be ensured in ground with intense yielding, ICE < 15.

To install bolts in this type of ground, self-drilling bolts have been developed. In this invention, the bolt bar acts as a drill rod and features a drill bit. Once the drilling process is complete, slurry of conventional cement is injected through an axial drill in the bar.

A self-drilling bolt and its installation are shown in Figure 9.15. Self-drilling bolts can be placed without major problems at lengths up to 15 m, by using extendible drill rods. With them, very high anchorage strengths are obtained, although obviously, these bolts are significantly more expensive than conventional bolts.

9.3.1.4 Bearing plates

The load capacity of a rock bolt is transmitted to the ground by means of a bearing plate (or face-plate), which is screwed to the bolt.

Under normal conditions, the forces on the faceplates are moderate and to have a good performance it is enough if either the nut or faceplate itself has a semi-spherical surface.

However, if the bolts are close to their maximum strength capacity, the faceplates must be designed to withstand these forces.

As shown in Photograph 9.13, a good solution is to insert a washer between the nut and the faceplate.

9.3.2 Fields of application of rock bolting

Bolting, including both conventional anchoring systems and self-drilling bolts, can be used to support underground works in any type of ground, although this statement requires two important remarks.

The first one is the need to combine the bolts with shotcrete to achieve radial confinement pressures of several MPa, due to the low bearing capacity of the bolts.

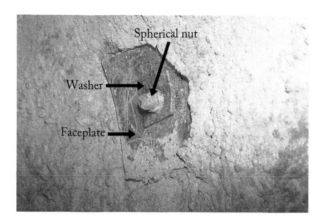

Photograph 9.13 Faceplate for high loads.

Source: Celada, 2011.

The second one refers to the impossibility to place conventional bolts in grounds where the bore hole is not stable during the time necessary to place the bolt.

In these cases, self-drilling bolts can be used, but due to their high cost and the need to have a support, in such grounds the use of combined supports composed of a shotcrete seal and steel arches is preferred.

Accordingly, it should be noted that it makes no sense to combine conventional bolts with steel arches because when it is possible to place conventional bolts, their high interaction with the ground and the radial pressure achieved with them will be greater than those provided by the steel arches.

9.3.3 Design calculations of rock bolting patterns

Except when bolting is used to support rock blocks, calculations of the bolt layouts are made though geomechanical models, solved with specific stress–strain calculation programs. To use them it is necessary to know the following properties:

- Resistant section of bolt bar (m²)
- Tensile strength of the bar (N/m²)
- Compression strength of the bar (N/m²), for steel bars it is equivalent to its tensile strength and it is negligible for fiberglass bars
- Modulus of elasticity of the bar (N/m²)
- Perimeter of the drill (m)
- Resin/grout bonding to the ground (N/m)
- Resin/grout stiffness (N/m/m)
- Cohesion (N/m²) and friction angle (°) of the interface with the ground

It must be kept in mind that the bonding between the anchor and the ground depends on the stress normal to the bolt, which is usually linked to the tangential stress around the tunnel and the type of rock in which the bolt is anchored, as verified in the tests performed by Geocontrol (1990) with the device shown in Figure 9.16.

Photograph 9.14 shows a detail of the cylindrical jack used to apply the confining pressure during the test.

Bonding (τ)

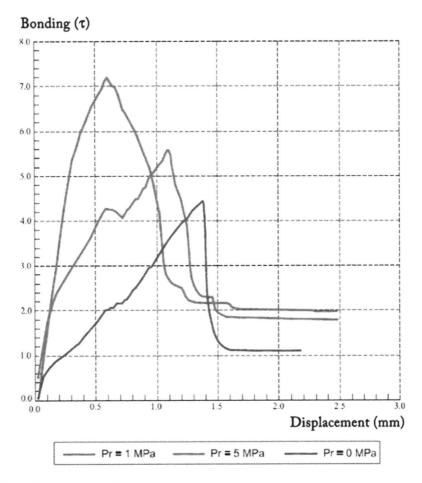

Displacement (mm)

Pr = 1 MPa   Pr = 5 MPa   Pr = 0 MPa

Figure 9.16 Tensile tests with bolt confinement.

Source: Celada, 2011.

Photograph 9.14 Cylindrical jack used to confine the bolts in the tensile tests.

Source: Celada, 2011.

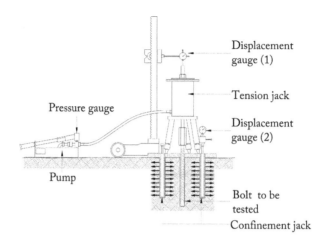

Figure 9.17 Results of the confined tensile tests performed in Carboniferous sandstones.
Source: Celada, 2011.

Table 9.1 Bonding of various carboniferous rocks as a function of the confining pressure of the bolt

Lithology	Sandstone			Slates			Limestones		
Confining pressure (MPa)	0	1	5	0	1	5	0	1	5
Maximum axial load (MN/m)	0.15	0.19	0.25	0.05	0.08	0.17	0.08	0.11	0.22

The curves shown in Figure 9.17 were obtained through these tests; whose peak values are summarized in Table 9.1.

Figure 9.18 shows the distribution of the axial forces acting on the bolts of a tunnel, calculated using FLAC software.

9.3.4 Site supervision of the bolting installation

When bolts are placed, only its free end can be seen, which is why they must be carefully supervised during their installation and have to be tested later.

In tunnels where it is planned to place bolts, there must be clear instructions for their correct installation. Critical aspects to achieve the correct anchorage must be specified: inflation time of the Swellex bolts, grout characteristics and injection pressures for self-drilling bolts, number of resin cartridges to be used and procedure to ensure the correct mixing in the bolts anchored with resin.

In all cases, the suitable diameter of the drill must be clearly specified.

Probably the most important aspect is to verify that the bonding strength of the bolts in the rocks to be excavated is equal to or greater than the minimum values established in the project.

To calculate the bonding force of a bolt, it is necessary to carry out tensile tests in bolts with a length shorter than 60 cm, otherwise the bonding strength can be greater than the axial strength of the bolt and, in that case, the bolt would break before sliding due to the bond failure.

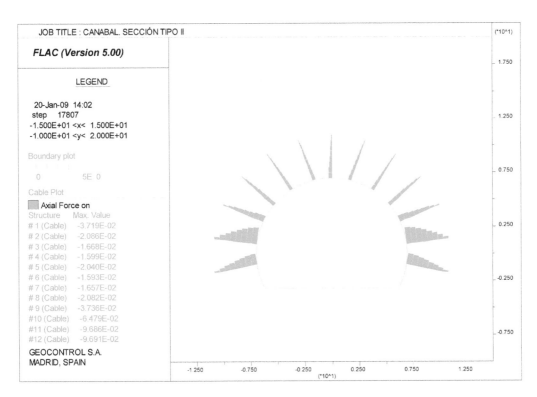

Figure 9.18 Axial forces acting on the bolts of a tunnel.

To carry out this test, a hollow hydraulic jack is required, to pass the bolt through it, as shown in Figure 9.19. The bonding force of the anchor is calculated with the following expression:

$$F = \Pi \cdot D \cdot \tau \cdot l$$

where
    F = maximum axial force of the anchorage
    D = diameter of the drill where the bolt is introduced
    τ = bonding stress of the bolt with the ground
    l = anchored bolt length

To correctly perform this test it is necessary to solve two problems; the first one is to attach the hydraulic jack to the installed bolt and the second one is to apply the load in the direction of the bolt axis.

To attach the tensioning system to the bolts, the usual method is to use a coupling, threaded to the free end of the bolt. When it is desired to test non-threaded bolts, such as Swellex bolts, it is necessary to use a coupling with steel claws.

To apply the tensile force exactly on the bolt axis, the best solution is to use a hemispherical washer, as shown in Figure 9.20.

Tensile tests on bolts placed in the ground are normally performed by applying an axial load of 15 tons so that it is a non-destructive test.

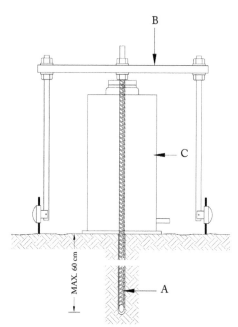

A. BOLT TO BE TESTED
B. STRUCTURE TO MEASURE GROUND
   DISPLACEMENTS
C. HYDRAULIC JACK

Figure 9.19 Tests to determine the bonding of the bolt.

Source: Celada, 2011.

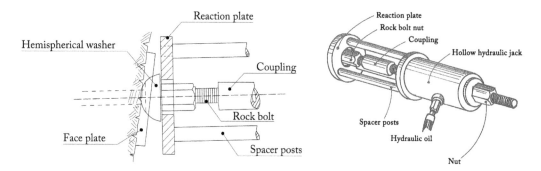

Figure 9.20 Hydraulic jack and loading device to perform tensile tests on bolts placed in the ground.

Source: Hoek and Brown, 1980.

In these cases the admissible failure rate must be very small, usually less than 1% of the tested bolts, because a bolt longer than 3 m must have been very badly installed to slide under an axial force of 15 tons.

9.4 SHOTCRETE

Shotcrete is a type of concrete made with a maximum aggregate size of 10 to 12 mm, with additives to accelerate the setting time without producing a significant decrease in strength.

At the beginning of the 20th century, a product similar to shotcrete was introduced, which was called gunite, derived from an American company that manufactured the machines for spraying; but gunite had two main differences when compared to shotcrete.

The most important one is that the aggregates and the cement were pneumatically driven through a hosepipe, and were mixed with water at its nozzle. Therefore, the strength of the gunite was extremely dependent on the gunner's skills, so the quality obtained often did not meet the design requirements, since its uniaxial compressive strength at 28 days hardly exceeded 20 MPa.

The second difference was in the performance rates obtained, which hardly exceeded 2 m³/h, with high rebound losses, which exceeded 30% when projected onto the tunnel vaults.

At the end of the 20th century the process of wet spraying was commercialized, which, combined with very complex additives, overcame all the shortcomings of gunite by turning it into shotcrete.

9.4.1 Basic components of shotcrete

The basic specifications of shotcrete refer to the following aspects:

- Characteristic strength to uniaxial compression at 28 days, between 30 and 45 MPa
- Maximum size of the aggregate, between 10 and 12 mm
- Quantity and type of cement in the dosage, exceeding 400 kg/m³
- Settlement of the concrete before projection, between 18 and 20 cm measured with the slump test

The following sections explain the characteristics of shotcrete components.

9.4.1.1 Aggregates

The aggregates used in shotcrete are obtained by selecting and classifying natural materials, crushed materials or a mixture of both. The finer sands contribute to improving shrinkage, while the coarser ones increase the rebound percentage.

Aggregates must be composed of clean, hard, strong particles, with uniform quality. The use of fine or coarse aggregates, or a mixture of both, is made according to the thickness of shotcrete layer to be applied. In general, aggregates larger than 15 mm should not be used, due to the limitations of the pumping equipment and also, to avoid the high losses caused by the rebound.

More important are fine aggregates, which are defined as a material, composed of hard and strong particles, of which at least 95% of its weight passes through the sieve number 4 ASTM. The sands with a sand equivalent lower than 80% produce segregation, poor lubrication and risk of clogging and make it very difficult to achieve the required strength. A deficiency of fine material can be compensated by using more cement or additives; to reduce the excess of fine particles, the dosage of water-reducing additives is increased.

The size of the fractions most commonly used in shotcrete are: 0–8 mm, 0–12 mm, and 0–15 mm, as included in the UNE Technical Standard 83607. Figure 9.21 shows the particle size distribution curve recommended by EFNARC for shotcrete aggregates.

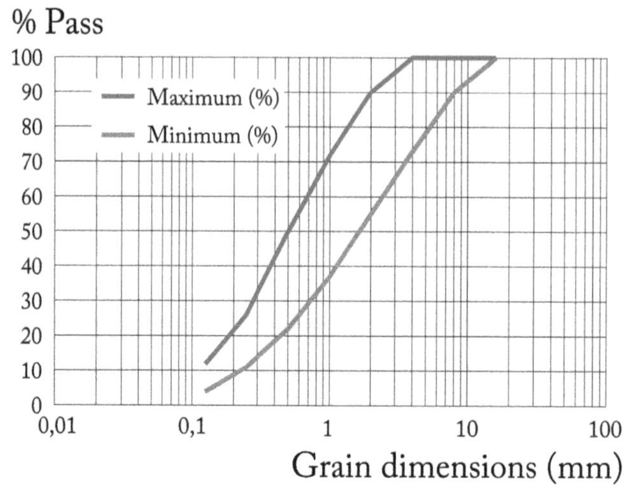

Figure 9.21 Characteristic particle size distribution curve for shotcrete aggregates.

Source: EFNARC.

9.4.1.2 Cement

The types of cement used in shotcrete production are indicated in the Spanish RC-08 instructions, although the most commonly used cements to manufacture shotcrete are Type I-42.5 R and II-52.5 R. In the presence of aggressive water, sulfate-resistant cements should be used.

Note that one of the elements that will have influence on the working period or "open time" of the shotcrete is the type of cement and its dosage. The quantity of cement used in the shotcrete dosage ranges usually between 400 and 500 kg per cubic meter of concrete, depending on the required concrete final strength.

9.4.1.3 Water

When determining the water amount required by the concrete, based on the water/cement ratio, the water absorption coefficient of the aggregate must also be taken into account, a parameter which is difficult to quantify. The typical water/cement ratio for shotcrete is between 0.35 and 0.45.

According to UNE Technical Standard 83-607-94, the mixing water must fulfill the requirements of the Code on Structural Concrete, known as the EHE-08, and must not contain any harmful component in quantities that affect the concrete properties.

9.4.2 Additives

Additives play an essential role in shotcrete to achieve the following objectives:

- Improve its strength and waterproofing
- Reduce rebound losses
- Improve the pumping of the concrete
- Check the service life of the mixture

The following sections explain the characteristics of the most common concrete additives.

9.4.2.1 Super plasticizers

Super plasticizers are the main additive in the shotcrete dosage because the following effects can be achieved:

- They confer cohesive properties to the shotcrete before its projection, maintaining a good workability of the mixture until its spraying and reduce the rebound during the projection.
- Reduction of the water content of the mixture, for the same consistency, between 30 and 40%.
- The final strength is significantly improved.
- Improvement of the durability.

Synthetic polymers such as naphthalene-sulphonates and melamine-sulphonates products are used as super plasticizers.

Recently, "hyper plasticizers" have been introduced resulting in a reduction in the water/cement ratio of more than 40%.

The reference dosage of super plasticizers is 1% of the cement weight and must be adjusted the following characteristics:

- Water/cement ratio around 0.40
- Fluid/liquid consistency of the mixture to enable pumping
- Workability during more than 60 or 90 minutes
- Homogeneous mixture and without particle disaggregation

9.4.2.2 Set-retarding additives

Set-retarding concrete additives are used to delay the concrete setting process up to 36 hours, without decreasing its final properties.

These additives control the hydration process in the concrete by creating a complex of calcium ions on the surface of the cement particles. The process can be reversed after adding another additive in the gun nozzle.

Set-retarding additives are very useful when the distance between the concrete manufacturing plant and the working site is large and when the outer temperature is very high, which demands larger water amounts in the mixture.

9.4.2.3 Set-accelerating additives

Set-accelerating additives produce two important effects:

- Change the shotcrete consistency to dry-plastic, to improve its bonding to the ground and to reduce rebound losses
- Achieve an initial strength as high as possible, so that the shotcrete can start to act as a support as soon as possible

The negative point is that set-accelerators reduce the final concrete strength, within the following limits:

- Approximately 50% with silicate accelerators
- Between 20 and 25% with aluminate accelerators
- Between 2 and 5% with alkali-free accelerators

The typical set-accelerator dosages, expressed as a percentage of the cement weight in the mixture, are usually as follows:

- Between 8 and 12% for silicates
- Between 4–6% for aluminates
- Between 4 and 8% for alkali-free accelerators

Nowadays, there is a tendency to use alkali-free accelerators for the following reasons:

- They are non-caustic products which do not contain soluble alkaline hydroxides, with a pH between 3 and 5, which contributes to health and safety at work.
- Their negative effect on the final strength is significantly lower, although providing high initial strength.
- They contribute better to environmental protection due to the reduction in the amount of aggressive components; both in the shotcrete placed in situ and in the bounced material that is transported to the landfill.

9.4.2.4 Increase of the compactness

The finest particles present in the concrete dosage are cement particles; which have a diameter of about 60 microns.

At the end of the 20th century, micronized silica, obtained as a sub-product of the semiconductor industry, began to be commercially produced to fill the existing gaps between the cement particles in the concrete and to increase its compactness. In this way the following advantages are achieved:

- A greater cohesion of the mixture, as well as an increase of the bonding with the ground and an improvement in the shotcrete projection performances in vault zones
- Reduction of the rebound, to levels lower than 10%
- Increased workability and reduction of the dust in the working site
- Increase of the initial and final compressive strength, thus reducing the accelerant dosage
- Decrease of the shotcrete's permeability, with water ingresses smaller than 30 mm, thereby increasing its resistance to carbonation and its durability

The minimum dosage of microsilica is 30 kg per m^3 of concrete.

Products which contain nanosilica, which are silica particles a thousand times smaller than those of microsilica, improve the characteristics of the concrete, as they create a reactive inorganic network and improve its compactness.

In addition, nanosilica particles play an active role in the concrete setting as they act as initial points in crystallization, reacting with the portlandite released during the cement hydration process. Thus, it acts as a binder that contributes to strengthening the concrete skeleton and improves the following properties:

- Cement hydration
- Gunite compactness, the process of setting and hardening
- Anti-segregation capacity
- Alkali-aggregate reaction in the concrete; because its acidic oxide effect neutralizes this risk
- Permeability

9.4.2.5 Increase of the ductility

Concrete is a material with very low tensile strength compared to its uniaxial compressive strength, which makes it very brittle at post-failure. This brittle behavior does not adapt to the ground displacements generated during the stress redistribution process after the excavation.

To eliminate this drawback, a technology to reinforce the concrete was developed at the end of the 20th century, which consisted in reinforcing the shotcrete with steel fibers and some years later, with synthetic fibers.

Steel fibers are pieces of wire with a diameter of between 0.25 and 1.0 mm, and a length of between 25 and 100 mm and their ends are bent to improve their interaction with the concrete, as is shown In Figure 9.22.

The A820 ASTM Standard indicates that fibers must be made, at least, of steels with 345 MPa of yield strength, but fibers of steels with yield strength up to 1,000 MPa are also manufactured.

The dosages currently used vary from 0.25% to 1.0%, expressed as a percentage of the concrete volume, which means 20 to 80 kg of fibers per m^3 of concrete.

These dosages are small and do not increase the concrete strength significantly, but there is a sharp decrease in its brittle behavior at post-failure, as illustrated in Figure 9.23.

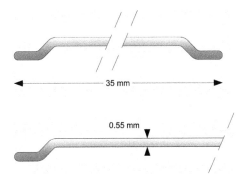

35 mm

0.55 mm

Figure 9.22 RC-65/35-BN Dramix fibers. (Adapted from image provided courtesy of Bekaert).

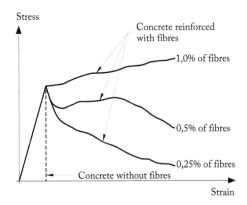

Stress

Concrete reinforced with fibres

1,0% of fibres

0,5% of fibres

0,25% of fibres

Concrete without fibres

Strain

Figure 9.23 Effect of steel fibers on the concrete behavior at post-failure.

Photograph 9.15 Concrete mixer truck and shotcrete robot.

It is thought that during the concrete manufacturing process, the fibers are evenly distributed throughout the concrete mass, so when it is loaded and the first cracks appear, the fibers will bind the two sides of the cracks.

Therefore, it is accepted that steel fibers delay the progression of the cracks in the concrete, so that it loses its brittle behavior and begins to have a ductile behavior which increases with the fiber amount.

9.4.3 Shotcreting

Shotcrete is manufactured in conventional concrete plants, where the additives explained in Section 9.4.2 are added to the mixture and transported in normal concrete mixer trucks.

Close to the tunnel face, the trucks or mixers will pour the concrete into the shotcrete robot, as illustrated in Photograph 9.15.

The two essential parts of the shotcrete robot are the pump that drives the concrete through a rubber hose and the telescopic structure that supports the gun nozzle, remotely controlled by the operator of the projection.

The accelerator is added into the gun nozzle, whose flow rate is regulated by a small pump located in the robot and compressed air, to speed up the shotcrete projection.

The compressed air must have a pressure of between 5 and 6 kp/cm² and can be supplied through the existing network in the tunnel or by a compressor located in the robot.

The following advantages are obtained with the existing robots for shotcreting compared to the old manual systems:

- No dust is produced during the projection.
- Rebound losses are smaller than 10%.
- Projection performance rates reach 30 m³/h.
- The shotcrete quality is excellent; because the gun nozzle can be placed near the surface and with the tilt necessary to do a perfect projection.
- The operator who makes the projection is far from the projection area; so the health and safety conditions are improved.

Photograph 9.16 shows shotcreting by means of a robot.

Photograph 9.16 Shotcreting with a robot.

9.4.4 Field of applications of the shotcrete

Shotcrete has three functions in tunnels:

- **Sealing:** preserves the ground from moisture changes and also acts as a light support in the short term.
- **Primary lining:** provides the radial pressure necessary to stabilize the excavations.
- **Secondary lining:** provides a functional lining to the tunnel perimeter.

The following sections discuss these functions.

9.4.4.1 Shotcrete sealing

After excavation and removing the loose ground pieces with mechanical equipment to detach unstable rock fragments, the ground must be stable during the time necessary to place the support.

During this time the moisture changes at the tunnel face, caused by the ventilation, have a negative impact on the stability as it contributes to opening the joints, with the risk of having rock falls.

Since concrete-spraying robots were marketed, this problem has been eliminated using a shotcrete layer of 5 cm minimum thickness, which is known as the seal.

During the seal spraying, the operator is located at the tunnel section which has been already supported, as shown in Photograph 9.17, so he is protected from possible ground falls at the tunnel face.

To reduce the time at which the seal reaches a minimum strength, between 1.0 and 1.2 MPa, the concrete is usually over-accelerated. In these cases, the seal thickness is not considered for the support calculation, due to the possible significant loss of strength in the long term.

9.4.4.2 Primary lining of shotcrete

Shotcrete provides good interaction between the ground and the support, although it is not as good as the interaction obtained with bolts. However, shotcrete has an advantage over

Photograph 9.17 Shotcreting without operators at the spraying area.

any other type of support, which is the high radial pressure achieved, which easily exceeds several MPa.

As already mentioned, a tunnel support made of shotcrete must be combined with bolts or steel arches. In the latter case, arch lattice girders, made of steel bars, are preferred to those made with steel profiles, because they create spray obstructions that, when shotcreting, become weakness zones in the support.

When shotcrete is used as a support element, the thickness of the layers reaches 80 cm in extreme cases, although it is recommended to shotcrete in layers up to 20 cm thick.

The uniaxial compressive strength required for shotcrete at 28 days is usually 35 MPa, but 45 MPa can be easily reached with careful manufacture.

9.4.4.3 Secondary lining of shotcrete

Shotcrete used in the secondary support has two advantages compared to cast-in-place concrete. One is that it incorporates part of the overbreaks into the resistant tunnel section; and the other one is that linings of less than 30 cm thick, which can't be achieved with cast-in-place concrete, are possible.

The savings associated with these two advantages are maximized in short tunnels, where the economic impact of using a formwork carriage is important and in tunnels with little water, where shotcreting can be done without problems.

Photograph 9.18 shows a gallery, built for skiers, lined with shotcrete.

The quality level that has been achieved in the manufacture and spraying of the concrete, allows the use of this material as a tunnel lining without any problems. Photograph 9.19, shows the cavern of the Roquetes Station, Line 3 Barcelona Subway (Spain), entirely lined with shotcrete.

9.4.5 Shotcrete design specifications

Like other support types, shotcrete is currently designed by solving geomechanical models, which enable knowing the distribution of the bending moments, axial and shear forces.

When the shotcrete thickness is less than 30 cm, it is modeled with a "shell" structural element; but when the thickness is greater, it may be better to model it using the elements

Photograph 9.18 Gallery built for skiers, lined with shotcrete.

Photograph 9.19 Cavern of Roquetes Station, Line 3 Barcelona Subway (Spain), lined with shotcrete.

from the model, assigning them the specific properties of the shotcrete and its variation during the setting process.

The properties required for the calculations are the following:

- Uniaxial compressive strength (N/m^2) and its evolution over time
- Modulus of deformation (N/m^2) and its evolution over time
- Poisson ratio
- Density (kg/m^3)
- Residual strength at post-failure (N/m^2)
- Interface properties between the structural element and the ground (normal and tangent stiffness, failure model, etc.)

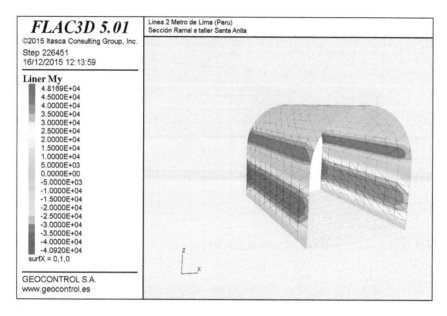

Figure 9.24 Distribution of the bending moments in the shotcrete.

Figure 9.24 shows the distribution of the bending moments in the shotcrete, obtained using FLAC3D software.

9.4.6 Shotcrete control

The following sections explain the principal tests that should be performed on shotcrete samples.

9.4.6.1 Sand equivalent test

The determination of the sand equivalent of the fine aggregate in the shotcrete is essential, because sand equivalents below 80% make it impossible to obtain good concrete strength. This test is regulated by the UNE-EN 933-8:2012 standard.

9.4.6.2 Strength at a tender age

The determination of the shotcrete uniaxial compressive strength at a tender age, below 1 hour, is essential to ensure the proper performance of the seal.

These strengths are measured with a needle penetrometer, which can be used within 15 to 120 minutes after the shotcreting, and enables estimation of the concrete strength in the range of 0.2 to 12 MPa. A needle penetrometer is shown in Photograph 9.20.

At ages between 1 and 24 hours, the most common method used to estimate the strength is a pull out test with steel nails, which are pierced into the shotcrete by shooting them with a nail-gun, like the one shown in Photograph 9.21.

To estimate the shotcrete uniaxial compressive strength using this method, it is necessary to measure the nail penetration length and the force necessary to pull it out.

Photograph 9.20 Needle penetrometer to estimate the shotcrete strength at ages below 1 hour.

Photograph 9.21 Nail-gun to pierce nails into the shotcrete for estimating its compressive strength.

All the existing methods to determine the shotcrete strength at a tender age require access to the site where the concrete has been sprayed, which is a risk when evaluating the shotcrete strength of the tunnel vault. To eliminate this problem, a method is being developed at the University of Cambridge (UK) to measure the shotcrete strength using infrared images.

This method, called strength monitoring using thermal imaging (SMUTI), is being tested during the project at the Bond Street Station of the London Underground (Owen, 2016).

A snapshot of the test performed with this new system is shown in Photograph 9.22.

Photograph 9.22 Determination of the shotcrete strength by thermal images.

Source: Owen, 2016.

9.4.6.3 Shotcrete standard strength

In shotcrete, as in cast-in-place concrete, the reference strength is reached at 28 days, although due to its accelerated setting, the uniaxial compressive strength is also achieved between three and seven days.

Unlike cast-in-place concrete, the test specimens to determine the shotcrete uniaxial compressive strength are 6 cm in diameter and 12 cm in height.

These specimens are extracted from square troughs 50 cm of side and 15 cm thick which are filled in situ, as shown in Photograph 9.23.

9.4.6.4 Tensile strength determination of shotcrete reinforced with fibers

One of the most relevant parameters to characterize the behavior of a fiber reinforced concrete is its tensile strength and there are two specific tests to determine this parameter.

Photograph 9.23 Troughs to test the shotcrete.

Photograph 9.24 Bending test.

Source: Albert de la Fuente, Estado normativo actual a nivel nacional y Europeo, Hormigón con fibras: propiedades y aplicaciones estructurales en la edificación, 2012 with permission

The first test consists in breaking fiber reinforced concrete prisms at bending, which are 550 mm long and have a square section with 150 mm each side. This test is regulated by several international standards: RILEM TC 162-TDF, EHE-08 and the Model Code 2010, among others.

Photograph 9.24 shows a snapshot of the test.

The second test is the Barcelona test, which is regulated by the UNE Standard 83515; it is performed on specimens which are 15 cm in diameter and 15 cm in height, and have the advantage of weighing only 6.5 kg and can be extracted from the already sprayed shotcrete.

Figure 9.25 shows the scheme of the Barcelona test and a specimen ready to be tested.

9.4.6.5 Control of the shotcrete thickness

The control of the shotcrete thickness has been one of the most difficult aspects in the application of this technique because it is impossible to determine by visual observation.

The most effective method to control the shotcrete thickness is to place nails on the surface to be sprayed, whose length must be completely covered by the shotcrete.

The second most effective method is to make drills, around 10 mm in diameter, with which the transition from concrete to rock is detected during its drilling and the real shotcrete thickness is then measured.

Recently, some robots equipped with a topographic scanner have made it possible to measure the cross profile of the sections to be sprayed before and after the projection, which enables to estimating the sprayed thickness.

Photograph 9.25 shows an example of a topographic control of the shotcrete thickness.

9.5 MICROPILES (FOREPOLING)

Micropiles are steel tubes of around 100 mm in diameter and 10 mm thick, which are bonded to the ground by cement slurry and have lengths between 9 and 25 m.

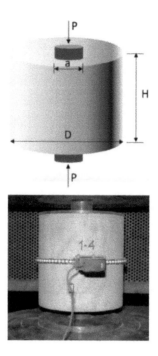

Figure 9.25 Scheme and performance of the Barcelona test. I. Scheme of the test. 2. Specimen ready to be tested.

Source: Aire, Molins and Aguado, 2013.

Photograph 9.25 Topographic control of the shotcrete thickness.

Source: Rivas, 2013.

Also called forepoling, micropiles have two very different applications in tunnel construction: (i) temporary support umbrella and (ii) structural underpinning of some tunnels, as explained in the following sections.

9.5.1 Temporary support umbrella

Temporary support umbrellas are constituted by a fan-like arrangement of micropiles that surrounds part of the tunnel crown and parallel to its axis.

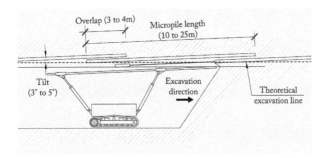

Figure 9.26 Longitudinal tunnel section with a micropile umbrella.

Figure 9.26 shows a longitudinal tunnel section with a micropile umbrella, where the overlap of two consecutive umbrellas is between 3 and 4 m.

Micropile umbrellas are constructed to control the development of overbreaks that appear in tunnel sections with very intense yielding, ICE <15, although they should be systematically used at the tunnel portals as a security measure.

In tunnels excavated at full section and in the tunnel portals, drilling rigs equipped with feeds up to 15 m in length can be used, which can execute micropiles up to 25 m in length without major deviations.

Photograph 9.26 shows an example of the construction of an umbrella of great length.

Drilling rigs with feeds of 15 m require very bulky machines and their use creates difficulties in many tunnels. For that reason, they were no longer used when micropile drilling systems were commercialized which could be placed on the standard jumbos.

In these systems, the steel tubes and the drill rods are introduced into the ground at the same time, without being subjected to any kind of force, joining them in successive sections up to lengths of 12 to 15 m, depending on the type of ground.

Once the drilling is completed, the central part of the drilling bit is removed, leaving the steel tubes ready to be injected.

These systems are much faster than the drilling rigs with long feeds; but, as indicated, with them, the umbrella's length cannot exceed 15 m.

Photograph 9.26 Construction of a heavy umbrella of great length.

Photograph 9.27 Execution of a heavy umbrella at a tunnel portal using a conventional drilling rig.

Photograph 9.28 Heavy umbrella executed with a standard jumbo.

Photograph 9.27 shows the execution of an umbrella with a conventional drilling rig at the tunnel portal, while Photograph 9.28 shows an umbrella made with this technique.

9.5.2 Underpinning

When tunnels are built in grounds of very poor quality, such as soft soils in shallow tunnels, there may be settlement problems with supporting the structure that constitutes the tunnel support.

These problems are easily controlled by underpinning the base of the tunnel support with micropiles, as shown in Figure 9.27.

In these cases, it is very important to design two brace beams containing the upper-ends of the two rows of micropiles, to prevent the micropiles from penetrating the support, which would make the use of micropiles ineffective.

9.5.3 Design of the micropile umbrellas

The micropiles in an umbrella work in a complex way that can be considered as an embedded beam in the ground, supported by steel arches.

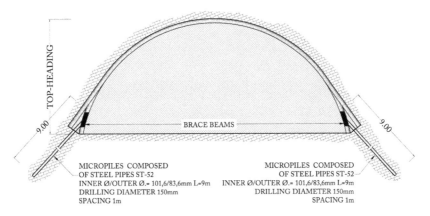

Figure 9.27 Micropile underpinning of a tunnel support.

It is also necessary to take into account an overlap between the umbrellas, between 3 to 4 m in length, as this has a significant impact on design modeling.

When micropiles are used to underpin the support, they work in a simpler way, because they work in compression by transferring the loads to the ground by the friction along the lateral surface of the micropiles and by compression, at the micropile tip.

In both cases, the best way to calculate micropiles is to integrate them into the geomechanical calculation models, which require knowing the following data:

- Drilling diameter (m)
- Uniaxial compressive strength of the slurry (N/m²)
- Inner and outer diameter of the steel tube (m)
- Tensile/compressive strength of steel (N/m²)
- Modulus of elasticity of the steel (N/m²)
- Steel Poisson ratio
- Properties of the interface between the structural element that models the tube and the ground; normal and tangential stiffness

9.6 COMPRESSIBLE SUPPORTS

In deep tunnel sections that have to be built in grounds of very poor quality, with ICE values clearly below 15 points, strong yielding will appear at the tunnel perimeter.

As a result of the strong yielding, significant displacements will appear at the perimeter of the excavation, which may not be controlled by conventional supports and the tunnel section might progressively decrease its size until the support is ineffective.

Figure 9.28 illustrates this situation in a tunnel section, excavated at a depth of 1,500 m, using the methodology of the characteristic curves.

The starting point of the excavation characteristic curve corresponds to the radial stress $\sigma_r = 40$ MPa at no strain.

The stress of 40 MPa is taken from a hydrostatic stress state, $K_0 = 1$, which is consistent with a highly yielded soil at a great depth.

The transition from elasticity to yielding takes places approximately at $\sigma_r = 20$ MPa with $\varepsilon = 0.5\%$. From this point the strains quickly grow until the radial stress reaches its minimum at a strain of $\varepsilon = 4.5\%$.

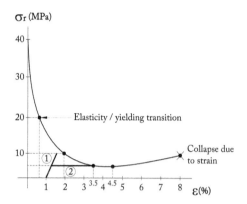

Figure 9.28 The concept of stabilization of a ground with very poor quality in a deep tunnel.

From the minimum radial stress, as the strains increase, so does the radial stress until the limit strain, $\varepsilon = 8\%$, is reached, at which the excavation collapses.

If a conventional support, marked with 1 in Figure 9.28, is placed, at a strain of 1%, it is verified that it is not possible to stabilize the excavation even if the radial stress provided by the support is 10 MPa, which is an extremely high value.

As an alternative, another support, marked with 2 in Figure 9.28, can be designed to stabilize the excavation, which, when a radial stress of 5 MPa is reached, it maintains this stress even though the strain increases, which can stabilize the excavation at a strain of 3.5%.

From a practical point of view, a support with a practically perfect elastic-plastic behavior, as the support 2 of Figure 9.28, can be obtained by placing longitudinal strips of compressible elements in the support.

In these cases, the most important problem to be solved is how to design a combined support, which combines the high deformability of the compressible elements with the reduced capacity of deformation of the conventional support elements, particularly the steel arches.

The following sections present the most relevant characteristics of some compressible elements that are commercialized.

9.6.1 Compressible coaxial cylinders

Between 1996 and 1999, the geotechnical group of the University of Graz, Austria, developed compressible elements, called lining stress controllers (LSC), which, in their most modern version, are made of three coaxial cylinders, as shown in Figure 9.29, which can be shortened 200 mm under a load of 200 kN.

Photograph 9.29 show the LSC cylinders placed in a tunnel and a detailed view of their deformation. The deformable coaxial cylinders have been used in the construction of the Gotthard Base Tunnel.

9.6.2 Compressible conglomerates

Compressible conglomerates, marketed under the name of high deformable concrete (hiDcon), are made of cement, steel fibers and hollow glass particles and are brick-shaped, as shown in Photograph 9.30.

HiDcon bricks can have a strain of 50% of their initial thickness, but this high strain depends on the other support elements; this happens with all compressible elements.

Figure 9.29 LSC Compressible cylinders.

Source: Moritz, 1999.

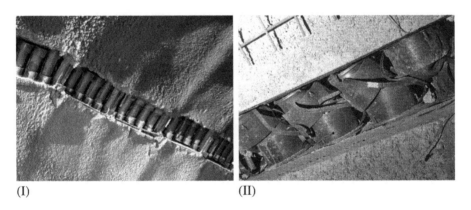

(I)   (II)

Photograph 9.29 LSC cylinders placed in a tunnel. I. LSC cylinders placed in a tunnel. II. Deformed LSC cylinders.

Source: Barla, 2001.

A hiDcon brick with a strain of 15%, placed between two steel arches, is shown in Photograph 9.31.

HiDcon bricks have been used in the access gallery to the San Martin de La Porte Tunnel from Lyon to Turin (France–Italy) that, once constructed, suffered convergences of almost two meters, necessitating a rebuild. Photograph 9.32 shows the reconstruction of San Martin de la Porte gallery.

9.6.3 Compressible tubes between steel plates

Another solution to construct elements of high deformability is to place tubes, which can be crushed in a controlled way, between steel plates. This is the case of the elements marketed as Stress Controller, which have been used in the construction of the second Tauern Tunnel (Austria).

Photograph 9.30 Compressible hiDcon bricks.

Image courtesy of Solexperts AG.

Photograph 9.31 HiDcon brick with a strain of 15%.

Source: Barla, 2001.

Photograph 9.33 shows a row of compressible elements, stress controller type, deformed together with the shotcrete and reinforced with arch lattice girders, which has cracked.

9.7 SEGMENTAL RINGS

Rings composed of reinforced concrete segments are used in shielded Tunnel Boring Machines, to act as tunnel support and lining.

Photograph 9.34 shows one of the tunnels of the M-30 South Bypass in Madrid (Spain), which has been supported and lined with segmental rings.

The following sections focus on the most important aspects that should be taken into account when designing the rings.

Photograph 9.32 Reconstruction of San Martin de la Porte gallery.

Photograph 9.33 Stress Controller elements placed together with cracked shotcrete due to the strains.

Image courtesy of Porr Bau GmbH.

9.7.1 Geometric characteristics

The geometric design of the rings must meet the following criteria:

- The rings must withstand all stresses generated during their manufacture, transportation and installation.
- In deep tunnels, the loads generated by the ground pressure on the rings must be properly considered.
- The installation of the rings must be easy and done in the shortest time possible.

Photograph 9.34 Tunnel of the M-30 South Bypass, supported and lined with concrete segmental rings.

In order to meet the design criteria above, the geometric characteristics of the rings, which are presented in the following sections, must be carefully selected.

9.7.1.1 Length and thickness of the lining rings

The length of the rings, which range between 1 and 2 m, depends on the curvature of the tunnel alignments to be traced by the TBM and the space available to handle the segments inside the Shield.

As a general rule, it can be stated that for excavation diameters between 8 and 10 meters, the rings have a length of 1.5 m; while for excavation diameters above 10 m, its length increases progressively until reaching 2 m for excavation diameters of 14 m.

Below 8 m the situation is the opposite, as the length of the rings decreases to 1 m for excavation diameters of 4 m.

The thickness of the rings depends on the tunnel diameter and the ground factors, mainly the presence of a water table above the tunnel.

Figure 9.30 shows the ratio between the segment thickness (E) and the inner diameter (Di) of several tunnels, expressed in %, and the inner diameter of each tunnel.

This figure shows that the minimum value of $E/D_i \times 100$, for tunnel diameters between 7 m and 13 m, is constant and almost equal to 3%.

For diameters smaller than 7 m, the minimum values of $E/D_i \times 100$ increase as the tunnel inner diameter decreases, because the segment thickness starts to be influenced by the handling and storage conditions.

In tunnels with an inner diameter greater than 13 m, the minimum values of $E/D_i \times 100$ also increase, because in these tunnels, structures are often designed to take advantage of the available space, as shown in Photograph 9.35, which depicts Line 9 in the Barcelona Subway.

9.7.1.2 Geometry of the segments

All rings have at least one segment which covers an angle much smaller than the others and has a trapezoidal shape, so that it plays the role of a "key" which can close the ring easily.

In the document. *The design, sizing and construction of precast concrete segments installed at the rear of a Tunnel Boring Machine*, published by AFTES in 1999, three

Ring thickness / Inner diameter (%)

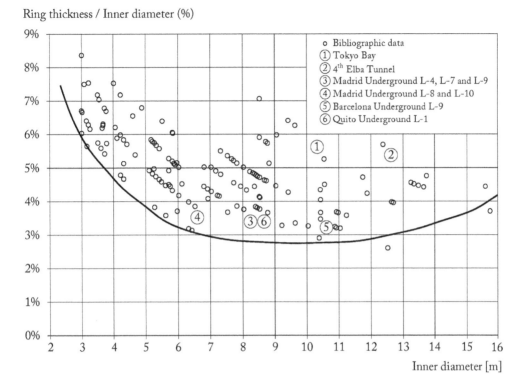

Figure 9.30 Distribution of the ratio between the segment thickness and the inner diameter of the tunnel (%) as a function of the inner diameter.

Adapted from Della Valle and Castellvi, 2016.

Photograph 9.35 Tunnel inner structure of Line 9 in the Barcelona Subway.

types of rings, with parallel transverse faces, are distinguished, depending on the segment shape:

- Rectangular segments
- Parallelogrammic segments
- Trapezoidal segments

In tunnels with a diameter smaller than 5 m, hexagonal-shaped segments of reduced thickness and large dimensions, featuring only four segments, have been used as shown in Photograph 9.36.

These hexagonal rings have difficulty in tracing curves with a radius smaller than 1,000 m, which together with the limitation in diameter of these rings restricts their use to hydraulic tunnels.

The following sections present the characteristics of the three types of rings defined by the AFTES.

9.7.1.2.1 Rings with parallel transverse faces and rectangular segments

In these rings, the segments have parallel transverse faces, except the key and the counter segments, as shown in Figure 9.31. In this figure, the segments are numbered in accordance with their order of placement.

Photograph 9.36 Ring with hexagonal segments.

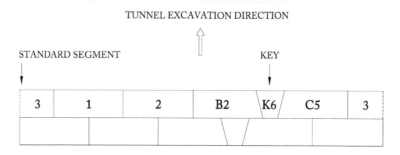

Figure 9.31 Ring with parallel transverse faces and rectangular segments.

This type of ring was commonly used when the joints between two adjacent rings were fitted with screws that were introduced in their housings when the segments were already placed.

9.7.1.2.2 Rings with parallel transverse faces and parallelogrammic segments

When the connection between two adjacent rings is made with plastic connectors, which are described in Section 9.7.3.1, rectangular segments greatly limit the displacements to install them, which increases the assembly time of the ring. To avoid this inconvenience, the parallelogrammic segments, shown in Figure 9.32, were developed.

With this type of ring two keys, placed in the reverse position, are required.

9.7.1.2.3 Rings with parallel transverse faces and trapezoidal segments

Rings with parallel transverse faces and trapezoidal segments were developed so that the excavation with the shield could continue before finishing the ring assembly. To do this, the rings have an even number of segments with the same shape, as shown in Figure 9.33.

To assemble the ring, the even segments are placed first, and the shield thrust jacks are supported on them, so that the excavation can continue while the odd segments are assembled.

The installation of the second half of the segments is very hard, since all these segments are placed as if they were "keys", which increases the assembly time of the rings. For this reason, this type of ring has not been widely used.

9.7.1.3 Rings to trace curves

Rings with segments of parallel faces cannot adapt to the curvature of the tunnel alignment and can only be used in tunnels with a straight layout.

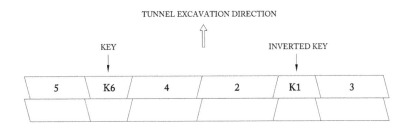

Figure 9.32 Rings with parallel transverse faces and parallelogrammic segments.

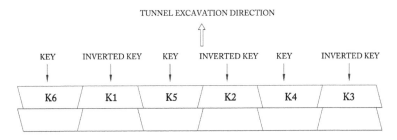

Figure 9.33 Rings with parallel transverse faces and trapezoidal segments.

For curved sections, it is necessary to design tapered segments, which leads to symmetric or asymmetric tapered segments, as shown in Figure 9.34.

The tapering (c) is defined as the difference between the maximum (L_{max}) and the minimum (L_{min}) length of the segment, and has to increase as the radius of the curve decreases.

Figure 9.35 shows a tunnel constructed with symmetrical tapered segments which describe a curve of radius R.

In this case, the necessary tapering (c) is calculated by the expression:

$$C = \frac{D \cdot L}{R}$$

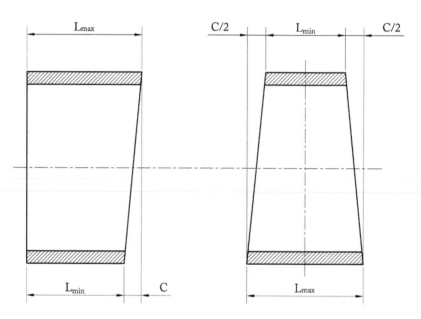

I.-ASYMMETRICAL TAPERED RING   II.-SYMMETRICAL TAPERED RING

Figure 9.34 Tapered segment.

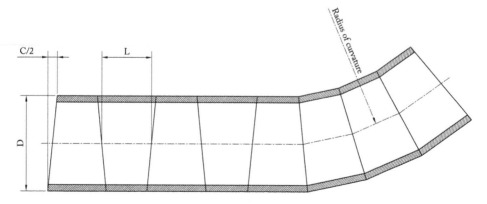

Figure 9.35 Curve described by a tunnel constructed with symmetric tapered segments.

where:

D = outer ring diameter
L = ring length
R = radius of the curve

To simplify the manufacture, storage and even the assembly of the rings, the current trend focuses on the use of a single type of ring for each tunnel.

The following sections present the characteristics of the two types of rings most used in tunnels, that are referred to as conventional and universal rings.

9.7.1.3.1 Conventional rings

Conventional rings are composed of tapered segments of different length, which have the same position in the three possible patterns: straight, curve to the right and curve to the left, so this ring is also known as right-left.

Photograph 9.37 shows a section built with conventional rings in Line 10 of the Madrid Subway. Conventional rings have the advantage that the bottom segment, as it has a fixed position, can have the appropriate size to accommodate the temporary tracks giving access to the tunnel face, as shown in Figure 9.36.

9.7.1.3.2 Universal rings

Universal rings are constituted by tapered segments which are all equal except the key, which covers half the angle compared to the others.

The n segments that compose a universal ring can be placed in 2n-1 positions, as illustrated in Figure 9.37, which shows the 13 possible combinations in a universal ring composed of seven segments. Universal rings can trace curves in both the vertical and horizontal plane and the great number of possible segment combinations enables them to trace curves of different radius.

The most important problem with universal rings is that it is not possible to keep the bottom segment in a fixed position to accommodate the tracks; because the segments change their position according to the selected configuration.

This problem is solved by placing an additional piece of concrete, as shown in Figure 9.38, which acts as a bottom segment to accommodate the tracks.

Photograph 9.37 Line 10 of the Madrid Subway, constructed with conventional rings.

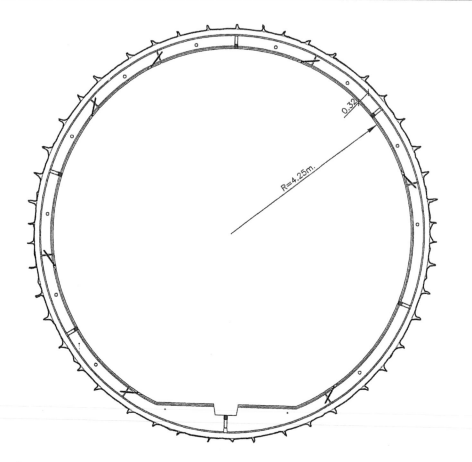

Figure 9.36 Conventional ring with the bottom segment prepared to accommodate the temporary tracks.

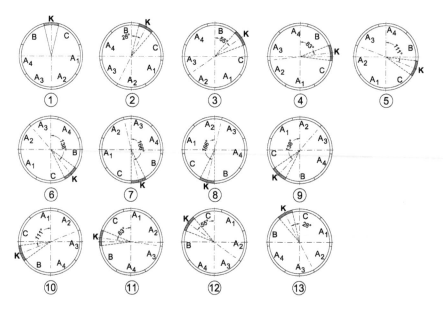

Figure 9.37 Detail of the 13 possible combinations of a universal ring with seven segments.

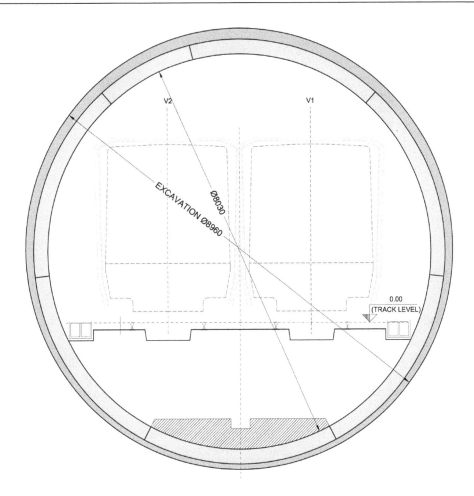

Figure 9.38 Universal ring with a concrete piece to accommodate the temporary tracks.

Photograph 9.38 shows a section of Line 4 of the Madrid Subway built with universal rings.

The use of universal rings is becoming more common in all types of tunnels, because these rings are very easily assembled when the TBM crew is properly trained.

Another advantage of the universal rings is that they can trace a very wide range of curves, and it is possible to quickly correct the small deviations of the TBM and to remove them, which eliminates the eccentricities in the cylinders' thrust.

9.7.1.4 Number of segments in each ring

The minimum number of segments per ring is four in those rings with hexagonal segments for tunnels whose excavation diameter is less than 5 m. From 5 m of diameter, the number of segments in a ring progressively increases; it being reasonable to find seven to nine segments in those tunnels whose diameter is between 8 and 12 m.

The two basic criteria to determine the number of segments in a ring are to make the ring assembly easy and to avoid damaging the segments during their storage and handling. For that reason it is necessary to ensure that segments length is less than 3.5 m in tunnels with a diameter smaller than 12 m.

Photograph 9.38 Line 4 of the Madrid Subway built with universal rings.

9.7.2 Completion of the segments

The rings must fulfill the functions required for both the support and the lining which involves not only stabilizing the tunnel section but also ensuring its long-term service.

To do this, the segments must have complements to enable their correct connection and to make them waterproof and, for deep tunnels, to adapt to large ground displacements.

9.7.2.1 Watertight joints

The rings must be functional as tunnel lining, so they should make the tunnel waterproof or watertight, when this condition is required and possible to achieve. To make rings waterproof, joints of an elastic material are used, which, until a few years ago, were fixed along the perimeter of the segment.

When assembling the rings, the elastic joints were compressed to prevent the water flow. The behavior of these joints is illustrated in Figure 9.39. The dimensions and characteristics of waterproofing joints must be carefully chosen to work effectively in the presence of water.

As the joints are fixed to the segments in specific housings, it is difficult to guarantee their integrity when the rings are assembled, since, in this operation, particularly when placing the key, the segments rub heavily between them and the joints usually detach.

To avoid these problems, joints integrated in the segments during manufacturing are available as shown in Figure 9.40.

If there are high water pressures, it is impossible for the rings, even those equipped with watertight joints, to achieve waterproofing of the tunnel, as shown in Photograph 9.39, which corresponds to a tunnel with segments in which the water pressure was 22 MPa and the water inflow was around 250 l/s.

In these cases, in order to achieve the waterproofing of the tunnel, there is no other solution but to attach a lining of reinforced concrete to the rings, such as the one shown in Figure 9.41.

With the lining in Figure 9.41, applied to the tunnel section corresponding to Photograph 9.39, it was possible to reduce the water flow from 250 l/s to only 7 l/s.

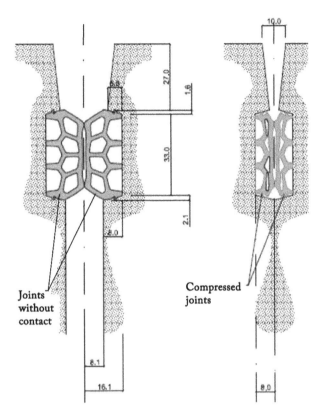

Figure 9.39 Waterproofing joints.

Figure 9.40 Waterproofing joint integrated during manufacturing of the segments.

Photograph 9.40 shows a tunnel section where a watertight ring of reinforced concrete was attached to the rings.

9.7.2.2 Connection elements

The segments must have connection elements to enable easy assembly of the ring and to avoid discontinuous joining between two adjacent rings; thus the connection elements, explained below, have been developed.

Photograph 9.39 Water inflow through the segments of a tunnel.

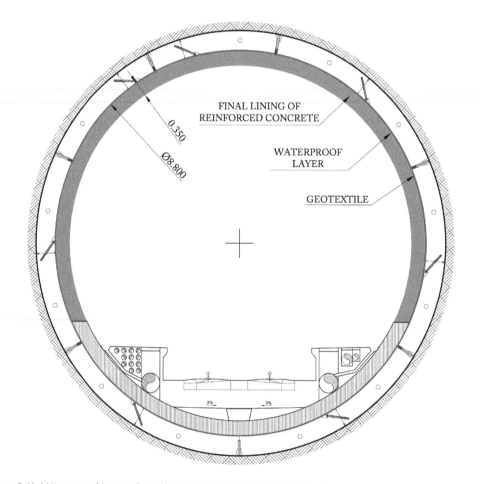

Figure 9.41 Waterproof lining of reinforced concrete attached to the ring.

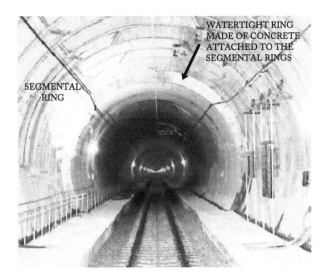

Photograph 9.40 Watertight ring attached to the rings.

9.7.2.2.1 Socket pin

Socket pins are used to enable the correct positioning of two adjacent segments in the same ring. They are plastic bars, about 60 cm in length and 30 mm in diameter, which are fixed into specific housings in the segment, to fit into the housings of the adjacent segment. A detail of a socket pin is shown in Photograph 9.41.

9.7.2.2.2 Longitudinal connectors

Longitudinal connectors fulfill a task similar to socket pins when connecting the segments of two adjacent rings. Figure 9.42 shows an exploded view of a longitudinal connector, which consists of a threaded rod, which fits into two housings made in the segments to join, and a centralizer.

Photograph 9.41 Socket pin.

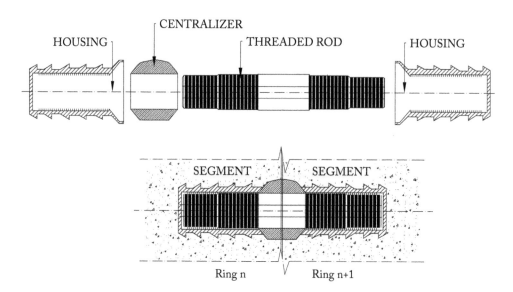

Figure 9.42 Elements of a longitudinal connector.

9.7.2.2.3 Socket bolts

Both socket pins and longitudinal connectors are designed to facilitate the ring assembly but to achieve stiffer rings it is necessary to use socket bolts.

Figure 9.43 shows a socket bolt used to join two segments of a ring and the pockets designed to accommodate the socket bolt in each segment.

Socket bolts are essential to ensure that the waterproofing gaskets work correctly. However, in some railway tunnels they are removed after the rings have been assembled to avoid leakage and in some hydraulic tunnels, if they are installed, it is required to cover the pockets to reduce the frictional losses when the water flows.

9.7.2.3 Compressible elements

If segments are used in tunnel sections where large strains are expected, corresponding to ICE values lower than 15 points, the rings must be designed to deal with large ground movements.

Although this problem is not definitively solved today, there are two techniques which attempt to achieve this goal.

The first possibility is to include in the rings the "stress controller" compressible elements, presented in Section 9.6.3, as illustrated in Figure 9.44.

This option requires a careful design of the ring and the manufacture of some segments with different dimensions than those used in tunnel sections where usual strains are expected.

The other option is to fill the gap between the rings and the ground with a compressible cement mortar, type Compex which, as shown in Figure 9.45, tolerates strains up to 65% for a stress of 5 MPa.

9.7.3 Manufacture, handling and storage of the lining segments

Tunnel lining segments are made of reinforced concrete in plants, following the process shown in Figure 9.46, which includes an oven for the accelerated and controlled setting of the concrete.

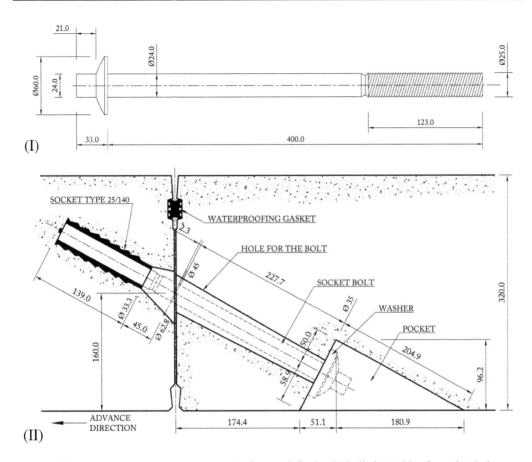

Figure 9.43 Socket bolt between two segments of a ring. I. Socket bolt. II. Assembly of a socket bolt.

Figure 9.44 Stress Controller elements attached to the ring.

Source: Bochumer Eisenhütte.

Concrete usually has a characteristic strength of 40 MPa, tested in cylindrical specimens, although it is common to use concrete with a characteristic strength of between 50 and 60 MPa and, in extreme cases, up to 100 MPa.

The steel quantity varies between 80 and 120 kg per cubic meter of concrete.

Since the late 20th century, the conventional steel reinforcement has been supplemented and sometimes fully replaced by steel fibers, which undoubtedly have the advantage of

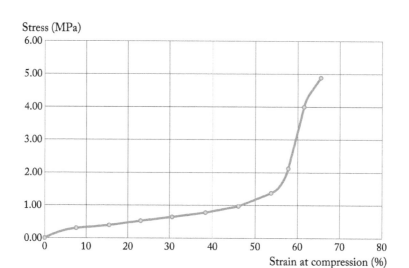

Figure 9.45 Strain of a Compex mortar in compression.

Source: Schneider & Cie.

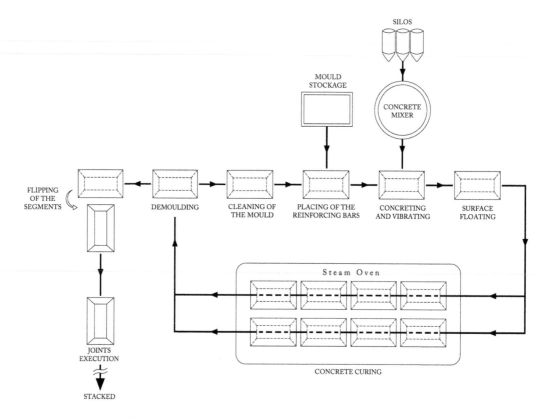

Figure 9.46 Concrete plant, with an oven, for manufacturing lining segments.

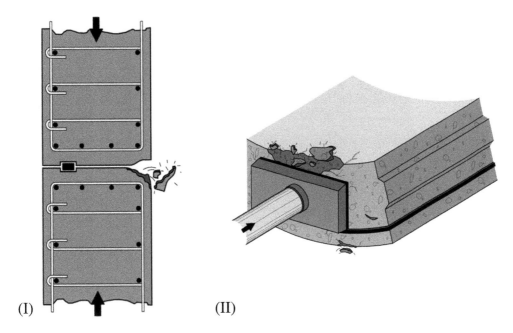

(I)   (II)

Figure 9.47 Breaks at the edges of the segments reinforced with conventional steel bars. I. By compression between two segments. II. By the compression produced by the thrust jacks.

preventing damage in the segment edges where no reinforcement exists. This effect is illustrated in Figure 9.47.

These flaws are avoided when steel fibers are included in the concrete, with a minimum amount of 30 kg of fibers per m^3 of concrete.

Given the advantages of steel fibers in controlling the development of concrete cracks, as discussed in Section 9.4.2.5, it is appropriate to replace part of the reinforcement with steel fibers or even, to replace it completely.

The first tunnels made of segments completely reinforced with steel fibers were the Sörenberg Tunnel in Switzerland and the Eurostar Tunnels between London and Castle Hill. Photograph 9.42 shows the rings, placed in the Sörenberg Tunnel, reinforced only with steel fibers.

In rings with a diameter larger than 10 m, the complete replacement of the conventional reinforcement by steel fibers has limitations depending on the segment thickness.

Nowadays it is possible to fully replace the reinforcement by fibers if the segment slenderness, defined as the ratio between its length and its thickness, is less than 10.

Recently, the practice of including polypropylene fibers in the concrete to manufacture the segments has been widespread. These fibers, with lengths between 6 and 12 mm and a diameter of 20 microns, are used to prevent damage to the segments during a fire.

The polypropylene melts at 160°C, so the volume filled by the fibers creates a network of microchannels in the concrete, which allow the dissipation of the steam pressure produced by the rise in the temperature, thus avoiding crack effects in the concrete.

When the molds for casting the segments are taken out of the oven, they are removed with a vacuum system as illustrated in Photograph 9.43.

Referred to as "demolding", this is the most demanding activity in terms of the segment strength, in the process from its manufacture to its assembly because to demold, the concrete must have a uniaxial compressive strength greater than 15 MPa. Once the segments

Photograph 9.42 Rings from the Sörenberg Tunnel in Switzerland reinforced with steel fibers.

Photograph 9.43 Demoulding of a segment with a vacuum system.

are demolded, they are transported to a storage yard by lifting devices, as illustrated in Photograph 9.44.

At the storage yard, the segments are grouped in sets which contain the segments necessary to assemble a complete ring, as shown in Photograph 9.45.

9.7.4 Segment design

To design the lining segments it is necessary to consider the stresses generated during the demolding and the storage process and those generated when assembled in the tunnel.

9.7.4.1 Loads during the demolding and the storage process

The stresses in the segments during the demolding process are usually calculated using simple models, which reproduce the real stresses reasonably well.

Photograph 9.44 Segments transportation for their storage.

Photograph 9.45 Storage of segments in sets that constitute one lining ring.

Figure 9.48 shows a model used to calculate the stresses during the stacking of the segments that compose a ring.

As already mentioned, demolding is usually the most critical operation that the lining segments must undergo, before being placed in the tunnel. To accurately determine these stresses, it is useful to perform calculations on three-dimensional models, as illustrated in Figure 9.49.

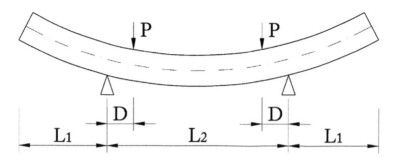

Figure 9.48 Model to calculate the stresses generated during the stacking of the segments.

This figure shows the demolding of the segment with a vacuum system, the three-dimensional model created to simulate the suction of the segment and, once the analysis has been performed, the distribution of the tensile stresses in the segment will be known.

9.7.4.2 Loads generated by the thrust of the shield cylinders

Shielded TBMs are able to move forward by the action of the hydraulic cylinders on the last ring. This generates compression stresses in the lining segments which they must withstand.

However, when the TBM deviates from its trajectory, corrections must be made to compensate for such mistakes, which result in situations when the cylinders act eccentrically and generate unfavorable stresses.

These situations can be analyzed by simulating the eccentric support of the cylinders on the segments with a three-dimensional model, as shown in Figure 9.50, which shows a snapshot of the cylinders pushing eccentrically, their modeling and the distribution of the tensile stresses generated.

In shallow tunnels, the rings do not have significant ground pressures because, at small depths, there are very small pressures, which are easily withstood by the rings.

However, the situation changes for tunnels built at several hundred meters below the surface, because in highly yielded grounds, i.e., ICE <15, the rings will deal with significant pressures.

To analyze this situation, three-dimensional geomechanical models have to be used to model the rings by considering each segment individually and the connection elements between them.

9.7.5 Quality control of the rings

Once the lining segments are completed, it is very difficult to detect any manufacturing defects due to poor installation of the reinforcement, lower fiber content than expected, inadequate concrete strength and small variations in the segment dimensions. When assembled on the corresponding rings, these defects will result in breaks that are complex and expensive to repair.

Therefore, during the manufacturing process, the segments must be subjected to the usual quality controls on prefabricated pieces of reinforced concrete.

To verify the dimensions of the segments, it is very useful to apply the technique of photogrammetry, which enables obtaining a three-dimensional model of the segment with an accuracy of 0.01 mm.

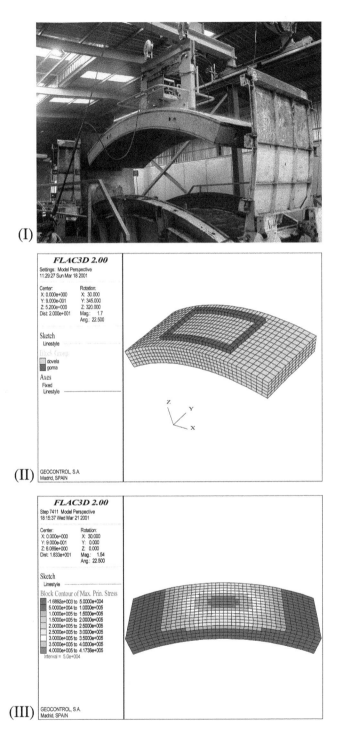

Figure 9.49 Distribution of the tensile stresses in the segment by the suction applied to demould them. I. Demoulding. II. Model of the demoulding process. III. Distribution of the tensile stresses during demoulding.

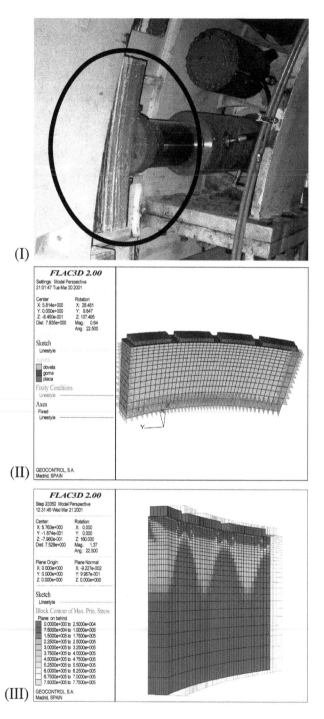

Figure 9.50 Distribution of the tensile stresses in a segment due to the eccentric thrusts. I. Detail of the eccentric support of the thrust cylinders on the segments. II. Modeling of the eccentric support of the thrust cylinders on the segments. III. Tensile stresses generated by the eccentric thrust on the segments.

Photograph 9.46 Targets for the photogrammetric control of the segment dimensions.

To apply this technique to the segments, targets must be glued in predefined positions, as shown in Photograph 9.46.

All manufactured segments must be identified with a bar code or something similar, to depict the manufacture date, the lot, their characteristics and their position in the ring.

9.8 THE BEHAVIOR OF THE DIFFERENT SUPPORT TYPES

The most common support types have different stress–strain behaviors reflecting their different characteristics when compared, as shown in Figure 9.51.

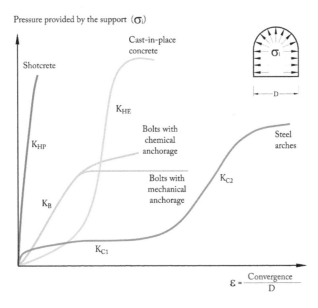

Figure 9.51 Stress–strain behavior of common support types. The K symbols are the equivalent stiffness of the supports displayed.

Steel arches have two different strain behaviors; one corresponds to the loading phase and the other to the displacement with the ground once the arch is in contact with the ground along the whole perimeter of the excavation.

As steel arches are always placed together with shotcrete, it is not easy to accurately determine their stress–strain characteristics but, in general, it can be stated that its strength is the lowest compared to other supports.

Both cast-in-place concrete and shotcrete, can reach very high and significantly similar strengths. However, shotcrete, due to its accelerated setting, has an initial stiffness larger than cast-in-place concrete.

Rock bolts have an intermediate behavior between concrete and steel arches, although it is necessary to distinguish between the ones with a chemical anchoring system, which have a brittle post-failure behavior, and the ones with a mechanical anchoring system, which have almost perfect elasto-plastic behavior.

To roughly estimate the strength of the different support elements, Figure 9.52 shows some formulas made for circular tunnels (Kaiser, 2000) which assume that the support is placed all around the tunnel perimeter.

Figure 9.53 shows, on the logarithmic scale, the strength of the different types of support defined in Figure 9.52, as a function of the tunnel diameter.

Support type	Flange width (mm)	Section depth (mm)	Weight (kg/m)	Curve number	Maximum support pressure P_{imax} (MPa) for a tunnel of diameter D (metres) and a set spacing of s (metres)
Wide flange rib	305	305	97	1	$P_{imax}=19{,}9D^{-1{,}23}/s$
	203	203	67	2	$P_{imax}=13{,}2D^{-1{,}3}/s$
	150	150	32	3	$P_{imax}=7{,}0D^{-1{,}4}/s$
I section rib	203	254	82	4	$P_{imax}=17{,}6D^{-1{,}29}/s$
	152	203	52	5	$P_{imax}=11{,}1D^{-1{,}33}/s$
TH section rib	171	138	38	6	$P_{imax}=15{,}5D^{-1{,}24}/s$
	124	108	21	7	$P_{imax}=8{,}8D^{-1{,}27}/s$
3 bar lattice girder	220	190	19	8	$P_{imax}=8{,}6D^{-1{,}03}/s$
	140	130	18		
4 bar lattice girder	220	280		9	$P_{imax}=18{,}3D^{-1{,}02}/s$
	140	200			

Support type	Flange width (mm)	Section depth (mm)	Weight (kg/m)	Curve number	Maximum support pressure P_{imax} (MPa) for a tunnel of diameter D (metres) and a set spacing of s (metres)
Concrete or shotcrete lining	1m	28	35	20	$P_{imax}=57{,}8D^{-0{,}92}$
	300	28	35	21	$P_{imax}=19{,}1D^{-0{,}92}$
	150	28	35	22	$P_{imax}=10{,}6D^{-0{,}97}$
	100	28	35	23	$P_{imax}=7{,}3D^{-0{,}98}$
	50	28	35	24	$P_{imax}=38D^{-0{,}99}$
	50	3	11	25	$P_{imax}=1{,}1D^{-0{,}97}$
	50	0,5	6	26	$P_{imax}=0{,}6D^{-1.0}$
Rockbolts or cables spaced on a grid of s x s metres	34 mm rock bolt			10	$P_{imax}=0{,}354/s^2$
	25 mm rockbolt			11	$P_{imax}=0{,}267/s^2$
	19 mm rockbolt			12	$P_{imax}=0{,}184/s^2$
	17 mm rockbolt			13	$P_{imax}=0{,}10/s^2$
	SS39 Split set			14	$P_{imax}=0{,}05/s^2$
	EXX Swellex			15	$P_{imax}=0{,}11/s^2$
	20 mm rebar			16	$P_{imax}=0{,}17/s^2$
	22 mm fibreglass			17	$P_{imax}=0{,}26/s^2$
	Plain cable			18	$P_{imax}=0{,}15/s^2$
	Birdcage cable			19	$P_{imax}=0{,}30/s^2$

Figure 9.52 Strengths of the different support elements in circular tunnels.

Source: Hoek and Brown, 1980.

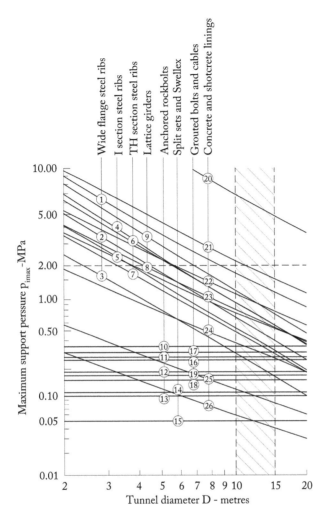

Figure 9.53 Strength of the support elements as a function of the tunnel diameter.

Source: Adapted from Kaiser, 2000.

This figure shows that for circular tunnels with a diameter ranging from 10 to 15 m it is only possible to achieve an initial support pressure greater than 2 MPa with a shotcrete layer of 30 cm in thickness and a uniaxial compressive strength at 28 days of 30 MPa.

BIBLIOGRAPHY

Aire, C., Molins, C., Aguado, A., "Ensayo de doble punzonamiento para concreto reforzado con fibras: efecto del tamaño y origen de la probeta", *Concreto y Cemento: Investigación y Desarrollo*. Vol. 5. No. 1. 2013, pp. 17–31.

Association Francaise des Tunnels et l'Espace Souterrain (AFTES), Work Group no 7. *Use of steel ribs in underground works*. AFTES Recommendations GT7R3A1. Paris, France. 1993.

Association Francaise des Tunnels et l'Espace Souterrain (AFTES), Work Group no 20. *Design of sprayed concrete for underground support*. AFTES Recommendations GT20R1A1. Paris, France. 2000.

Association Francaise des Tunnels et l'Espace Souterrain (AFTES), *The design, sizing and construction of precast concrete segments installed at the rear of a tunnel boring machine (TBM)*. Paris, France. 1999.

Barla, G., Bonini, M.C., Debernardi, D., "Modelling of tunnels in squeezing rock", *Proceedings of the 1st International Conference on Computational Methods in Tunnelling*. Vienna, Austria. 2001, pp. 1–18.

Bieniawski, Z.T., Aguado, D., Celada, B., Rodríguez, A., "Forecasting tunnelling behaviour", *Tunnels & Tunnelling Internatioanal*. August. 2011, pp. 39–42.

British Tunnelling Society and The Institution of Civil Engineers, *Specifications for Tunnelling*. Thomas Telford, London. 2010.

Celada, B., *Tecnología del bulonaje. Estudio particular del bulonaje a la resina*. ETS de Ingenieros de Minas, Oviedo, Spain. 1970.

Celada, B., *Delimitación de las condiciones de utilización para la aplicación de las técnicas de sostenimientos activos en la minería española*. Instituto Geológico y Minero de España, Madrid, Spain. 1983.

Celada, B., "Concepto del sostenimiento en minería y obras subterráneas", *Canteras y Explotaciones*. No. 2215. Madrid, Spain. 1985.

Celada, B., *Sostenimiento de excavaciones subterráneas*. Serie Geotecnia Minera. Instituto Geológico y Minero de España, Madrid, Spain. 1988.

Celada, B., "Concepto y diseño del sostenimiento de túneles", *Manual de Túneles y Obras Subterráneas*. ETSI Minas y Energia, Madrid, Spain. 2011.

Della Valle, N., Castellvi, H., "Diseño de túneles con tuneladora", *40 Simposio Internacional sobre Túneles y Lumbreras en Suelos y Rocas*. Amitos, Mexico City, Mexico. 2016.

Dolsak, W., "Technical advances of self-drilling rock reinforcement and ground control systems", *World Tunnel Congress*. Agra, India. 2008.

Ehrbar, H., Graber, L.R., Sala, A., *Tunnelling the Gotthard*. Swiss Tunnelling Society. 2016.

Gullón, A., "Sostenimiento con cerchas metálicas", *Manual de Túneles y Obras Subterráneas*. ETSI Minas y Energia, Madrid, Spain. 2011.

Habenicht, H., "The anchoring effects – our present knowledge and its shortcomings – A key note lecture", *Proceedings of International Symposium on Rock Bolting*. Abisko, Sweden. 1983.

Heilegger, R., Beil, A., "Full automated tunnel segment production system. A case study: Tunnel Boring Machines", *Trends in Design and Construction of Mechanized Tunnelling*. A.A. Balkema, Rotterdam, the Netherlands. 1996.

Heuer, R.E., "Selection design of shotcrete for temporary support", *Use of Shotcrete for Underground Structural Support*. Special Publication ACI-ASCE, Detroit, MI. 1974.

Hoek, E., Brown, E.T., *Underground Excavations in Rock*. Institution of Mining and Metallurgy, London. 1980.

Kovari, K., Fechtig, R., *Percements historiques de tunnels alpins en Suisse: St. Gothard, Simplon*. Societé pour l'art de l'Ingenier Civil, Zurich. 2000.

Melbye, T.A., Dimmock, R.H., "Modern advances and applications of sprayed concrete", *International Conference on Engineering Developments in Shotcrete*. Hobart, Australia. 2001.

Moritz, A.B., *Ductile Support System for Tunnels in Squeezing Rock*. Gruppe Geotechnik Graz series, Vol. 5. Technische Universitët, Graz, Austria. 1999.

Owen, R., "Test of strength", *Tunnel & Tunnelling International*. August. 2016.

Rey, A., "Hormigón Proyectado: Dosificación, Fabricación y Puesta en Obra", I Jornada sobre Hormigón Proyectado. Madrid, Spain. 2006.

Rey, A., "Sostenimiento con hormigón proyectado", *Manual de Túneles y Obras Subterráneas*. ETSI Minas y Energia, Madrid, Spain. 2011.

Rivas, J.L., "Gunita: Colocación en obra y control", *VIII Master Universitario de Túneles y Obras Subterráneas*. Asociación Española de Túneles y Obras Subterráneas, Madrid, Spain. 2013.

Scott, J.J., "Split–set rock stabilizers", *Proceedings of 18th U.S. Symposium on Rock Mechanics*. Golden, CO. 1977.

Stillborg, B., *Professional Users Handbook for Rock Bolting*, 2nd Edition. Trans Tech Publications, Zurich, Switzerland. 1994.

Chapter 10

Stress–strain analysis

Benjamín Celada Tamames and Pedro Varona Eraso

Success can only be achieved by those who do something while waiting for the success.

Thomas Alva Edison

10.1 INTRODUCTION

Within the DEA, a tunnel design methodology, Phase II is dedicated to the structural design of underground excavations and is based on stress–strain analysis of the tunnel support sections. Selected to excavate and stabilize an underground excavation, the sections are designed according to the activity diagram presented in Chapter 2, Figure 2.22.

This chapter presents the basic principles to be taken into account for the application of the stress–strain analysis methodology.

10.2 CONCEPT OF THE STRESS–STRAIN ANALYSIS

Closed analytical solutions have many limitations blocking the successful solving of the problems of rock mechanics, due to the great difficulty in solving theoretical models which take into account the complex mechanical behavior of rock masses.

For example, a problem as simple as the determination of the characteristic curve of a circular tunnel, in a stratified ground and subjected to a non-hydrostatic stress field ($K_0 \neq 0$), does not currently have an analytical solution.

Stress–strain analysis was initially developed in the second decade of the 20th century by Zienkiewicz and Cheung (1967), introducing numerical calculation techniques to the resolution of the geomechanics models featuring elements of small size (called finite elements).

Development of the stress–strain analysis and personal computers has been parallel since, with low performance computers, it is impossible to perform the calculations necessary to solve the geomechanics models based on finite elements.

In the 1970s the geomechanics models had only a few hundred elements, in the 1990s there were thousands of elements in a model, and at the beginning of the 21st century, the most complex models of underground excavations have about one million elements.

Despite the impressive development of the stress-strain analysis during the last decades, this methodology has its limitations, so it is important to know its conceptual foundations to choose the most suitable option in each case.

10.2.1 Solving the algorithms

The algorithms most commonly used in stress–strain analysis are the boundary elements, finite elements and finite difference elements algorithms, whose most important characteristics are presented in the following sections.

10.2.1.1 Boundary elements

The boundary element algorithm considers that, after excavating, the tangential and radial stresses at the perimeter are zero, which is accurate if the excavated ground behaves elastically, as shown in Figure 10.1.

Accordingly, the stress state after the excavation is obtained by combining two stress states: the initial state and the one produced in the ground after applying the stresses, as shown in Figure 10.2.

To know the initial stress state at the perimeter of the excavation, it is discretized in segments or elements where distributed dummy forces, F_X and F_Y, are applied, as shown in Figure 10.3.

The set of dummy forces, F_X and F_Y, is determined by a system of analytical equations.

The boundary element algorithm has the advantage of requiring very short calculation times but it can only be used in excavations with an elastic behavior.

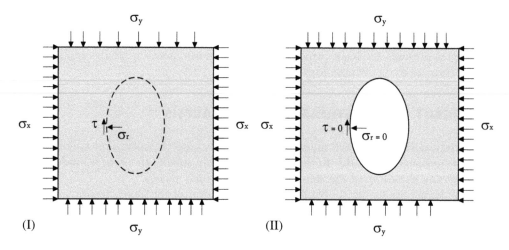

Figure 10.1 Basic hypothesis of the boundary element algorithm. I. Stress state before the excavation. II. Stress state after the excavation.

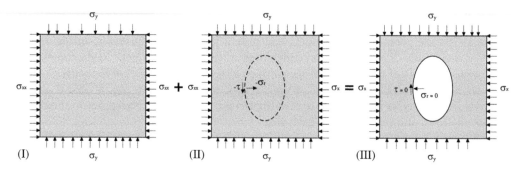

Figure 10.2 Determination of the stress state after an excavation according to the boundary element algorithm. I. Initial stress state. II. Application of $-\sigma_r$ and $-\tau$. III. Stress state after the excavation.

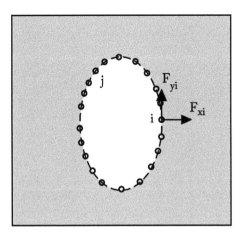

Figure 10.3 Dummy forces at the boundary of the excavation.

10.2.1.2 Finite elements

The finite element algorithm shares its name with the stress–strain calculation methodology, since when this methodology began being developed in the 1960s, there was only one algorithm to solve the equations, and therefore, the mathematical algorithm used kept this name.

The finite element algorithm, initially developed for continuous media with an elastic behavior, is based on the principle of virtual work, which postulates that if a solid in equilibrium moves under the effect of external forces, the work of the elements of the solid is the same as the work produced by the external forces.

For a single elastic element, the principle of virtual work assumes that:

$$\{q^e\} = [K^e]\{u^e\} + \{f^e\} \tag{10.1}$$

$\{q^e\}$ = vector of nodal forces
$[K^e]$ = stiffness matrix, that depends on the shape functions and the elastic parameters of the material
$\{u^e\}$ = vector of nodal displacements
$\{f^e\}$ = vector of initial stresses + mass forces

Figure 10.4 illustrates the meaning of the terms in Equation 10.1, applied to a triangular element in a two-dimensional finite element model.

When all the elements from the discretized model are considered, Equation 10.1 becomes:

$$\{r^G\} = [K^G]\{u^G\} + \{f^G\} \tag{10.2}$$

where:
$\{r^G\}$ = vector of the reaction forces at the nodes
$[K^G]$ = stiffness matrix
$\{u^G\}$ = vector of nodal displacements
$\{f^G\}$ = vector of initial stresses + mass forces

By making the nodal displacements, from adjacent elements, compatible and taking into account the constitutive model of the ground, which defines the stress–strain relations in the elastic domain and in yielding, a system of 2N equations and 2N unknown is obtained, where N is the number of nodes in the finite element model.

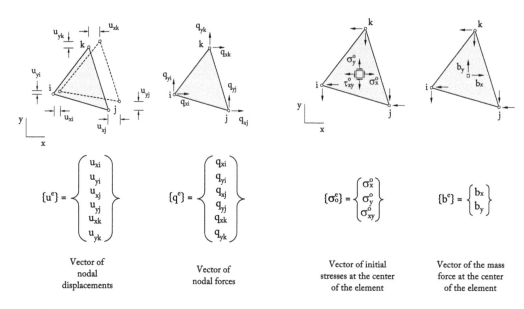

$$\{u^e\} = \begin{Bmatrix} u_{xi} \\ u_{yi} \\ u_{xj} \\ u_{yj} \\ u_{xk} \\ u_{yk} \end{Bmatrix}$$

Vector of nodal displacements

$$\{q^e\} = \begin{Bmatrix} q_{xi} \\ q_{yi} \\ q_{xj} \\ q_{yj} \\ q_{xk} \\ q_{yk} \end{Bmatrix}$$

Vector of nodal forces

$$\{\sigma_0^e\} = \begin{Bmatrix} \sigma_x^o \\ \sigma_y^o \\ \sigma_{xy}^o \end{Bmatrix}$$

Vector of initial stresses at the center of the element

$$\{b^e\} = \begin{Bmatrix} b_x \\ b_y \end{Bmatrix}$$

Vector of the mass force at the center of the element

Figure 10.4 Terms that define the virtual work of an element.

This system of equations is solved using the direct method, the iterative method or algebraic matrixes, which enables knowing the nodal displacements and the resulting stresses in the centers of the elements.

The information obtained with the finite element algorithm is much more complete than that obtained with boundary elements, and, more importantly, it can refer to excavations with elastic behavior or with yielding. In contrast, the calculation times with the finite element algorithm are much longer than those required with the boundary element algorithm.

The finite element algorithm has another important disadvantage, namely, the way in which the models with intense yielding are solved.

The stress–strain calculation software based on the finite element algorithm solves the yielding states by modifying the components of the stiffness matrix in an iterative way so that, at any calculation step, the balance, compatibility and constitutive model equations are satisfied.

To do this, the sequence of application of the loads in the model is done in increments, small enough so that the solution, of the system of equations, could numerically converge.

This way of solving the yielding states, significantly increases the calculation times with the finite element algorithm

10.2.1.3 Finite differences

The finite difference algorithm was created to solve stress–strain problems in a continuous medium with elasto-plastic behavior.

The finite difference algorithm is based on Newton's first law:

$$F = m \cdot a \tag{10.3}$$

where:

  F = applied force
  m = mass of the solid
  a = acceleration

This law applied to a solid divided into finite elements is expressed as:

$$\rho \frac{d\dot{u}}{dt} = \frac{\partial \sigma_{ij}}{\partial x_j} + \rho g \tag{10.4}$$

where:

ρ = density
x_j = position vector (x, y)
σ_{ij} = components of the stress tensor
g = gravitational acceleration

Accordingly, the nodal forces acting on an element in the model will produce an acceleration, which will relate to a velocity and a displacement.

The nodal displacements of an element in the model produce a translation, a rotation and a strain in that element.

Depending on the constitutive model of the ground, for a given strain in an element there is an associated specific stress, as shown in Figure 10.5.

The main advantage of the finite difference algorithm is that it was specifically developed to solve stress–strain problems in grounds with yielding, so it is very effective in analyzing excavation problems in highly yielded grounds, even being able to simulate the excavation collapse.

10.2.1.4 Distinct elements

The distinct element algorithm is an evolution of the finite difference algorithm, which has been created to solve stress–strain calculation problems in a discontinuous medium with elasto-plastic behavior.

This algorithm is much more complex than those described before, since it takes into account the potential rock blocks that can be individualized as a result of the existing discontinuities in the ground and, in addition, identifies those that behave as rigid blocks from those which behave as deformable blocks.

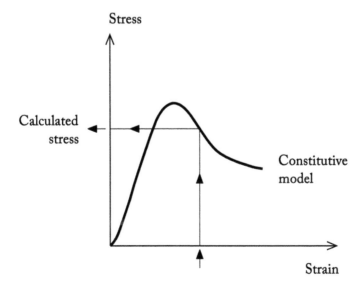

Figure 10.5 Calculation of the stress as a function of the strain.

The procedures used by the distinct element algorithm to identify the behavior of the blocks require a calculation time proportional to N^2, where N is the number of blocks in the model.

The advantages of the distinct element algorithm are as follows:

- Allows analyzing the stress-strain behavior of the blocks in the ground, provided that their discontinuities are properly characterized.
- As it uses dynamic equations, the simulation of large block displacements is done in a realistic way and without creating instabilities in the calculation process.
- Like the finite difference algorithm, it enables directly estimating the stresses from the constitutive models.
- The yielding phenomena of the blocks in the model are solved with great realism, since the algorithm of distinct elements enables identifying the possible failures in the contacts between the blocks and also, the softening of the rock matrix in the block.

10.2.2 Ground modeling

The grounds can be classified in three groups depending on their joint sets, taking into account the criteria contained in Table 10.1 according to Kaiser et al. (2000).

Massive and highly fractured grounds can be modeled as a continuous media; massive grounds as they barely have joints and highly fractured grounds, because as they have more than four joints sets, the size of the resulting blocks is very small.

10.2.3 Most common calculation softwares

Table 10.2 contains the characteristics of the most used calculation softwares in stress–strain analyses, which can be synthesized in the following criteria:

- Software that use the boundary element algorithm give very limited results and should be only used in the preliminary studies of a project.
- Software that use the finite element algorithm can be used, at any phase of the project, for underground works in grounds which can be considered as a continuous medium, although some software can include joint sets in the model and its field of application extends to grounds with $RMR < 65$.
- Software that use the finite element algorithm are not suitable to solve problems with high yielding, so it is recommended to use them in areas with $ICE > 15$.
- Software that use the finite difference algorithm can represent any type of yielding phenomenon in grounds with a continuous behavior or with a joint set and are applied in grounds with $RMR < 65$.
- Software that apply the distinct element algorithm are complicated to use, particularly the 3-DEC, and should only be used for stress–strain analysis in clearly discontinuous grounds with $40 < RMR < 85$.

Table 10.1 Classification of the ground according to their discontinuities

Type of ground	Joint sets in the ground	RMR_{89}
Massive	1 or none	> 75
Moderately fractured	2 or 3	50–74
Highly fractured	4 or more	< 50

Table 10.2 Characteristics of the most used calculation software for stress–strain analysis

Calculation algorithm	Commercial software	Model dimensions	Reference of the supplier (www)	Field of application		
				Project phase	Rock mass quality	Excavation behavior
Boundary elements	Examine 2D	2	rocscience.com	Preliminary studies	RMR>75	ICE>130
	Examine 3D	3		Feasibility		
Finite elements	RS2 (Phase 2)	2		Detailed engineering	RMR<65	ICE>15
	RS3 (Phase 3)	3				
	PLAXIS	2 or 3 (Extruded)	plaxis.nl			
	Abaqus	2 or 3	principia.es			
	MIDAS	2 or 3	midasoffice.com			
	ANSYS-CIVIL FEM	2 or 3	civilfem.com			
Finite differences	FLAC	2	itascacg.com	Feasibility		Any type of yielding
	FLAC 3D	2 or 3		Detailed engineering		
Distinct elements	U DEC	2		Feasibility	40<RMR<85	
	3 DEC	2 or 3		Detailed engineering		

10.3 BASIC REQUIREMENTS IN STRESS–STRAIN ANALYSES

Stress–strain calculation software programs are very powerful tools to accurately analyze the redistribution process of the stresses in the ground when an underground excavation is constructed.

These programs have benefited from the great progress in computer-aided calculation in the last decades and, above all, their use has become so friendly that, in many cases, they can be used by professionals with very limited knowledge in tunnel engineering. This is associated with the risk of accepting results that lack physical meaning and have low representation.

In this context, the comment from Alan Muir Wood (2004) can be highlighted:

> Unlike empirical models, which require laboratory tests, and theoretical models, which require a thorough process of simplification and analysis, numerical models are available on any computer in any office and give results by just pressing a button.

Another aspect that cannot be forgotten is the objective of the stress–strain calculations, focused on obtaining the stresses in the support elements of an underground excavation and the ground displacements originated during its construction.

According to DEA methodology, the ground stresses obtained in the calculations should be used to check the safety factors of the support elements and the expected ground displacements must be used, during construction, to check the quality of the calculations.

The following sections present the basic requirements that must be met to obtain representative results using stress–strain analysis methodology.

10.3.1 Ground data collection

To build a tunnel without significant deviations in time and budget, it is essential to characterize the ground rigorously, for which it is recommended to follow the methodologies presented in Chapters 3, 4, 5 and 6.

The results from the laboratory and in-situ tests must be consistent with each other, with the data obtained in geological mapping and with those obtained in the monitoring stations.

Discrepancies between the results obtained with the different site investigation and characterization techniques must be carefully analyzed, especially in tunnel sections with strong yielding.

10.3.2 Representativeness of the ground data

The data used in the models to characterize the ground must be as representative as possible, since the better the real ground behavior is modeled, the more realistic the results obtained in the calculations are.

The degree of representation must be higher for those ground parameters with more influence on the calculations:

- Uniaxial compressive strength
- Triaxial compressive strength
- Modulus of deformation
- Tensile strength

To obtain representative data, the following recommendations should be taken into account:

I. **The data to characterize the ground has to be obtained from tests** specifically performed for the tunnel to be constructed, and should *not* be taken from bibliographic references.

II. **The data that characterizes the ground behavior cannot be represented by a single value;** instead each set of the results obtained from the tests must be characterized by the maximum, the minimum and the average value.

- If the average values are selected in the design, there is a risk of collapse during construction in the long term.
- On the other hand, it must be kept in mind that the design with minimum values leads to very conservative construction solutions.
- Accordingly, the properties of the ground must be characterized by a range and the value used in the calculations must be justified.

III. **The values that represent the properties of the ground cannot be affected by any safety factor,** otherwise, the calculated displacements could not be compared with those measured during construction.

IV. **The distribution of the ground quality in the geomechanics profile must be accurately defined** by actual values (i.e., RMR) based on site investigations carried out. Therefore, it is not admissible to simply list percentage proportions of the different types of ground; since this will mean that the site investigation might not have been enough.

10.3.3 Identification of the construction process

The construction process should be modeled including all intermediate stages necessary to complete the final section of the underground excavation.

This definition should include the geometry of the different construction stages, the distance between the tunnel faces and the excavation advance length in each of them.

10.3.4 Characterization of the support

The support types to be used must be modeled with structural elements, whose characteristics must correspond to the support placed in the works.

In the case of using shotcrete, the law of strength increase must be specified, according to the age of the concrete.

If the shotcrete thickness exceeds 60 cm, it can be modeled with elements of the model by assigning them the properties of the shotcrete, taking into account their evolution over time.

The distance to the excavation face, when the support is installed, should be modeled for each type of support.

10.4 STRESS–STRAIN MODELING PROCESS

Stress–strain modeling enables the approach of more complex stabilization problems of underground excavations. This is because there are no limitations in the excavation shapes, nor in the number or characteristics of the ground to be considered. It is possible to model the most complex constructive methods as well as any support elements, as stated by Starfield and Cundall (1988).

In contrast, the stress–strain modeling process requires much more time than the application of the analytical calculation methodologies.

Figure 10.6 shows the diagram of activities of the stress–strain analysis process, which can be grouped into the following five phases:

I. Preliminary activities
II. Preparation of the model

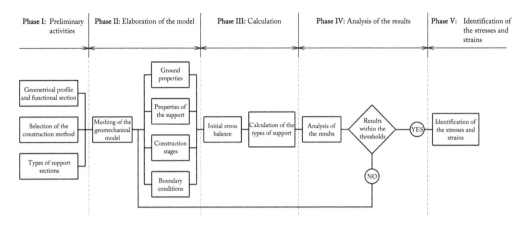

Figure 10.6 Stress–strain analysis.

III. Calculation
IV. Analysis of the results
V. Identification of the stresses and strains

It is important to note that stress–strain analysis is an iterative process, since the resolution of a model essentially requires finding the strains in the ground and the stresses on the support for a given radial pressure applied in the perimeter of the excavation.

In fact, the resolution of a geomechanics model is equivalent to calculating one of the infinite equilibrium states that define the characteristic curve of an underground excavation, as illustrated in Figure 10.7.

It is possible that, after finishing the calculations of a geomechanics model, the calculated stresses could require optimization of the support.

In such cases there is no choice but to do the calculation process again and, in order to perform the minimum number of calculations, the best alternative is to define, in as much detail as possible, the modeling process.

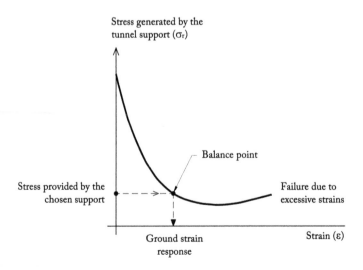

Figure 10.7 Characteristic curve of an underground excavation obtained from the geomechanics model.

10.4.1 Activities before modeling

The activities prior to modeling refer to the collection of the information necessary to correctly prepare the relevant geomechanics model.

10.4.1.1 Representative tunnel section and geomechanics profile

A representative section of an underground excavation project is essential to define the geometry to be excavated. The section must include the construction tolerances, to deal with possible ground displacements until stabilization or correction of some constructive defects.

The geomechanics profile contains the distribution of the sections to be excavated, having similar stress–strain behavior, to be analyzed in terms of stress–strain.

The geomechanics profile must contain enough information about the ground behavior, as specified in Sections 10.3.1 and 10.3.2.

10.4.1.2 Selection of the construction process

Several construction methods can be selected to construct a tunnel but only one of them will minimize the construction time, the cost and, in the case of tunnels in urban environments, the displacements of the surface and the buildings close to the tunnel.

Therefore, the construction method must be carefully selected, applying the criteria presented in Chapter 7.

In the detail engineering phase, it is very important to define, as accurately as possible, the construction aspects associated with temporary situations during construction, since they are of great importance in the stabilization process and must be taken into account in stress–strain modeling.

It is necessary to keep in mind that changing any detail in the construction process will not only require repeating the calculations, but also modifying the mesh of the model, which is usually a very tedious task.

10.4.1.3 Initial definition of the tunnel support sections

As indicated in Section 10.4, to solve a geomechanic model in terms of stress–strain it is necessary to know the characteristics of the support selected to stabilize the excavation because before making a geomechanics model, it is essential to define the tunnel support section to be modeled.

Initially, the tunnel support sections can be defined using the recommendations based on the ICE, presented in Section 2.4.3.3, or those from the Q index, indicated in Section 2.4.3.2.

With ICE values < 39, the recommendations provided by ICE will not be accurate enough and it will be necessary to analyze the characteristic curve of the excavation.

In these cases, the characteristic curve of the excavation must be calculated by points, using a two-dimensional geomechanics model, as illustrated in Figure 10.8.

The characteristic curve in the above figure corresponds to a road tunnel, located at a depth of 120 m, which is 12 m wide and is excavated in tertiary grounds, which means that the ICE of the excavation is 32 points.

This characteristic curve was obtained with a two-dimensional geomechanics model which was solved for the following five radial stresses: $\sigma_r = 1.1$ MPa; $\sigma_r = 0.7$ MPa; $\sigma_r = 0.4$ MPa; $\sigma_r = 0.2$ MPa and $\sigma_r = 0$ MPa.

The total time required to obtain this characteristic curve was around 2,5 hours.

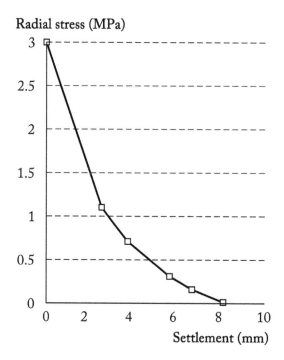

Figure 10.8 Characteristic curve obtained with a two-dimensional geomechanics model.

With this characteristic curve it is possible to analyse the stabilization of the excavation by placing a support, which transmits a radial stress of 0.5 MPa and generates a settlement of 5 cm, to achieve the balance of the excavation, as illustrated in Figure 10.9.

According to the recommendations presented in Table 2.2, for ICE = 32, the support should be composed of TH-29 steel arches with a spacing between axes of 1 m and 25 cm of shotcrete.

In this case, to estimate the required strength of the support using the characteristic curve, a safety factor of 2.5 can be considered, so the support must provide a radial pressure of 0.5 MPa × 2.5 = 1.25 MPa.

If a shotcrete layer of 15 cm and 35 MPa at 28 days is considered, the radial pressure can be calculated by the formula 22 in Figure 9.52:

$$P_i = 10.6 \cdot D^{-0.97}$$

For D = 12 m, the radial pressure provided by the shotcrete is $P_{iHP} = 0.93$ MPa.

The pressure provided by the TH-29 steel arches, placed with a spacing between axes of 1 m, can be estimated by interpolating the values obtained from the expressions 6 and 7 from Figure 9.52, obtaining a value of $P_{iTH} = 0.53$ MPa.

As a first approach, the total radial pressure provided by the support can be obtained as the sum of the pressures provided by the steel arches and the shotcrete, therefore the maximum radial pressure achieved in this case will be $P_i = 0.93$ MPa + 0.53 MPa = 1.46 MPa, which exceeds the estimated value of 1.25 MPa necessary to stabilize the excavation.

This support, composed of TH-29 steel arches with spacing between axes of 1 m and 15 cm of shotcrete, is lighter than the support recommended by the ICE and, if the stress–strain calculation validates it, it will be a more cost-effective solution.

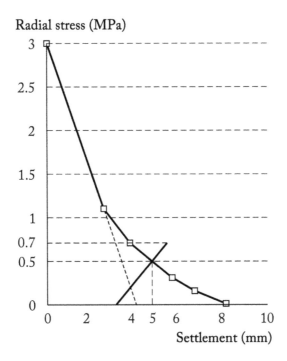

Radial stress (MPa)

Settlement (mm)

Figure 10.9 Estimation of the necessary support using the characteristic curve in Figure 10.8.

10.4.2 Elaboration of the geomechanics model

On the one hand, to successfully apply the stress–strain analysis, it is very important to prepare an efficient geomechanics model, which is not an easy task. This is because the geomechanics model must include all the information that affects the stabilization of the underground excavation and, on the other hand, the model must be divided into finite elements, which have to be reduced as the stresses on them increase.

These two constraints, applied to three-dimensional models of underground excavations with bifurcations or changes in their cross sections, lead to models with an extraordinary complexity, due to the difficulty in dividing them into finite elements.

Nowadays, all stress–strain calculation programs have meshing routines that greatly simplify the preparation of the geomechanics models.

Meshing routines are very easy to use and work very well on two-dimensional models, but their application to complex three-dimensional models requires significant work.

To simplify the preparation of a geomechanics model it is convenient to analyze if it is possible to introduce any type of symmetry in it, since this will reduce drastically the number of elements in the model.

Figure 10.10 shows an example of a two-dimensional model with symmetry in the vertical plane that passes through the center of the tunnel.

The following sections present the most important aspects to be considered when preparing geomechanics models.

10.4.2.1 Model size

The most important requirement to be met by the geomechanics model is to reproduce as accurately as possible the tunnel environment conditions and the construction process.

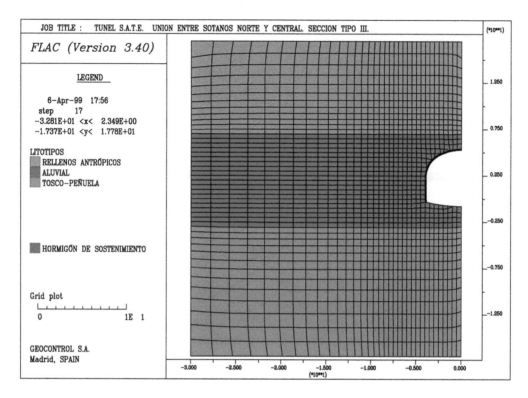

Figure 10.10 Symmetrical two-dimensional model.

This objective should be compatible with obtaining a model as simple as possible, especially in the three-dimensional analysis, to have admissible calculation times.

In two-dimensional models, the model width and height is usually about six times the excavation width, depending on the type of project.

Figure 10.11 shows a two-dimensional model of a shallow tunnel.

Three-dimensional models include both the tunnel cross section and its construction steps along the tunnel axis, so it is possible to simulate the interaction between the ground and the support as it is constructed.

In the direction of the tunnel axis, three-dimensional models usually have lengths between 60 and 150 m, as illustrated in Figure 10.12.

This model is 100 m wide, 80 m long, 47 m high and is composed of 271,840 elements.

When special excavations are modeled, the models are constrained by the dimensions of the excavations, and therefore, do not follow the previous pattern. Figure 10.13 shows the three-dimensional geomechanics model prepared to analyse the North Portal of the Montegordo Tunnel (Chile).

This model has a width of 170 m, a maximum height of 148 m, its length is 163 m and is composed of 227,450 elements.

10.4.2.2 Two-dimensional model in a continuous medium

When compared to other models, two-dimensional geomechanics models require less work to make them and also shorter calculation times.

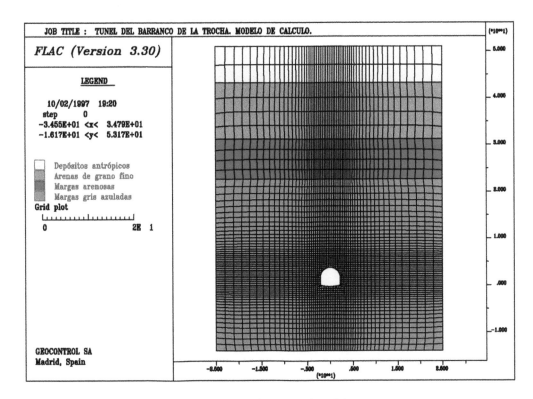

Figure 10.11 Shallow tunnel represented in a two-dimensional model.

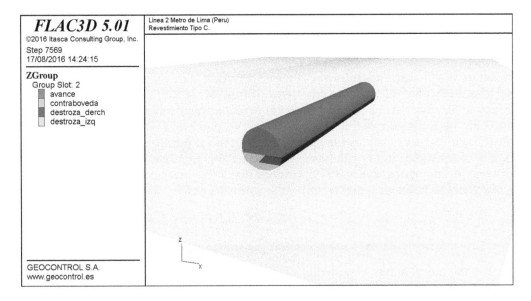

Figure 10.12 Three-dimensional geomechanics model of a tunnel.

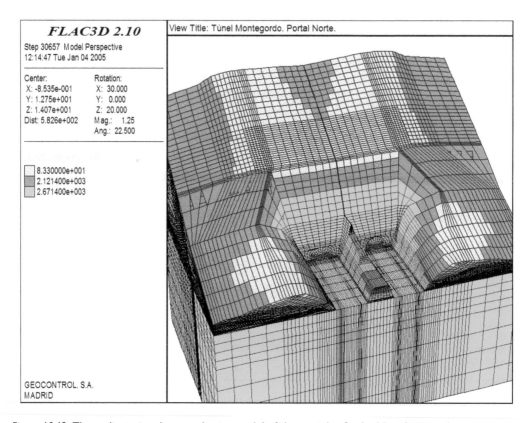

Figure 10.13 Three-dimensional geomechanics model of the portals of a double-tube tunnel.

However, this has a significant drawback because two-dimensional models cannot explain the stress balance produced in the unsupported span close to the tunnel face, as shown in Figure 10.14.

In section AA', located ahead of the tunnel excavation face, the pressure σ_r acts in the radial direction with a value equal to the vertical stress, for $K_0 = 1$.

In section BB', located in a section which has been excavated and supported, the support pressure (σ_p) balances the radial pressure (σ_r).

However, in the area close to the tunnel face, the radial pressure acting on the ground cannot be compensated by any other pressure, since the support has not been placed yet.

To avoid this drawback, Panet (1979) assumed that at the unsupported span, the excavation face creates a fictitious radial pressure (σ_f), which is given by the expression:

$$\sigma_f = 0.75 \cdot \sigma_o \left(\frac{0.75 \cdot R}{0.75 \cdot R + X} \right) \tag{10.5}$$

where R is the tunnel radius, considering a circular tunnel, and X the distance between the support and the tunnel face.

If the support is placed at the tunnel face, then X = 0 and $\sigma_f = 0.75 \cdot \sigma_o$ and if the support is placed at a distance to the tunnel face equal to its diameter, then X = 2·R and $\sigma_{f2R} = 0.2 \cdot \sigma_o$.

Afterward, Panet (1995) introduced a variation in Equation 10.5 to take into account the ground yielding effect at the tunnel face.

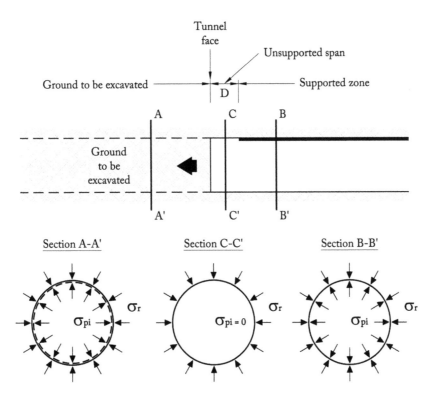

Figure 10.14 Sections of interest in the area close to the tunnel face.

Figure 10.15 shows four graphs to estimate the contribution of the excavation face in a tunnel with a radius of 6 m, constructed with excavation advance lengths of 2 m, under moderate yielding conditions (ICE = 52).

Three out of the four graphs have been calculated applying Panet's theory; two with the elastic models from 1979 and 1995 and the third by applying the plastic model from 1995. The fourth graph has been obtained with FLAC3D software.

This figure shows that both Panet formulations in the elastic domain give similar results, but are very optimistic, in the analyzed case, due to the yielding at the tunnel face.

It is also appreciated, in Figure 10.15, that the Panet formulation which considers ground yielding fits quite well with the calculations made with FLAC3D software.

In the analyzed case, considering that the support is placed 2 m away from the excavation face, its contribution to stabilize the unsupported span is 18% of the in situ stress.

10.4.2.3 Axisymmetric models in a continuous media

Axisymmetric models have symmetry with respect to the excavation axis, which enables accurately analyzing the vertical shaft behaviors, as long as the ground is homogeneous and composed of horizontal layers.

In tunnels, axisymmetric models can be used whenever the ground is homogeneous with a lithostatic initial stress state ($K_0 = 1$), if the tunnel has a circular cross section and is located at a depth enough so the differential stress between the tunnel crown and the invert is negligible when compared to in situ stresses.

Figure 10.16 shows an axisymmetric model of a tunnel, which has axial symmetry along the OY axis.

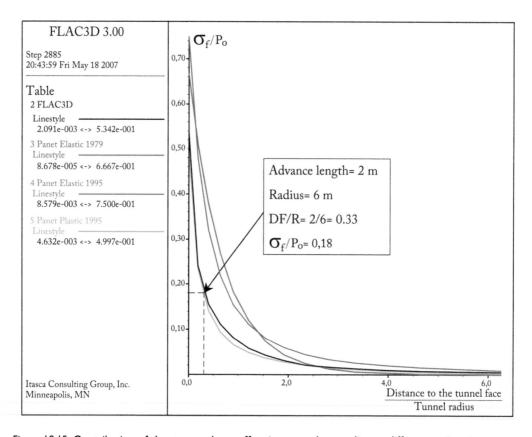

Itasca Consulting Group, Inc.
Minneapolis, MN

Figure 10.15 Contribution of the stress release effect in a tunnel, according to different estimations.

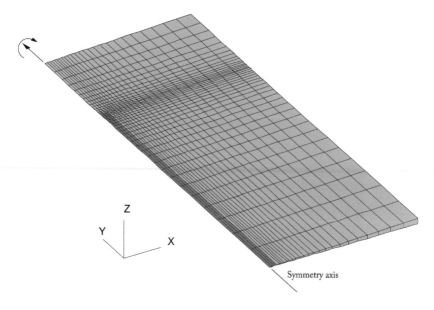

Figure 10.16 Axisymmetric model of a tunnel.

10.4.2.4 Three-dimensional models in a continuous medium

Three-dimensional models enable analyzing, with great accuracy, the interaction between the ground and the support at the different construction stages, provided that the ground is correctly characterized from the geomechanics point of view.

Usually, one of the axes from the three-dimensional model is aligned with the tunnel axis, so that the different construction stages can be simulated with great realism and the interaction between the excavation faces can be analyzed.

As the tunnel axis coincides with one of the model axes, it is possible to sequentially analyze the development of the excavation faces, which requires a thickness of the elements, in the direction of the excavation, equal to the excavation advance round. This increases considerably the number of elements in the model.

The most complex three-dimensional geomechanics models correspond to underground works that must be built close to existing excavations, because in these cases, as well as modeling the new structures to be constructed, it is necessary to evaluate the incidence of the existing excavation.

This means that the model has to include two areas of interest, which significantly increases the difficulty to elaborate it, as happened in the model for the Vall d'Hebron Station, Line 5 of the Barcelona Subway.

The crown from Vall d'Hebron Station had to be designed at a distance of only 8 m from the bottom slab of San Genís engine sheds, as shown in Figure 10.17.

Figure 10.18 shows the model prepared to analyze Vall d'Hebron Station that incorporates symmetry with respect to the vertical plane that passes through the axis of the station, in order to reduce the model dimensions.

This model is 80 m high, its dimensions in a horizontal plane are 120 m and 100 m and, it is composed of 158,080 elements. Due to the symmetry, this model is equivalent to another, without symmetry, of 316,160 elements.

To minimize the effects generated by the construction of Vall d'Hebron Station on San Genís trains depot, the station was excavated with foundation galleries, as described in Section 7.2.3.1.2.

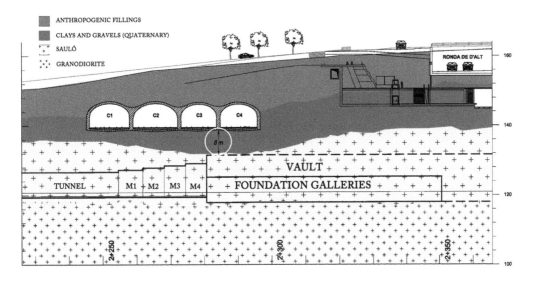

Figure 10.17 Longitudinal section of Vall d'Hebron Station, Line 5 of the Barcelona Subway (Spain).

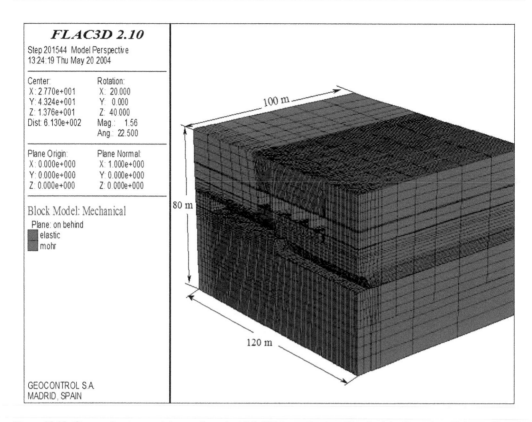

Figure 10.18 Geomechanics model to analyze the Vall d'Hebron Station, Line 5 of the Barcelona Subway (Spain).

Figure 10.19 shows the detail of the geomechanics model prepared to perform the stress–strain analysis of this station.

This figure shows that the station support, which was designed with shotcrete and cast-in-place concrete, has been modeled using the finite elements of the model.

As an example of a highly complex geomechanics model, Figure 10.20 shows the model made to perform the stress–strain analysis of Ñuñoa Station, Line 6 of the Santiago de Chile Subway.

Ñuñoa Station consists of a vertical shaft and two horizontal galleries connected to the tunnel station, whose length is 120 m and its maximum width is 17 m.

This model has a length of 240 m, a width of 168 m, a height of 80 m and consists of 660,000 elements.

10.4.2.5 Models in a discontinuous medium

The models for performing stress–strain calculations in discontinuous media are much more complex to compute than models in continuous media, because it is necessary to consider the difficulties in defining the rock blocks that can appear in the excavation surroundings.

The following sections explain the most relevant characteristics of two- and three-dimensional models in discontinuous media.

10.4.2.5.1 Two-dimensional models in discontinuous media

Two-dimensional models in discontinuous media are of limited application, because, with them, it is impossible to model non-prismatic rock blocks.

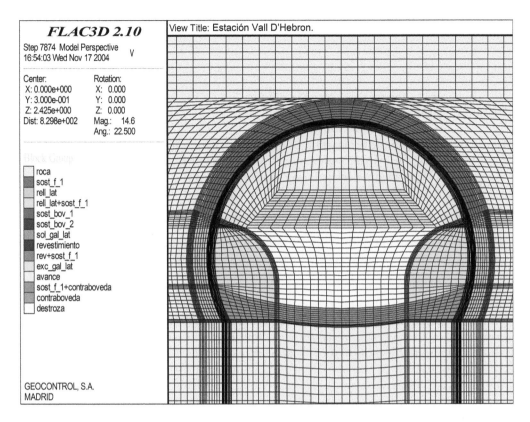

Figure 10.19 Detail of Vall d'Hebron Station model, Line 5 of the Barcelona Metro (Spain).

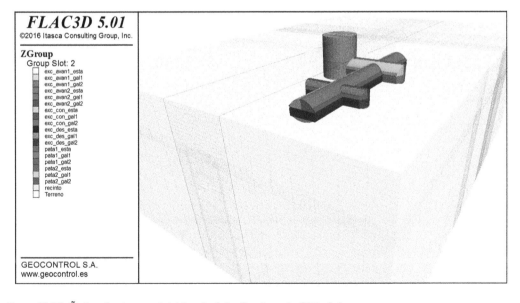

Figure 10.20 Ñuñoa Station model, Line 6 of the Santiago de Chile Subway.

These models are useful to analyze, in terms of stress–strain, stratified grounds affected not only by the stratification, but also by individual joints of great persistence; since, under these conditions, the ground displacements will be concentrated at the contact surface of the layers and in the joints, without producing three-dimensional displacements.

Assuming that the blocks modeled in two-dimensions are prismatic and that the dimension perpendicular to the modeled plane is infinite, discontinuous two-dimensional models can be used to perform rough calculations.

In these cases, the calculation software has the routines necessary to individualize each rock block depending on the orientation of the joint sets considered.

Figure 10.21 shows a discontinuous model to perform the stress–strain analysis of the Cavern of Venda Nova Hydroelectric Power Station (Portugal).

10.4.2.5.2 Three-dimensional model in discontinuous media

The methodology to compute three-dimensional models in continuous media is very specific, because the elements in the model must adapt to the potential rock blocks that could appear.

On the other hand, it should be noted that 3DEC software, which uses the distinct element algorithm, is almost the only program available to perform this type of calculations.

3DEC software has a sophisticated meshing routine which, based on the joint sets orientation identified in the ground, enables the definition of the rock blocks that can be individualized and also to distinguish those that behave as rigid solids from those that deform. Deformable blocks must, in turn, be divided into other elements to calculate the strains and stresses produced as they deform.

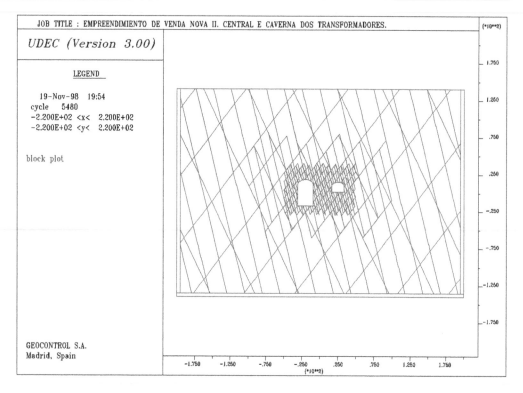

Figure 10.21 Discontinuous two-dimensional model of Venda Nova Hydroelectric Power Station (Portugal).

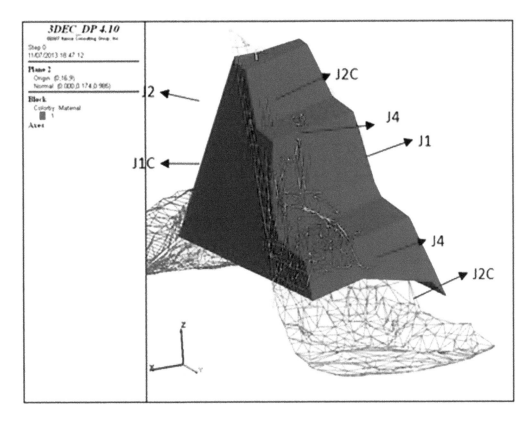

Figure 10.22 Rock block with seven discontinuities, computed with 3DEC software.

As an example of 3DEC software ability to manage the existing joint sets in the ground, a rock block with seven discontinuities that has been modeled with this program is shown in Figure 10.22. Figure 10.23 shows a complex three-dimensional model computed with the 3DEC software, which is 2 km wide, 1 km high and 1 km deep.

This model was prepared to analyze the stabilization of a cavern to contain crushers inside the slope shown in Figure 10.24.

In the model shown in Figure 10.23, more than 200,000 blocks have been distinguished and each block has been divided into 5 to 20 elements, so the number of elements in the model approaches a million.

Figure 10.25 shows the detail of the mesh of the cavern and the access tunnels.

10.4.3 Calculation of the geomechanics models

The following sections contain the most relevant aspects to keep in mind in the calculation process of the geomechanics models.

10.4.3.1 Boundary conditions

If the ground surface is included in the geomechanics models, they must reproduce the topography and the existing structures.

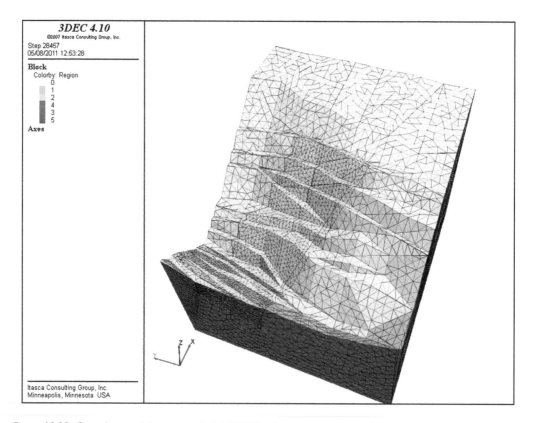

Figure 10.23 Complex model computed with 3DEC software.

To introduce the boundary condition of the displacements, the base of the model is usually constrained in all directions; the horizontal displacements are constrained at the lateral sides and the nodes in the upper face can move freely.

For in situ stresses, a lithostatic state is usually assumed for the vertical stress with the corresponding stress ratio, K_{0H} and K_{0h}, in the horizontal plane.

In models for shallow tunnels, when the overburden is smaller than twice the excavation width, loads should be applied at the model surface depending on the existing structures.

10.4.3.2 Stress distribution in the model before calculations

As indicated in the previous section, the stresses in the ground which correspond to the in situ stress state before the excavation are applied at the boundaries of the model. Therefore, before starting the calculations, it is necessary to solve a previous stage, called initial stress adjustment, so all the elements of the model will have the desired in situ stresses.

Figure 10.26 shows the distribution of the vertical stresses in a model after the initial stress adjustment.

To reduce the calculation time of this phase, it is usual to increase the cohesion of the elements in the model, with respect to the values that they will have in later stages, to avoid the yielding of the elements in the model which increases the calculation time.

Once the initial stress adjustment phase is completed, the next step is to analyze the results obtained to verify that the stress state in the model is the desired one.

Figure 10.24 Slope analyzed with the model shown in Figure 10.23.

10.4.3.3 Calculation of the construction stages

Once the initial stress equilibrium is reached, the construction phases are calculated, one after the other.

To simulate an excavation stage, the resistant properties of the elements that are going to be excavated in that stage are set to zero producing an imbalance in the model which enables calculating the new distribution of the nodal displacements and stresses in the center of each element.

Once a construction phase is balanced, the next phase is calculated, until the complete construction sequence has been calculated.

In sequential calculations, each excavation advance, usually between 0.5 and 5 m, produces an imbalance in the model and, therefore, a new calculation is required. However, the stress imbalance induced in the model by simulating the excavation is small so the calculation times to reach the new equilibrium are also small.

In any case, sequential calculations significantly increase the calculation time.

10.4.3.4 Calculation time

The calculation time necessary in a stress–strain model depends on the following parameters:

- Number of elements in the model
- Simulated construction phases
- Type of calculation algorithm
- Degree of yielding in the elements of the model
- Equipment available to perform the calculations

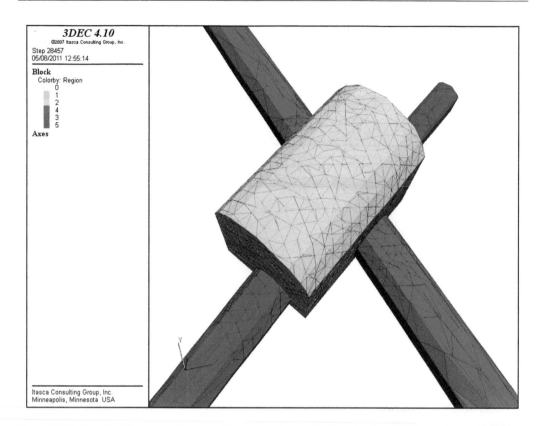

Figure 10.25 Detail of the mesh of the crushing cavern located inside the slope shown in Figure 10.24.

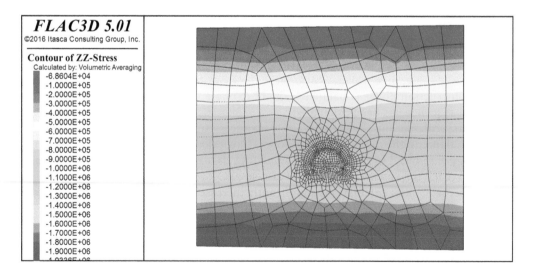

Figure 10.26 Distribution of the vertical stresses before calculating the interaction between the ground and the support.

Table 10.3 Calculation times estimated to solve the most frequent stress-strain models, using an Intel Core i7 4960×3.66 Ghz processor

Type of model	No. of elements	Estimated calculation time (hours)
Two-dimensional model	5,000/10,000	0.25
Non-sequential three-dimensional model	40,000/400,000	1 to 4
Sequential three-dimensional model	100,000/1,000,0000	24–240
Three-dimensional model in discontinuous media	50,000/500,000	48–480

Table 10.3 presents an estimation of the calculation times necessary to solve the most frequent stress–strain models, using an Intel Core i7 4960×3.66 GHz processor.

The data contained in Table 10.3 highlight two aspects that have to be considered; one is the speed in solving two-dimensional models and the other is the relation between the number of elements in a three-dimensional model and the calculation time.

Two-dimensional models have many limitations, but by applying Panet theory to simulate the support effect of the excavation face and choosing the most suitable curve to simulate this effect, as indicated in Section 10.4.2.2, reasonably representative results can be obtained.

Accordingly, two-dimensional models combined with Panet theory are very useful in stress–strain analysis of underground works with moderate yielding, that is, for ICE > 40.

Table 10.3 highlights the high calculation time required by three-dimensional models in discontinuous media, in particular those made with 3DEC software, because this program, before starting the calculations, distinguishes the blocks with a rigid behavior from those that are deformed plastically in the contacts between the blocks.

10.4.4 Analysis of the results

The analysis of the results obtained after calculating a geomechanics model should provide the following information:

- Verification of the excavation stability
- Details of the stresses on the support
- Details of the ground displacements

As a result of calculating a geomechanics model, a large amount of information is obtained, such as the displacements in each node of the model, the pressures applied in the centers of each element and the stresses on the structural elements that simulate the elements of support.

All stress–strain calculation software have graphical outputs based on isolines or color codes that allow representing, in a very effective way, the distribution of the values calculated in the model or in a part of it.

The following sections present the methodology for analyzing the results obtained in a geomechanics model.

10.4.4.1 Stabilization of the excavation

Once the geomechanics model has been calculated, it is necessary to verify that the adopted design enables reaching the stabilization in conditions compatible with the functionality required in the tunnel.

This operation is particularly important when calculations are done using programs with the finite difference algorithm, since this algorithm gives results without warning, even when unacceptable displacements are obtained for the tunnel to function as planned.

The verification of the excavation stability can be made by analyzing the distribution of the elements in the model that have yielded during the calculation process or evaluating the increase in the shear strains that have occurred, as presented in the following sections.

10.4.4.1.1 Yielded elements

An element of a geomechanics model is considered as having yielded, if during the calculation process or in the final situation it has reached the stress limit imposed by the constitutive model.

If an element of a geomechanics model has yielded, it does not mean that the excavation has collapsed but simply that the ground that constitutes that element is in the post-failure phase.

To verify the effect of local yielding on the excavation perimeter, it is necessary to analyze the factors of safety of the support elements.

In any case, generalized yielding around an excavation indicates that the design must be modified.

Figure 10.27 shows an example of intense yielding around a tunnel, which clearly indicates that the analyzed design is not adequate.

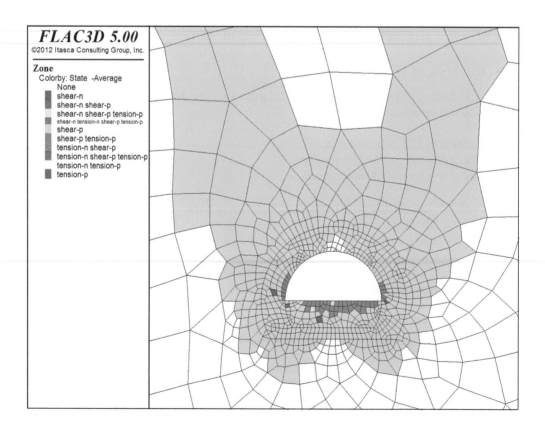

Figure 10.27 Intense yielding around a tunnel.

Two modes of yielding are distinguished in this figure; one at shear and the other at tension. In addition, it is possible to distinguish if the element has yielded in a calculation phase, which is identified with the letter p, or in the final stage, identified with the letter n.

In this case, both – the elements which have yielded in the final stage or those in an intermediate phase of the calculations – are the elements loaded to the limit of the selected constitutive model.

Figure 10.28 shows a tunnel where the yielded elements are minor and local, which indicates that the adopted design is correct.

10.4.4.1.2 Increments in the shear strain

The stability analysis of a design, based on the yielded elements after the stress–strain calculations, is fully qualitative and therefore of limited application.

Much more useful is the analysis of the shear strain increments, produced on the elements during the simulation of the construction process.

The increase of the shear strain can be measured in laboratory tests with controlled post-failure and, therefore, it is an objective index of the safety margin until the collapse.

In addition, the shear strain is an intrinsic parameter of the ground which has a limit value and, therefore, is in itself a failure criterion.

Figure 10.29 shows the distribution of the shear strain increments in the tunnel of Figure 10.27.

At first sight, the situation shown in Figure 10.29 seems relatively normal; but if the color scale is observed in detail, it is verified that the increase of the shear strain in most of the

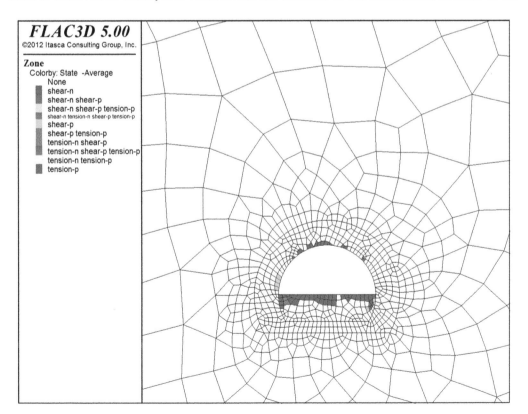

Figure 10.28 Moderate yielding around a tunnel.

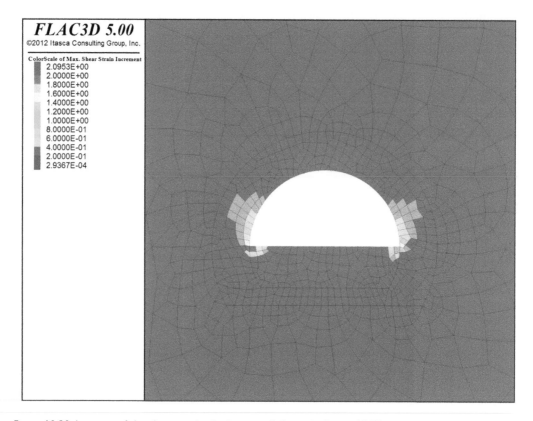

Figure 10.29 Increase of the shear strains in the tunnel shown in Figure 10.27.

model is larger than $2 \cdot 10^{-1}$, that is of 20%. Most grounds cannot reach this value, which clearly indicates that the analyzed design is not correct.

In addition, if the shear strain increase at the support of the steel arches is observed, it is larger than 1.8, that is 180% which has no physical meaning in rock mechanics.

Figure 10.30 shows the distribution of the shear strain increments for the yielded elements shown in Figure 10.28.

The distribution of the shear strain increments in Figure 10.30 seems to show a more problematic situation than the one shown in Figure 10.29, but if the color scale is observed in detail, it can be verified that is not so.

On the one hand, the largest value of the shear strain shown in Figure 10.30 is $1.2 \cdot 10^{-3}$ (0.12%), which is acceptable in many types of ground.

On the other hand, at the perimeter of the tunnel crown the shear strain is less than $0.9 \cdot 10^{-4}$ (0.09%) which is a very typical value.

The two examples above highlight the great potential of analyzing the distribution of the shear strain increments in the ground to evaluate the degree of stability of an underground excavation.

10.4.4.2 Ground displacements

The distribution of the nodal displacements in the model is obtained from the calculation of the three-dimensional geomechanics model and can be represented in three-dimensions as shown in Figure 10.31.

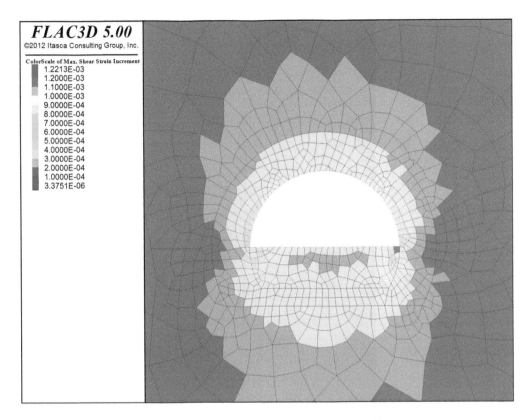

Figure 10.30 Increase of the shear strains in the tunnel shown in Figure 10.28.

In this figure, it is verified that the ground displacements around the tunnel are in millimeters, which can be associated to an almost elastic behavior in the analyzed tunnel.

In addition, in the above figure, it can be seen that the displacements induced in the already constructed tunnel are smaller than $1.4 \cdot 10^{-3}$ (1.4 mm), so the influence of the new tunnel is very small.

In three-dimensional calculations, it is often more convenient to represent the displacements in the selected planes since the analysis is much more intuitive.

Figure 10.32 shows the distribution of the ground displacements in a tunnel cross section where it is clear that the major displacements appear at the tunnel crown, 5.7 cm, and at the center of the invert slab, 3.6 cm. These centimetric displacements may be associated with moderate yielding.

However, the displacements calculated in the geomechanics models represent the displacements generated in the ground after the stress readjustment that follows the excavation and that the origin of the displacements is several meters ahead of the tunnel face.

It should therefore be considered that the ground displacements measured in the tunnel during its construction will be much smaller than those calculated.

Figure 10.33 shows that the convergence, relative displacement between two points at the tunnel perimeter, which can be measured is much lower than the one calculated.

To calculate the convergence or the vertical displacement of the tunnel crown that can be measured during the tunnel construction, the best option is to define the position of the reference points in the model, so that the displacement of these points can be recorded during the calculation, as shown in Figure 10.34.

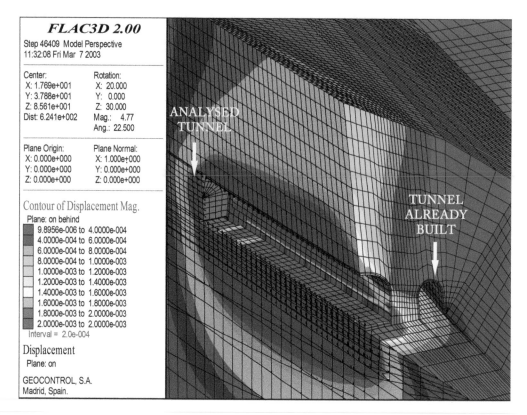

Figure 10.31 Nodal displacements in a tunnel constructed next to another tunnel already built.

The calculation of three-dimensional geomechanics models also enable defining the subsidence trough that will occur at the ground surface after tunnel construction. Figure 10.35 shows the subsidence trough calculated for Cerrillos Station, Line 6 of the Santiago de Chile Subway.

10.4.4.3 Loads on the support elements

The loads on the support elements can be used to calculate the minimum factor of safety provided by the steel arches, bolts and shotcrete placed as support, but are not very useful as control parameters, since their measurement during construction is difficult, not very accurate and very expensive.

The stresses calculated on the bolts are obtained as the distribution of the axial loads along the rock bolt length, which thus allows analyzing their contribution. Figure 10.36 shows the distribution of the axial loads in the bolts of a tunnel. It is usually considered that the steel arches in the tunnels work at bending-compression and, therefore, the data necessary to structurally verify them are the bending moments and the axial loads, which are provided by the structural elements used to model them.

Figure 10.37 shows the distribution of the bending moments and axial loads acting on a steel arch.

Shotcrete can also be modeled with structural elements or, when its thickness is large enough, with the elements of the model. In the first case, the bending moments and axial

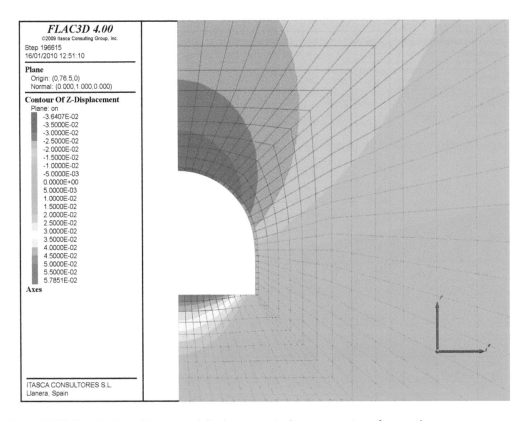

Figure 10.32 Distribution of the ground displacements in the cross section of a tunnel.

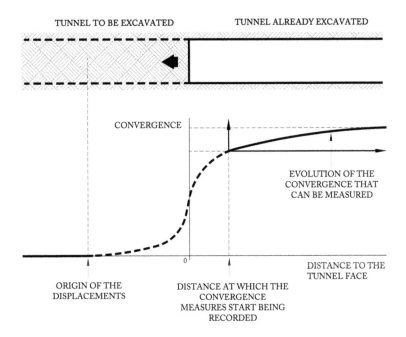

Figure 10.33 Convergence that can be measured in a tunnel.

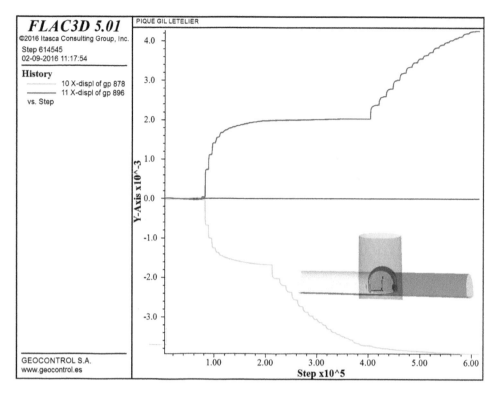

Figure 10.34 Semi-convergence measured at a reference point in the tunnel.

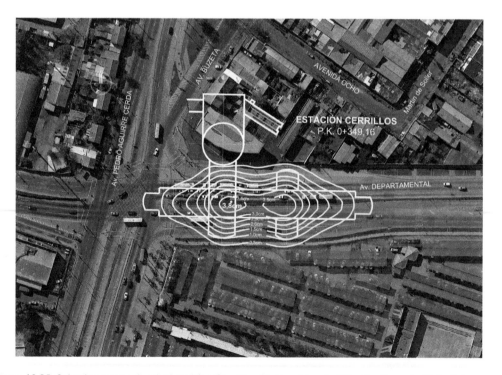

Figure 10.35 Subsidence trough calculated for Cerrillos Station, Line 6 of the Santiago de Chile Subway.

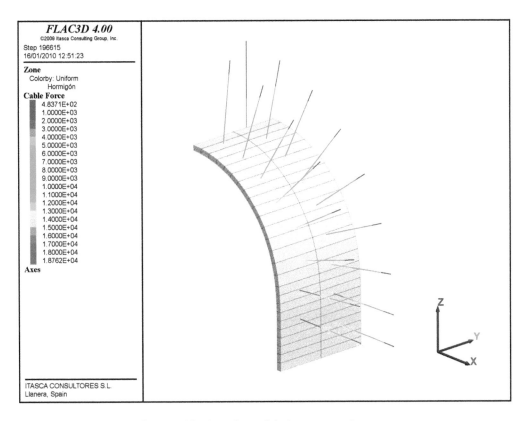

Figure 10.36 Distribution of the axial loads in the rock bolts in a tunnel.

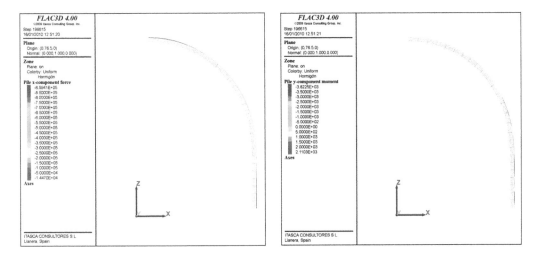

Figure 10.37 Axial loads and bending moments calculated in a steel arch. 1- Axial loads. 2. Bending moments.

forces are obtained, whereas the second case gives the results in tensile and compressive stresses, as shown in Figure 10.38.

Figure 10.39 shows the results of a model, where the shotcrete has been modeled with finite elements and the concrete of the invert slab with a structural element.

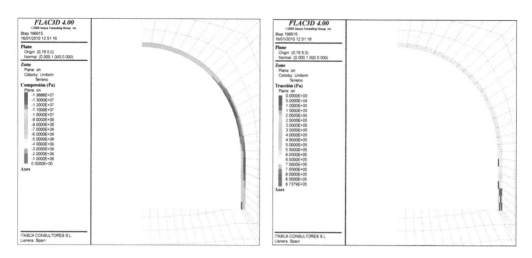

Figure 10.38 Compressive and tensile stresses in shotcrete. 1. Compressive stresses. 2. Tensile stresses.

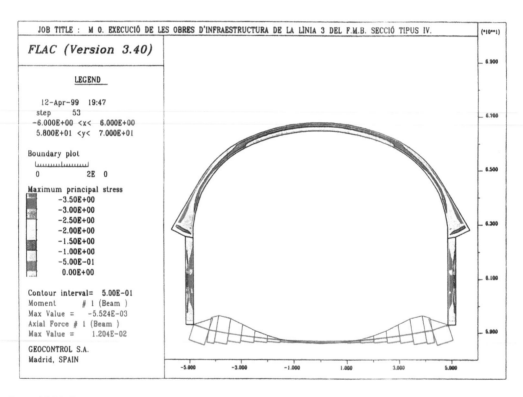

Figure 10.39 Stresses in the shotcrete, calculated with finite elements and structural elements.

The stresses on the support are used to calculate the factors of safety and the amount of steel reinforcement required in the concrete.

Figure 10.40 shows the structural verification of a steel arch, using the bending moment-axial load diagram. In this case, the resulting minimum factor of safety of the steel arch is $FS_{min} = 1.11$.

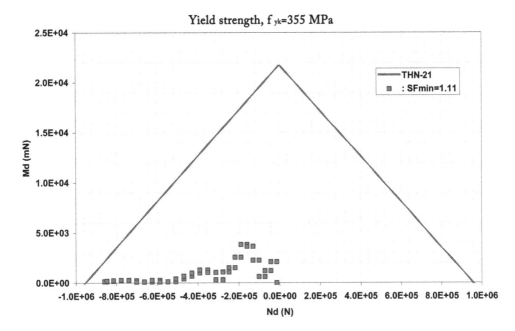

Figure 10.40 Structural verification of a steel arch.

10.5 CASE STUDIES

The following sections present three particular cases of tunnel design involving modeling.

10.5.1 Stress calculation of rock blocks stability

The stress–strain calculation of underground excavations, whose stability can be simulated as a block mechanism, is performed with programs based on the distinct elements algorithm.

As indicated in Table 10.2, these programs can be applied to grounds with many discontinuities, defined by the RMR between 40 and 85 points.

Table 10.2 also indicates that programs based on finite element and finite difference algorithms can be applied to grounds with an RMR up to 65 points.

In the upper range of application of these programs, which corresponds to $45 < RMR < 65$, it is always recommended to verify that the designed support ensures the stability of potentially unstable rock blocks.

To verify this, the UnWedge software, marketed by Rocscience (Vancouver, Canada) is very useful. It features analyses of the stability of rock blocks which can fall as a result of the gravitational force or slide as a rigid block along one or two of the joints in the block.

UnWedge software considers the effect of the in situ stress state under conditions of elastic behavior, but is limited to blocks formed by three discontinuities. When four or more joint sets are identified in the ground, it is necessary to analyze them several times, considering, in each analysis, only three out of the total number of joints identified in the ground.

As an example of an application of UnWedge software, Figure 10.41 shows the data of a tunnel, with N-S alignment, to be constructed in a ground where three joint sets have been identified.

Figure 10.42 shows the results obtained with UnWedge software, after analyzing the three joint sets defined in Figure 10.41.

Joint set	dip°	dip direction°
J1	70 ± 5	036 ± 12
J2	85 ±8	144 ± 10
J3	55 ± 6	262 ± 15

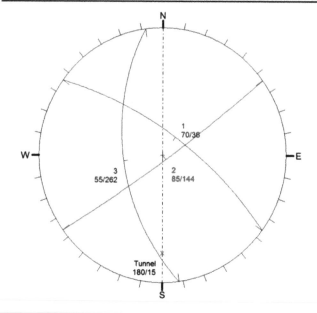

Figure 10.41 Description and representation of three joint sets in the ground where a tunnel must be excavated, N–S direction.

Figure 10.42 shows that the three joint sets identified in the ground produce four rock blocks. Three of these blocks are unstable because their factor of safety is lower than 1.3 but the block at the bottom slab of the tunnel is stable as it cannot fall.

After this first analysis, others should be done, considering the following:

- Potential reduction of the joints length, which will reduce the volume of the blocks, improving their stability
- The supporting effect of the rock bolts
- The effect of the in situ stress state around the tunnel.

Of these three actions, the most reliable is the supporting effect of the rock bolts because the reduction of the length of the joints is very subjective and the knowledge of the natural stress state does not always correspond to reality.

Nevertheless, the minimum factor of safety of the wedge stability should be 1.3 because it is a temporary situation.

10.5.2 Stress–strain calculation in underground works under the water table

When an underground excavation has to be constructed below the water table, the water will fill the joints in the ground creating hydrostatic pressure, also known as the pore pressure.

Wedge	Weight - tonnes	Failure mode	Factor of Safety
Roof wedge	44.2	Falls	0
Right side wedge	5.2	Slides on J1/J2	0.36
Left side wedge	3.6	Slides on J3	0.40
Floor wedge	182	Stable	∞

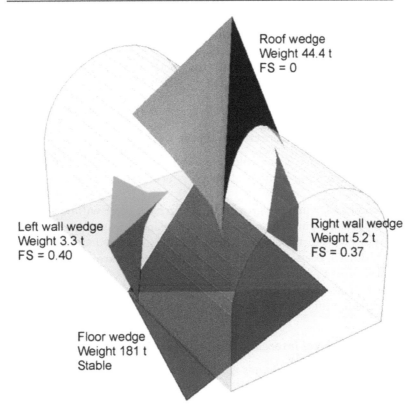

Roof wedge
Weight 44.4 t
FS = 0

Left wall wedge
Weight 3.3 t
FS = 0.40

Right wall wedge
Weight 5.2 t
FS = 0.37

Floor wedge
Weight 181 t
Stable

Figure 10.42 Results obtained with UnWedge software.

If the pore pressure has the same order of magnitude as the ground strength, it will create a significant mechanical effect on the ground displacements, which normally occurs in weathered rocks or soils and, to a lesser extent, in soft rocks.

In addition, an excavation under the water table will create a significant draining effect, so if the ground permeability and the aquifer flow are high enough, a permanent water infiltration regime will be created in the excavation.

If the ground permeability is low enough, it will not be possible to control the initial infiltrated water flow, so the pore pressure around the excavation will be zero.

Once the support and lining are placed, if the latter is watertight, there will be no water infiltrations and the hydrostatic pressures will be restored to the initial values, making it necessary to reinforce the lining with steel bars.

Designing a watertight lining is an expensive solution, so, alternatively, the design of a drained tunnel can be considered. In addition, drained tunnels have the advantage that they do not require a reinforced lining. As a counterpart, it should be kept in mind that, for environmental reasons, the draining effect of a tunnel is usually limited to flows of less than 1 l/s per km of tunnel.

To consider the effect of an excavation under the water table in the stress–strain calculations, the models described below can be adopted.

10.5.2.1 Model without a hydro-mechanical interaction

The simplest way to model the effect of the water on an underground excavation is to assume that the flow does not vary during the excavation and that the effect of the water generates stresses only on the elements of the model.

This model is suitable for grounds with low permeability and medium to high strength, because in these cases the pore pressure variations will be small and negligible with respect to the ground strength.

10.5.2.2 Hydro-mechanical interaction without water flow

This model assumes that the water volume in the ground remains constant during the excavation of an underground structure, which is acceptable in grounds with very low permeability.

Under these conditions, when the ground is compressed, there will be an increase in the pore pressures and when decompressed, they will be reduced.

Figure 10.43 shows the distribution of the pore pressures around a tunnel constructed with a TBM, which have been calculated with FLAC3D software assuming no water flow around the tunnel.

In these cases it is assumed that pore pressure variations exert an instant mechanical effect, so the calculation times are similar to those required to solve a model without water.

10.5.2.3 Hydro-mechanical interaction with water flow

This model is recommended for grounds with low strength and medium to high permeability, where an underground excavation will be constructed under a water table which generates a significant water flow in the tunnel.

The calculation with hydro-mechanical interaction and water flow requires calculation times much longer than for the two previous models, because when the pore pressures are varied, it is necessary to update its distribution during the calculation process and to calculate the mechanical effect produced by these variations on the ground.

Figure 10.44 shows the distribution of the pore pressures after excavating a tunnel, with a TBM, calculated assuming hydro-mechanical interaction with variations in the water flow.

10.5.3 Seismic effect

Underground excavations are fairly safe against earthquakes, although the tunnel portals, shallow tunnels and sections excavated in grounds of very poor quality, RMR < 20, should be specifically designed to withstand the seismic effects.

One of the methodologies most used in the seismic design of underground excavations consists in calculating the strains that the ground could suffer as a result of an earthquake before constructing the tunnel and applying these strains to the structure composed of support and lining to calculate the induced stresses.

Usually, earthquake-induced displacements in the direction of the tunnel axis have a smaller effect than the displacements induced in the cross section and therefore, seismic

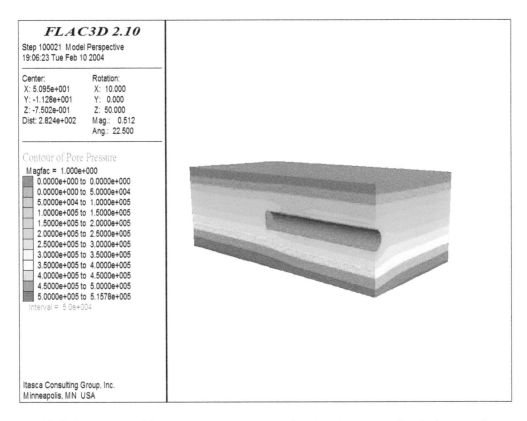

Figure 10.43 Distribution of the pore pressures assuming that there is no water flow in the ground.

analyses focus on defining the shear strain produced by the earthquake in a plane transverse to the underground excavation. This shear strain is often called seismic shear strain (θ_s).

The shear strain was calculated with the methodology used by Kuesel (1969) in the San Francisco Subway, which was also used in the Santiago de Chile Subway since 1971. This methodology was improved by Ortigosa and Musante (1991) and is included in Chapter 3.1003.5 of the *Road Manual of the Ministry of Public Works* in Chile, where the shear strain is calculated with the model shown in Figure 10.45.

The shear strain of the ground in the layer "i" is calculated by the expression:

$$\theta_{si} = \frac{(1 - 0.0167) \cdot a_o \cdot \gamma \cdot Z_i}{\delta_{ci}} \tag{10.6}$$

θ_{si} = Shear strain induced by an acceleration, a_o, at the center of the layer i, expressed in radians

Z_i = Depth to the center of layer "i", expressed in meters.

a_0 = Peak horizontal acceleration at the ground surface, expressed as a fraction of the gravity

γ = Soil unit weight, expressed in ton.f./m³

δ_{ci} = Ground shear modulus for seismic loads generated at the center of layer "i", expressed in ton.f./m³

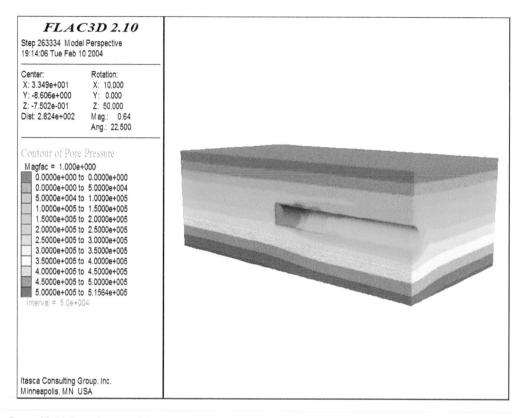

Figure 10.44 Distribution of the pore pressures assuming water flow in the ground.

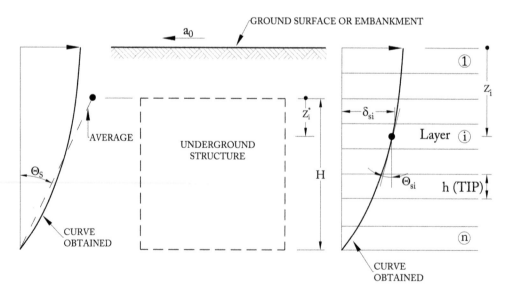

Figure 10.45 Model to calculate the shear strain.

Z_i = Vertical distance from the ground surface or top of the embankment to the center of layer "i", expressed in meters

Z_i^* = Vertical distance from the top of the structure to the center of layer "i", expressed in meters

δ_{si} = Free-field seismic displacement, relative to the base of the structure, that is generated at the top of layer "i"

θ_{si} = Free-field shear strain at the center of layer "i"

θ_s = Average free-field shear strain

h = Thickness of the layers in which the ground is divided

n = Number of layers in which the ground is divided

The δ_{ci} value is not a ground constant; but depends on the amplitude of the seismic displacements, that is, the shear strain, and can be estimated by the expression:

$$\delta_{si} = 53 \cdot K_{2i} \sqrt{\overline{\sigma}_{vi}} \qquad (10.7)$$

where:

K_{2i} = Dimensionless factor that depends on the shear strain, that can be calculated using the graphs in Figure 10.46

$\overline{\sigma}_{vi}$ = Vertical effective stress at the center of layer "i", expressed in ton.f./m²

δ_{si} = Free-field seismic displacement, relative to the base of the structure, that is generated at the top of layer "i"

Since the coefficient K_2 depends on the shear strain, it has to be calculated by successive iterations, starting from an arbitrary value, until the error between the initial and the calculated shear strain is small enough.

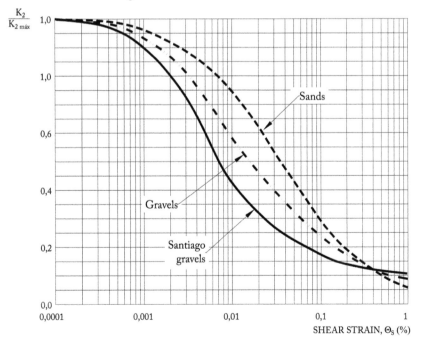

Figure 10.46 Evolution of the K_2 coefficient in relation with the shear strain.

Source: Road Manual. Ministry of Public Works. Chile.

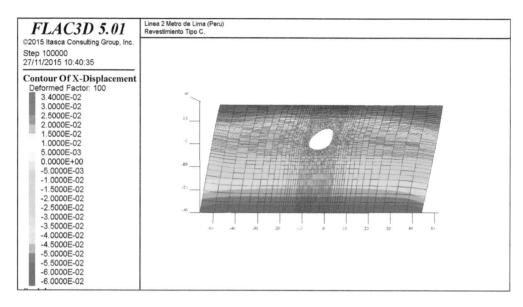

Figure 10.47 Tunnel cross section deformed by a shear strain.

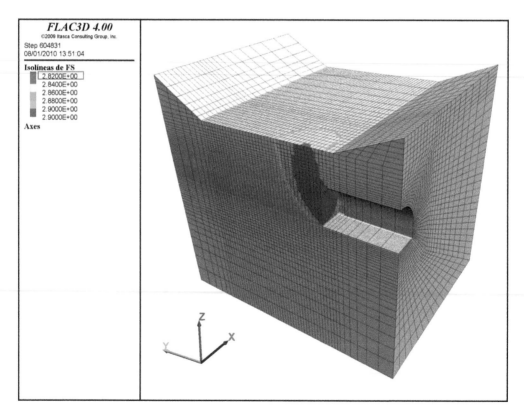

Figure 10.48 Distribution of factor of safety isolines of the ground around a tunnel being designed.

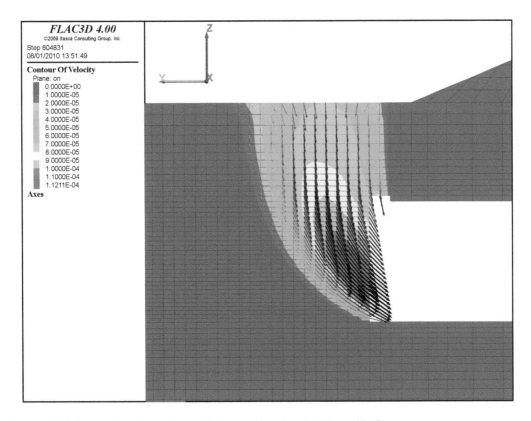

Figure 10.49 Potential surface collapse of the tunnel modeled in Figure 10.48.

Once the shear strains at each layer (θ_{si}) are obtained, the induced displacements in each layer are calculated by the expression:

$$\delta_{si} = \sum_{i=1}^{i=n} \theta_{si} \cdot h_i \tag{10.8}$$

where:

δ_{si} = Horizontal displacement in the center of the layer i, expressed in meters
θ_{si} = Shear strain in the center of the layer i, expressed in radians

To calculate the earthquake-induced stresses, the displacements are introduced in the geomechanics model and the corresponding stresses are calculated. Figure 10.47 shows a tunnel cross section deformed by the effects of a shear strain.

The loads generated by the shear strain must be combined with the static loads to verify that they can be withstood by the tunnel support and lining.

According to Appendix C of the ACI Standard 318-11 the combination of the loads for the seismic calculations is 1.3E + 1.0S, where E are the loads obtained in the static calculation and S are the ones obtained in the seismic calculation.

10.5.4 Ground safety factor

Calculation software based on the finite difference algorithm enable the modeling of underground excavation collapses since they are able to solve problems with strong yielding until collapse.

With these software programs it is possible to calculate the ground minimum safety factor for a specific construction method, using the methodology which consists in successively reducing the parameters of the ground, usually cohesion and friction angle, Varona and Ferrer (1998). Figure 10.48 shows the distribution of the safety factor isolines in the ground where a tunnel will be constructed.

The factor of safety isolines in a three-dimensional model enable identification of the potential failure zones and the specific spot where an excavation collapse would begin.

Figure 10.49 shows a vertical section of the model, shown in Figure 10.48, where it is seen that the excavation collapse starts at the lower part of the excavation face.

BIBLIOGRAPHY

Association Française des Tunnels et de l'Espace Souterrain, *The Convergence-Confinement Method.* AFTES, Paris. 2001.

Kaser, P. K., Diederichs, M.S., Martin, C.D., Sharp, J., Steiner, W., "Underground works in hard rock tunnelling and mining", *GeoEng 2000* Melbourne, Australia. 2000, pp. 841–926.

Kuesel, T.R., "Earthquake design criteria for subways", *Journal of the Structural Division, ASCE.* Vol. 95. No. ST6. 1969, pp. 1213–1231.

Ortigosa, P., Musante, H., "Seismic earth pressures against structures with restrained displacements", *Proceedings of the International Conference on Recent Advances in Geotechnical Earthquake Engineering and Soil Dynamics.* University of Missouri, Columbia, MO. 1991.

Panet, M., "Les déformations différées dans les ouvrages souterrains", *Proceedings of the 4th International Congress ISRM.* A. A. Balkema, Montreux, Suisse. Vol. 3. 1979, pp. 291–301.

Panet, M., *Le calcul des tunnels par la méthode convergence-confinement.* École Nationale de Pontes et Chausses, París. 1995.

Starfield, A.M., Cundall, P.A., "Towards a methodology for rock mechanics modelling", *International Journal of Rock Mechanics and Mining Sciences.* Vol. 25. No. 3. 1988, pp. 99–106.

Varona, P., *Diseño de Túneles.* Universidad Pontificia Católica de Chile, Santiago, Chile. 2014.

Varona, P., Ferrer, M., "Cálculo de factores de seguridad con Flac", *Ingeopres*, Vol. 58. 1998, pp. 38–41.

Wood, D.M., *Geotechnical Modelling.* Taylor & Francis, London. 2004.

Zienkiewicz, O.C., Cheung, Y.K., *The Finite Element Method in Structural and Continuum Mechanics.* McGraw-Hill, Berkshire, UK. 1967.

Index